Gopinath Kallianpur
Rajeeva L. Karandikar

Introduction to Option Pricing Theory

Birkhäuser
Boston • Basel • Berlin

Gopinath Kallianpur
Department of Statistics
University of North Carolina
Chapel Hill, NC 27599

Rajeeva L. Karandikar
Department of Mathematics & Statistics
Indian Statistical Institute
110016 New Dehli, India

Library of Congress Cataloging-in-Publication Data
Kallianpur, G.
Introduction to option pricing theory / Gopinath Kallianpur,
Rajeeva L. Karandikar.
p. cm
Includes bibliographical references and index.
ISBN 0-8176-4108-4 (alk. paper), — ISBN 3-7643-4108-4 (pbk. : alk. paper)
1. Options (Finance)–Prices–Mathematical models.
I. Karandikar, R. L. (Rajeeva L.) 1956– . II. Title.
HG6042.K35 1999
332.64'5–dc21
99-38324
CIP

AMS Subject Classifications: 60XX, 60Gxx, 60G40, 62P05, 90A05, 90A12

Printed on acid-free paper.

ISBN 0-8176-4108-4
ISBN 3-7643-4108-4

Formatted from authors' files by TEXniques, Inc., Cambridge, MA.
Printed and bound by Hamilton Printing Company, Rensselaer, NY.
Printed in the United States of America.

9 8 7 6 5 4 3 2 1

Contents

Preface

Since the appearance of the seminal works of Robert Merton, Fischer Black and Myron Scholes, stochastic processes have assumed an increasingly important role in the development of the mathematical theory of finance. Our aim in this monograph is to present, in some detail, that part of stochastic finance which pertains to option pricing theory.

Although the work is intended for probabilists, five introductory chapters on stochastic analysis are included to make the book self-contained and accessible to graduate students who are making their first acquaintance with the subject. These chapters cover the essentials of Itô's theory of stochastic integration, integration with respect to semimartingales, Girsanov's theorem and a brief introduction to stochastic differential equations (Chapters 1–5). Chapter 6 is devoted to discrete time problems with a view to preparing the reader for the development of continuous time finance.

Trading strategies and arbitrage opportunities are introduced in Chapter 7, as well as contingent claims and complete markets. One feature of the book worth mentioning is the importance attached to the concept of arbitrage. An entire chapter (Chapter 8) is devoted to the relationship between arbitrage and equivalent martingale measures (EMM) and the derivation of necessary and sufficient conditions for no arbitrage (NA).

The Black and Scholes theory is discussed in detail in Chapter 10. Approximations to the Black–Scholes PDE and to the Black–Scholes formula are derived in Chapter 11.

Chapter 12 is devoted to American options. Asset pricing when the volatility is random is the subject of Chapter 13. The Russian options—both the put and the call options—are studied in detail in Chapter 14 because of their independent interest as stopping time problems.

Some of the chapters, especially Chapters 8 and 13, contain new results. Chapter 13, in particular, presents joint work which one of us (GK) did with Jie Xiong of the University of Tennessee. We thank him for permission to include it in this book.

The treatment is kept throughout at a theoretical level, our concern being primarily to introduce the basic concepts of option pricing theory and their interrelationships.

Most of the material in the book has been presented since 1995 by the first author in a graduate course at the University of North Carolina.

G. Kallianpur

R.L. Karandikar

1
Stochastic Integration

This chapter is a self-contained introduction to the theory of stochastic integration with respect to continuous semimartingales. Apart from some simple facts on the predictable σ-field (which are included with proofs in Section 1.2), we do not need any other result from the general theory of processes. The Doob–Meyer decomposition of the square of a continuous (local) martingale is given by constructing the quadratic variation process directly. Once this is done, the stochastic integral is defined and its properties are obtained in a natural fashion, using arguments common in the theory of the Lebesgue integral, namely, the monotone class theorem and dominated convergence theorem. We first give the definition of the stochastic integral with respect to Brownian motion and then deal with the case of continuous semimartingales in stages. In the last section, we define the integral with respect to a general (r.c.l.l.) semimartingale. However, this time we state without proof some key results, including the Doob–Meyer decomposition. We show that these results, three to be precise, enable us to define the integral in the general case and derive the properties of the integral following the route adopted earlier.

1.1 Notation and definitions

$(\Omega, \mathcal{F}, P)$ will denote a complete probability space equipped with a filtration $\{\mathcal{F}_t\}$, i.e., an increasing family of sub σ-fields of $\mathcal{F}$ indexed by $[0, \infty)$. We will also assume that all P-null sets belong to $\mathcal{F}_0$. $\mathcal{F}_\infty$ is the σ-field generated by $\cup_{0 \leq t \leq \infty} \mathcal{F}_t$. The σ-fields will usually (but not always) be assumed to be right continuous, i.e., $\mathcal{F}_t = \mathcal{F}_{t+} := \cap_{s>t}\mathcal{F}_s$. These two conditions are together referred to as the *usual conditions* in the literature. When $(\mathcal{F}_t)$ satisfies the usual conditions, $(\Omega, \mathcal{F}, (\mathcal{F}_t), P)$ will be referred to as a probability basis. When right continuity of the σ-fields is not assumed, this fact will be explicitly stated.

A mapping $X : [0, \infty) \times \Omega \to \mathbb{R}^d$ is a stochastic process if, as a function of (t, ω), it is measurable with respect to the σ-field $\mathcal{B} \times \mathcal{F}$. Here $\mathcal{B}$ is the σ-field of Borel sets of $[0, \infty)$. When $d > 1$, we may call it an $\mathbb{R}^d$-valued stochastic process for emphasis. For $\omega \in \Omega$, the function $t \mapsto X(t, \omega)$ is the *path* of the process corresponding to ω. X_t is the random variable (r.v.) $\omega \mapsto X(t, \omega)$. Two stochastic processes X, Y are said to be equal if $P(X_t = Y_t) = 1$ for all t.

A process X is said to be *adapted* w.r.t. $(\mathcal{F}_t)$ (or simply adapted if there is only one filtration under consideration) if, $\forall t$, X_t is $\mathcal{F}_t$-measurable.

For a stochastic process X, let $\mathcal{G}_t^X = \sigma(X_u : u \le t)$ and

$$\mathcal{F}_t^X = \bigcap_{\varepsilon>0} \sigma\big(\mathcal{G}_{t+\varepsilon}^X \cup \{P - \text{null-sets }\}\big).$$

Thus $(\mathcal{F}_t^X)$ is the smallest filtration satisfying the usual conditions with respect to which X is adapted.

Definition A process M is said to be a *martingale* w.r.t $(\mathcal{F}_t)$ (also written as $(M_t, \mathcal{F}_t)$ is a martingale or M is a martingale if $(\mathcal{F}_t)$ is clear from the context) if
(i) M is $(\mathcal{F}_t)$ adapted,
(ii) $\mathbb{E}|M_t| < \infty \quad \forall t$,
(iii) $\mathbb{E}(M_t|\mathcal{F}_s) = M_s$ (a.s.) $\quad \forall 0 \le s \le t$.
M is called a super(sub)-martingale if (i), (ii) above hold, and if for $s \le t$,

$$E(M_t|\mathcal{F}_s) \le (\ge) M_s \ \ a.s.$$

If (M_t) is a martingale, then it can be deduced from Jensen's inequality that for $1 \le p < \infty$, $|M_t|^p$ is a sub-martingale.

A process X is said to be a continuous process if its paths are continuous. A process X is called an *r.c.l.l.* or a *cadlag* process if it is right continuous and has left limits, that is, if $X_t := \lim_{s \searrow t} X_s$ for all $t \ge 0$, and $\lim_{s \nearrow t} X_s := X_{t-}$ exists for all $t > 0$. It is easy to see that if X, Y are r.c.l.l. processes, then $P(X_t = Y_t) = 1$ for all t implies that $P(X_t = Y_t \ \ \forall t) = 1$. It is known that if M is a martingale, there exists a cadlag version of M, i.e., there exists a process $\tilde{M}$ such that $\tilde{M}$ is cadlag and $\tilde{M}_t = M_t$ a.s. for each t. We shall assume from now on that our martingales are r.c.l.l.

A martingale M is said to be a *square integrable martingale* (or an L^2-martingale) if $\mathbb{E}M_t^2 < \infty \quad \forall\, t < \infty$. Let $\mathcal{M}^2 = \{M : M$ is an L^2-martingale with $M_0 = 0\}$.

Definition τ is a stopping time for $(\mathcal{F}_t)$ if $\tau\colon \Omega \to [0, \infty]$ is a r.v. such that $\{\omega : \tau(\omega) \le t\} \in \mathcal{F}_t \ \ \forall\, t < \infty$. τ is a finite stopping time if $P\{\tau < \infty\} = 1$.

Definition For a stopping time τ, its associated σ-field $\mathcal{F}_\tau$ is defined as follows:

$$\mathcal{F}_\tau := \{A \in \mathcal{F}_\infty : A \cap \{\tau \leq t\} \in \mathcal{F}_t \quad \forall\, t\}.$$

It is easy to see that for a right continuous adapted process X, X_τ is $\mathcal{F}_\tau$-measurable.

Let $M_t = E(Y|\mathcal{F}_t)$, $E\,|Y| < \infty$. It is easy to see that $\{M_t\}$ is uniformly integrable. It follows by Doob's optional sampling theorem that $E(M_{\tau_2}|\mathcal{F}_{\tau_1}) = M_{\tau_1}$ a.s. for any two stopping times τ_1, τ_2, $\tau_1 \leq \tau_2$. Taking $\tau_1 = s \wedge \tau$, $\tau_2 = t \wedge \tau$, $s \leq t$, τ a stopping time, we have $E(M_{t\wedge\tau}|\mathcal{F}_{s\wedge\tau}) = M_{s\wedge\tau}$.

Hence $(M_{t\wedge\tau}, \mathcal{F}_{t\wedge\tau})$ is a martingale. It can then be shown that $(M_{t\wedge\tau}, \mathcal{F}_t)$ is a martingale. We will now state a very important result about martingales. Its proof is essentially based on the corresponding result for discrete martingales. This result and other results on martingales stated in this book without proof can be found in (Meyer, 1966; Neveu, 1965; Karatzas and Shreve, 1988).

Theorem 1.1 [Doob's maximal inequality]. *Let M be a martingale such that $\forall t \geq 0$, $\mathbb{E}|M_t|^p < \infty$, $1 < p < \infty$. Then*

$$\mathbb{E} \sup_{0\leq t\leq T} |M_t|^p \leq \left(\frac{p}{p-1}\right)^p \mathbb{E}|M_T|^p. \tag{1.1}$$

Definition A process M is said to be a *local martingale* if it is adapted and if there exists a sequence of stopping times $\tau_n \uparrow \infty$ (and $\tau_n \leq \tau_{n+1}$) such that, for each n, the process $M_t^n := M_{t\wedge\tau_n}$ is a martingale. A sequence $\{\tau_n\}$ with the above property is called a *localizing* sequence (for the local martingale M). Further, a local martingale is said to be a *local L^2-martingale* if $\{\tau_n\}$ above can be chosen so that M^n are square integrable martingales.

It is easy to see that every martingale M is a local martingale (take $\tau_n = n$) and that every continuous local martingale M (a continuous process that is a local martingale) is also a local L^2-martingale—take τ_n to be the first time that $|M_t|$ exceeds n. Let us note that if M is a local martingale such that the family $\{M_{t\wedge\tau} :$ τ a stopping time, $\tau \leq T\}$ is uniformly integrable or

$$\mathbb{E}[\sup_{t\leq T} |M_t|] < \infty$$

for all $T < \infty$, then M is a martingale.

Let $\mathcal{V}^+$ denote the class of all $(\mathcal{F}_t)$ adapted increasing processes A with $A_0 \geq 0$ and let

$$\mathcal{V} = \{B : B \text{ is written as } B_t = A_t^1 - A_t^2 \text{ for some } A^1, A^2 \in \mathcal{V}^+\}.$$

If $B \in \mathcal{V}$, its path $t \longrightarrow B_t(\omega)$ has bounded variation on $[0, T]$ for every T, denoted by $|B|_T(\omega)$. Clearly, $|B| \in \mathcal{V}^+$ if $B \in \mathcal{V}$. We will refer to elements in $\mathcal{V}$ as processes with locally bounded variation paths. Elements in $\mathcal{V}^+$ are called increasing processes.

Definition A process X is said to be a *semimartingale* if it can be written as $X_t = X_0 + M_t + A_t$ where M is a local martingale and $A \in \mathcal{V}$, i.e., A is a process whose paths are of locally bounded variation.

Definition A process X is said to be a *locally bounded (locally integrable)* process if there exists a sequence of stopping times $\tau_n \uparrow \infty$ (and $\tau_n \leq \tau_{n+1}$) such that, for each n, the process X^n, defined by $X_t^n = X_{t \wedge \tau_n}$, is bounded, respectively $\mathbb{E}(\sup_t |X_t^n|) < \infty$.

In this work, we concentrate on Brownian motion, also known as the Wiener process.

Definition A Brownian motion is a continuous process (W_t) such that

(i) $W_0 = 0$,

(ii) for $0 \leq s \leq t < \infty$, the distribution of $W_t - W_s$ is normal (i.e., Gaussian) with mean 0 and variance $t - s$,

(iii) for $0 = s_0 < s_1 < \ldots < s_k$, the random variables

$$(W_{s_1} - W_{s_0}),\ (W_{s_2} - W_{s_1}),\ \ldots,\ (W_{s_k} - W_{s_{k-1}})$$

are independent.

From the definition it is easy to see that W_t and $W_t^2 - t$ are $(\mathcal{F}_t^W)$ martingales. We will also consider d-dimensional Brownian motion defined below.

Definition An $\mathbb{R}^d$-valued process $W_t = (W_t^1, \ldots, W_t^d)$ is said to be a $d-$ dimensional Brownian motion (or Wiener process) if

(i) (W_t^j) is a Brownian motion for $1 \leq j \leq d$,

(ii) $W^1, W^2, \ldots, W^d$ are independent processes.

Black-Scholes a functional version of the usual monotone class theorem. Its proof is easy and can be found in (Dellacherie and Meyer, 1978).

Theorem 1.2 *Let $\mathcal{H}$ be a class of bounded functions on a set Ω_0. Suppose that $\mathcal{H}$ is closed under bounded pointwise convergence (i.e., $f_n \in \mathcal{H}$ for all n, $|f_n| \leq K$ for some constant K, $f_n \longrightarrow f$ pointwise implies $f \in \mathcal{H}$). Suppose $\mathcal{G} \subseteq \mathcal{H}$ and $\mathcal{G}$ is an algebra of functions. Further, suppose that either the constant function $1 \in \mathcal{G}$ or there exists $f_n \in \mathcal{G}$, $f_n \longrightarrow 1$ pointwise. Then $\mathcal{H}$ contains all $\mathcal{G}$ measurable bounded functions.*

1.2 The predictable σ field

As we will see later, the natural class of simple functions for the purpose of stochastic integration is the class $\mathcal{S}$ consisting of processes f of the form

$$f_t(\omega) = U_0(\omega)1_{\{0\}}(t) + \sum_{j=1}^{m-1} U_j(\omega)1_{(s_j,s_{j+1}]}(t), \tag{1.2}$$

where $0 = s_0 < s_1 < \cdots < s_m$, U_j is a bounded $\mathcal{F}_{s_j}$-measurable random variable for $0 \leq j < m$ and $m \geq 1$.

Let $\overline{\Omega} = [0,\infty) \times \Omega$. Elements of $\mathcal{S}$ (or more generally, any process (X_t)) can be considered as functions on $\overline{\Omega}$. The predictable σ-field $\mathcal{P}$ is defined to be the smallest σ-field on $\bar{\Omega}$ with respect to which every element of $\mathcal{S}$ is measurable. If there is more than one family of σ-fields under consideration, we will write $\mathcal{P}((\mathcal{F}_t))$ to denote the predictable σ-field corresponding to $(\mathcal{F}_t)$. The usefulness of $\mathcal{P}$ from the point of view of stochastic integration is evident from the next theorem.

Theorem 1.3 *Let μ be any finite positive measure on $(\bar{\Omega}, \mathcal{P})$. Then $\mathcal{S}$ is dense in $L^2(\bar{\Omega}, \mathcal{P}, \mu)$.*

Proof It is easy to check that $\mathcal{S}$ is an algebra, $f_t^n(\omega) = 1_{[0,n]\times\Omega}(t,\omega)$ belongs to $\mathcal{S}$ and f^n converges to the function $1_{\bar{\Omega}}$. The result now follows from Theorem 1.2. □

A process (g_t) is called a *predictable process* if g, considered as a function on $\bar{\Omega}$, is $\mathcal{P}$-measurable.

Let us begin by observing that every left continuous adapted process (g_t) is predictable. This is so because g is the pointwise limit of g^n where g^n is defined by

$$g_t^n(\omega) = \psi_n\big(g_0(\omega)\big)1_{\{0\}}(t) + \textstyle\sum_{j=1}^{n^2} \psi_n\big(g_{\frac{j}{n}}(\omega)\big)1_{(\frac{j}{n},\frac{j+1}{n}]}(t), \tag{1.3}$$

where $\psi_n(x) = (x \wedge n) \vee (-n)$, and clearly $g^n \in \mathcal{S}$. In particular, every adapted continuous process (g_t) is predictable. The next result shows that $\mathcal{P}$ can be equivalently defined as the σ-field generated by continuous adapted processes.

Lemma 1.4 *Let $\mathcal{P}_1 = \sigma(g : g$ continuous $(\mathcal{F}_t)$-adapted process$)$. Then $\mathcal{P} = \mathcal{P}_1$.*

Proof We have already noted that $\mathcal{P}_1 \subseteq \mathcal{P}$. For the other part, let us note that for $s < t$, we can get bounded continuous functions φ on $[0,\infty)$ such that $\varphi_n(u) \longrightarrow 1_{(s,t]}(u)$ pointwise. Thus it follows that every $f \in \mathcal{S}$ can be written as the pointwise limit of continuous adapted processes, and hence $\mathcal{S} \subseteq \mathcal{P}_1$. This implies $\mathcal{P} \subseteq \mathcal{P}_1$, completing the proof. □

Lemma 1.5 *Let σ, τ be stopping times such that $\sigma \leq \tau$ and let X be an r.c.l.l. adapted process. Then f, g, h defined by*

$$f_t(\omega) = 1_{[0,\tau(\omega)]}(t) \tag{1.4}$$

$$g_t(\omega) = 1_{(\sigma(\omega),\tau(\omega)]}(t) \tag{1.5}$$

$$h_t(\omega) = X_{\sigma(\omega)}(\omega)1_{(\sigma(\omega),\tau(\omega)]}(t) \tag{1.6}$$

are predictable.

Proof Since f, g, h are left continuous, it suffices to show that they are adapted. Now $\{f_t = 1\} = \{t \leq \tau\}$ and $\{g_t = 1\} = \{\sigma < t\} \cap \{t \leq \tau\}$. Since σ, τ are stopping times, it follows that these sets belong to $\mathcal{F}_t$ and hence f, g are adapted. To show that h is adapted, fix t. Let $\sigma_n(\omega) = (2^{-n}[2^n\sigma(\omega) + 1]) \wedge t$. Then

$$h_t(\omega) = \lim_n X_{\sigma_n(\omega)}(\omega)1_{\{\sigma(\omega)<t\}}1_{\{\tau(\omega)\geq t\}}.$$

It is easy to see that X_{σ_n} is $\mathcal{F}_t$ measurable and hence h_t is $\mathcal{F}_t$ measurable. □

1.3 The Itô integral

The Itô integral is the name given to the object

$$\int_0^t f_s dW_s, \tag{1.7}$$

where (W_t) is a Brownian motion and f is a stochastic process. The difficulty and the interest in defining this is due to the fact that the paths $s \to W_s(\omega)$ of the integrator W are of unbounded variation, and hence the integral cannot be interpreted as a Riemann–Stieltjes integral.

We begin by giving a proof of the fact that the paths of Brownian motion are of unbounded variation.

Theorem 1.6 *Let (W_t) be a Brownian motion. Fix T. Let $t_i^n = iT2^{-n}$, $0 \leq i \leq 2^n$,*

$$V^n = \sum_{i=0}^{2^n-1} |W_{t_{i+1}^n} - W_{t_i}^n| \quad Q^n = \sum_{i=0}^{2^n-1} (W_{t_{i+1}^n} - W_{t_i^n})^2.$$

Then (a) $V^n \to \infty$ a.s. and (b) $Q^n \to T$ a.s.

Proof We will first prove (b). Let

$$X_i^n = W_{t_{i+1}^n} - W_{t_i^n}, \quad Z_i^n = (W_{t_{i+1}^n} - W_{t_i^n})^2 - T \cdot 2^{-n}.$$

Then $\{X_i^n,\ i \geq 0\}$ are independent random variables with normal distribution and $\mathbb{E}(X_i^n) = 0, \mathbb{E}(X_i^n)^2 = T \cdot 2^{-n}$. So $\{Z_i^n,\ i \geq 0\}$ are independent random variables with $\mathbb{E}(Z_i^n) = 0$ and $\mathbb{E}(Z_i^n)^2 = 2(T2^{-n})^2$. Now

$$\begin{aligned} \mathbb{E}(Q^n - T)^2 &= \mathbb{E}(\sum_{i=0}^{2^n-1} Z_i^n)^2 \\ &= \sum_{i=0}^{2^n-1} \mathbb{E}(Z_i^n)^2 \\ &= 2^n \cdot 2 \cdot (T2^{-n})^2 = 2T^2 2^{-n}. \end{aligned}$$

Writing $\| X \|_2 = [\mathbb{E}X^2]^{\frac{1}{2}}$ for the L^2-norm, we have

$$\begin{aligned} \| \sum_{n=1}^{\infty} |Q^n - T| \|_2 &\leq \sum_{n=1}^{\infty} \| Q^n - T \|_2 \\ &\leq \sqrt{2}T \sum_{n=1}^{\infty} 2^{\frac{-n}{2}}. \end{aligned}$$

Thus $\sum_{n=1}^{\infty} |Q^n - T| < \infty$ *a.s.* and as a consequence, $Q^n \to T$ *a.s.*

For (a), note that for any ω,

$$\begin{aligned} Q^n(\omega) &\leq \left(\max_{0 \leq i \leq 2^n-1} |W_{t_{i+1}^n}(\omega) - W_{t_i^n}(\omega)| \right) \cdot \sum_{i=0}^{2^n-1} |W_{t_{i+1}^n} - W_{t_i}^n| \\ &= \left(\max_{0 \leq i \leq 2^n-1} |W_{t_{i+1}^n}(\omega) - W_{t_i^n}(\omega)| \right) \cdot V^n(\omega). \end{aligned} \tag{1.8}$$

Continuity of paths of W implies that

$$\max_{i<2^n} |W_{t_{i+1}^n}(\omega) - W_{t_i^n}(\omega)| \to 0 \quad \text{for all } \omega. \tag{1.9}$$

So if $\liminf_n V^n(\omega) < \infty$ for some ω, then $\liminf_n Q^n(\omega) = 0$ in view of (1.8) and (1.9). Since $Q^n \to T$ *a.s.*, we must have $V^n \to \infty$ *a.s.* □

Remark It is well known that the paths of Brownian motion are nowhere differentiable. For this and other path properties of Brownian motion, see (Brieman, 1968; McKean, 1969; Karatzas and Shreve, 1988).

Let (W_t) be a Wiener process and let $(\mathcal{F}_t)$ be a filtration such that W is a martingale w.r.t. $(\mathcal{F}_t)$. Further, suppose that for s fixed,

$$\{W_t - W_s\ :\ s \leq t\} \text{ is independent of } \mathcal{F}_s. \tag{1.10}$$

It follows that $(W_t^2 - t)$ is also a martingale w.r.t. $(\mathcal{F}_t)$.

Remark We will prove later that it is unnecessary to assume (1.10) in the sense that W, being a Wiener process and a martingale w.r.t $\mathcal{F}_t$, implies (1.10). See Theorem 2.10.

Let $\mathcal{S}$ be the class of simple processes f of the form

$$f_t(\omega) = \sum_{j=1}^{k} U_j(\omega) 1_{(s_{j-1}, s_j]}(t), \tag{1.11}$$

where $0 = s_0 < s_1 < \ldots < s_k$ and U_j is a $\mathcal{F}_{s_{j-1}}$ measurable bounded random variable. For an f given by (1.11), it is natural to define

$$\int_0^t f dW = \sum_{j=1}^{k} U_j (W_{s_j \wedge t} - W_{s_{j-1} \wedge t}). \tag{1.12}$$

If we can obtain an estimate on the growth of the integral defined above for $f \in \mathcal{S}$, we can then extend the integral to an appropriate class of integrands.

We now note some properties of $\int f dW$ for $f \in \mathcal{S}$ and obtain an estimate.

Theorem 1.7 *(a) For $f, g \in \mathcal{S}$*

$$\int_0^t (f + g) dW = \int_0^t f dW + \int_0^t g dW. \tag{1.13}$$

(b) For $f \in \mathcal{S}$ and $Y_t = \int_0^t f dW$, Y_t and $Y_t^2 - \int_0^t f_s^2 ds$ are $(\mathcal{F}_t)$-martingales.
(c) Let $f \in \mathcal{S}$. Then for any T,

$$\mathbb{E} \sup_{t \le T} | \int_0^t f dW|^2 \le 4\mathbb{E}(\int_0^t f_s^2 ds). \tag{1.14}$$

Proof It is easy to check (1.13) from the definition of $\int f dW$ for $f \in \mathcal{S}$. For (b), let f be given by (1.11). Then for $s < t$

$$\mathbb{E}(Y_t | \mathcal{F}_s) = \sum_{j=1}^{k} \mathbb{E}\left[U_j (W_{s_j \wedge t} - W_{s_{j-1} \wedge t}) | \mathcal{F}_s\right]. \tag{1.15}$$

Now if $s_{j-1} \le s$, then U_j is $\mathcal{F}_s$-measurable and hence

$$\begin{aligned} \mathbb{E}\left[U_j (W_{s_j \wedge t} - W_{s_{j-1} \wedge t}) | \mathcal{F}_s)\right] &= U_j \mathbb{E}(W_{s_j \wedge t} - W_{s_{j-1} \wedge t} | \mathcal{F}_s) \\ &= U_j (W_{s_j \wedge s} - W_{s_{j-1} \wedge s}). \end{aligned} \tag{1.16}$$

If $s < s_{j-1}$, then, noting that

$$\mathbb{E}(U_j (W_{s_j \wedge t} - W_{s_{j-1} \wedge t}) | \mathcal{F}_{s_{j-1}}) = 0, \tag{1.17}$$

we have

$$\begin{aligned}\mathbb{E}(U_j(W_{s_j\wedge t} - W_{s_{j-1}\wedge t})|\mathcal{F}_s) &= \mathbb{E}\big[\mathbb{E}(U_j(W_{s_j\wedge t} - W_{s_{j-1}\wedge t})|\mathcal{F}_{s_{j-1}})|\mathcal{F}_s\big] \\ &= 0 \qquad (1.18)\end{aligned}$$

For $s < s_{j-1}$, the right hand side of (1.16) is zero. We thus conclude that (1.16) is true for all s. From this it follows that $(Y_t, \mathcal{F}_t)$ is a martingale.

Note that for $i < j$, $s_i \le s_{j-1}$, $U_iU_j(W_{s_i\wedge t} - W_{s_{i-1}\wedge t})$ is $\mathcal{F}_{s_{j-1}}$ measurable and hence

$$\mathbb{E}(U_iU_j(W_{s_i\wedge t} - W_{s_{i-1}\wedge t})(W_{s_j\wedge t} - W_{s_{j-1}\wedge t})|\mathcal{F}_{s_{j-1}}) = 0. \qquad (1.19)$$

We can now prove that for $i < j$,

$$\begin{aligned}&\mathbb{E}(U_iU_j(W_{s_i\wedge t} - W_{s_{i-1}\wedge t})(W_{s_j\wedge t} - W_{s_{j-1}\wedge t})|\mathcal{F}_s) \\ &\quad = U_iU_j(W_{s_i\wedge s} - W_{s_{i-1}\wedge s})(W_{s_j\wedge s} - W_{s_{j-1}\wedge s}). \qquad (1.20)\end{aligned}$$

For $s < s_{j-1}$, we can first take conditional expectation given $\mathcal{F}_{s_{j-1}}$ as in (1.18) and then given $\mathcal{F}_s$, use (1.19) to conclude that the left hand side in (1.20) is zero and the right hand side is zero, as $W_{s_j\wedge s} = W_{s_{j-1}\wedge s} = W_s$. For $s_{j-1} \le s$, U_i is $\mathcal{F}_s$ measurable, and so using (1.16) with

$$U'_j = U_iU_j(W_{s_i\wedge t} - W_{s_{i-1}\wedge t})$$

and noting that $s_{i-1} < s_i \le s_{j-1} \le s < t$, we conclude that (1.20) is true. We have thus far proved (1.16) and (1.20). We will now prove

$$\begin{aligned}&\mathbb{E}\big[U_j^2(W_{s_j\wedge t} - W_{s_{j-1}\wedge t})^2 - U_j^2(s_j\wedge t - s_{j-1}\wedge t)|\mathcal{F}_s\big] \\ &\quad = U_j^2(W_{s_j\wedge s} - W_{s_{j-1}\wedge s})^2 - U_j^2(s_j\wedge t - s_{j-1}\wedge s). \qquad (1.21)\end{aligned}$$

Once this is done, (1.20) and (1.21) together imply that $Y_t^2 - t$ is an $\mathcal{F}_t$-martingale.

For $a < b$ and U, an $\mathcal{F}_a$-measurable bounded random variable, if $s \le a < b$, one has

$$\begin{aligned}\mathbb{E}[U^2(W_b - W_a)^2|\mathcal{F}_s] &= \mathbb{E}[\mathbb{E}(U^2(W_b - W_a)^2|\mathcal{F}_a)|\mathcal{F}_s] \\ &= \mathbb{E}[U^2\mathbb{E}[(W_b - W_a)^2|\mathcal{F}_a]|\mathcal{F}_s] \\ &= \mathbb{E}[U^2(b-a)|\mathcal{F}_s], \qquad (1.22)\end{aligned}$$

since $(W_b - W_a)$ is independent of $\mathcal{F}_a$. For $b > s > a$,

$$\begin{aligned}\mathbb{E}[U^2(W_b - W_a)^2|\mathcal{F}_s] &= \mathbb{E}[U^2(W_b^2 - 2W_bW_a + W_a^2)|\mathcal{F}_s] \\ &= \mathbb{E}[U^2(W_s^2 + (b-s) - 2W_sW_a + W_a^2)|\mathcal{F}_s] \\ &= \mathbb{E}[U^2(W_s - W_a)^2 + (b-s)|\mathcal{F}_s]. \qquad (1.23)\end{aligned}$$

For $s \ge b > a$,

$$\mathbb{E}[U^2(W_b - W_a)^2|\mathcal{F}_s] = U^2(W_b - W_a)^2. \qquad (1.24)$$

Together (1.22), (1.23) and (1.24) can be recast as

$$\mathbb{E}[U^2(W_b - W_a)^2|\mathcal{F}_s] = \mathbb{E}[U^2\{(W_{b\wedge s} - W_{a\wedge s})^2 + b \wedge s - a \wedge s\}|\mathcal{F}_s]. \quad (1.25)$$

(1.21) is an easy consequence of (1.25).

By Doob's maximal inequality, Theorem 1.1 and the fact that Y_t is a martingale, we get

$$\mathbb{E} \sup_{t\leq T} |Y_t|^2 \leq 4\mathbb{E}[Y_T^2]. \quad (1.26)$$

Since $Y_t^2 - \int_0^t f_s^2 ds$ is a martringale (with mean zero as $Y_0 = 0$), we also have

$$\mathbb{E}[Y_T^2] = \mathbb{E}\left[\int_0^T f_s^2 ds\right]. \quad (1.27)$$

The relations (1.26) and (1.27) together give (1.14). □

We can use the estimate (1.14) to extend the stochastic integral $\int f dW$ to a larger class of f's defined below. Let

$$\mathcal{L}^2(W) = \left\{f : f \ \text{ predictable}, \ \mathbb{E}\int_0^T f_s^2 ds < \infty \quad \forall T < \infty\right\}.$$

For $N \geq 1$, let μ_N be the measure on $\mathcal{P}$ defined by

$$\mu_N(C) = \mathbb{E}\int_0^N 1_C(s,\omega)ds, \quad C \in \mathcal{P}.$$

By Theorem 1.3, $\mathcal{S}$ is dense in $L^2(\overline{\Omega}, \mathcal{P}, \mu_N)$. So we can get $f_N \in \mathcal{S}$ such that

$$\mathbb{E}\int_0^N |f_s^N - f_s|^2 ds. \leq 2^{-2N}. \quad (1.28)$$

Let $Y_t^N = \int_0^t f^N dW$. We will prove that Y^N converges in an appropriate sense to a process Y. For $N \geq T$ using (1.14) for $f^{N+1} - f^N$, we get

$$\begin{aligned}
\mathbb{E}\sup_{t\leq T} |Y_t^{N+1} - Y_t^N|^2 &\leq 4\mathbb{E}\int_0^T |f_s^{N+1} - f_s^N|^2 ds \\
&\leq 4 \cdot 2\mathbb{E}\int_0^T \left\{|f_s^{N+1} - f_s)^2 + |f_s^N - f_s|^2\right\} ds \\
&\leq 4 \cdot 2\{2^{-2(N+1)} + 2^{-N}\} \\
&\leq 16 \cdot 2^{-2N}.
\end{aligned}$$

Hence the $L^2(P)$ norm of $\sum_{N\geq T}\{\sup_{t\leq T} |Y_t^{N+1} - Y_t^N|\}$ is finite. As a consequence,

$$\sum_{N\geq T}\left\{\sup_{t\leq T} |Y_t^{N+1} - Y_t^N|\right\} < \infty \quad a.s. \quad (1.29)$$

Let Ω^T be the null set where the series in (1.29) does not converge. Let $\Omega^0 = \cup_{T=1}^{\infty}\Omega^T$. Then $P(\Omega^0) = 0$. For $\omega \notin \Omega^0$, $Y_t^N(\omega)$ converges to, say $Y_t(\omega)$, uniformly on $[0, T]$ for every T. For $\omega \in \Omega^0$, define $Y_t(\omega) = 0$. We then have that Y_t is a continuous process, and also that

$$\mathbb{E}(\sup_{t\leq T} |Y_t^N - Y_t|^2) \longrightarrow 0.$$

We would like to define Y_t to be the stochastic integral $\int_0^t f dW$. We must verify that the (Y_t) constructed above does not depend on the sequence (f^N) chosen. For this, note that for any $g \in \mathcal{S}$,

$$\mathbb{E}\sup_{t\leq T} |Y_t^N - \int_0^t g dW|^2 \leq 4 \cdot \mathbb{E}\int_0^T |f_s^N - g_s|^2 ds.$$

Taking lim inf as $N \to \infty$ and using Fatou's lemma for the left hand side and (1.28) for the right hand side, we conclude that

$$\mathbb{E}\sup_{t\leq T} |Y_t - \int_0^t g dW|^2 \leq 4\mathbb{E}\int_0^T |f_s - g_s|^2 ds \quad \forall g \in \mathcal{S}. \tag{1.30}$$

It follows easily from this estimate that Y does not depend on the choice of (f^N) satisfying (1.28). Indeed, there is a unique process (Y_t) that satisfies (1.30).

Definition For $f \in \mathcal{L}^2(W)$, the unique process (Y_t) satisfying (1.30) is defined as the stochastic integral of f with respect to W and is written as $Y_t = \int_0^t f dW$.

Theorem 1.8 *The conclusions in Theorem 1.7 are valid for* $f, g \in \mathcal{L}^2(W)$.

Proof Fix $f \in \mathcal{L}^2(W)$ and let $f^N \in \mathcal{S}$ satisfy (1.28). Let $Y_t = \int_0^t f dW$, $Y_t^N = \int_0^t f^N dW$. We have seen earlier that $(Y_t^N, \mathcal{F}_t)$ is a martingale and that $E(Y_t^N - Y_t)^2 \to 0$. So $E|Y_t^N - Y_t| \to 0$. Now if $s < t, C \in \mathcal{F}_s$, then we have

$$\mathbb{E}1_C Y_t^N = \mathbb{E}1_C Y_s^N,$$

since Y^N is a martingale. Taking the limit as $N \to \infty$ and using $\mathbb{E}|Y_t^N - Y_t| \to 0$, $\mathbb{E}|Y_s^N - Y_s| \to 0$, we conclude that

$$\mathbb{E}1_C Y_t = \mathbb{E}1_C Y_s.$$

and so Y_t is a martingale.

Now $\mathbb{E}(Y_t^N - Y_t)^2 \to 0$ also gives $\mathbb{E}|(Y_t^N)^2 - (Y_t)^2| \to 0$ and (1.28) implies that

$$\mathbb{E}|\int_0^t (f_s^N)^2 ds - \int_0^t (f_s)^2 ds| \to 0.$$

We then have

$$\mathbb{E}|(Y_t^N)^2 - \int_0^t (f_s^N)^2 ds - (Y_t)^2 + \int_0^t (f_s)^2 ds| \to 0.$$

The L^1 convergence of the martingales $(Y_t^N)^2 - \int_0^t (f_s^N)^2 ds$ to $(Y_t)^2 - \int_0^t (f_s)^2 ds$ for every t implies (as noted above) that $Y_t^2 - \int_0^t (f_s)^2 ds$ is a martingale. The estimate (1.14) follows from this, or by taking $g = 0$ in (1.30). The linearity property (1.13) is easy to verify. □

The following result displays the interplay between stochastic integrals and stopping times.

Theorem 1.9 *Let $f \in \mathcal{L}^2(W)$ and let $Z_t = \int_0^t f dW$. Let σ be a stopping time. Then*

$$\int_0^t f 1_{[0,\sigma]} dW = Z_{t\wedge\sigma}. \tag{1.31}$$

Proof It is clear that we need to consider only bounded stopping times σ. We can verify (1.31) for simple f and a stopping time taking finitely many values from the definition of the integral for simple functions. If σ is a bounded stopping time, then we can approximate σ from above by stopping times $\sigma = [2^m\sigma]2^{-m}$ taking finitely many values to conclude that (1.31) is true for simple f and bounded σ. Now we can approximate $f \in \mathcal{L}^2(W)$ by simple f to conclude the result. □

In analogy with the Riemann–Stieltjes integral, we define, for a bounded stopping time σ (bounded by T),

$$\int_0^\sigma f dW = \int_0^T f 1_{[0,\sigma]} dW.$$

Remark If $f, g \in \mathcal{L}^2(W)$ are such that

$$\hat{P}\{(t,\omega) : f_t(\omega) \neq g_t(\omega)\} = 0 \tag{1.32}$$

where $\hat{P}$ is the product of the Lebesgue measure on $[0,\infty)$ and P, then it follows that

$$\int_0^t f dW = \int_0^t g dW. \tag{1.33}$$

Thus if $f \in \mathcal{L}^2(W)$ and g is any measurable process (need not be predictable) such that (1.32) is true, then we can define $\int g dW$ by (1.33). It can be shown that the class of processes g, such that there exists $f \in \mathcal{L}^2(W)$ with (1.32), consists of measurable adapted processes g such that

$$\mathbb{E}\int_0^T g_s^2 ds < \infty \quad \forall T. \tag{1.34}$$

In Itô's original definition, the integral was defined directly for this class instead of going via predictable processes. See (Itô, 1961; Kallianpur, 1980). We have chosen to deal with predictable integrands since this is the class of integrands for general semimartingale integrals.

Let us make a simple observation. Suppose $f, g \in \mathcal{L}^2(W)$ are such that for a stopping time σ,

$$P(f_t(\omega) = g_t(\omega) \quad \forall t \leq \sigma(\omega)) = 1. \tag{1.35}$$

Let $Y_t = \int_0^t f dW$ and $Z_t = \int_0^t g dW$. Then

$$(Y_t - Z_t)^2 - \int_0^t (f_s - g_s)^2 ds$$

is a martingale. As a consequence,

$$(Y_{t\wedge s} - Z_{t\wedge\sigma})^2 - \int_0^{t\wedge\sigma} (f_s - g_s)^2 ds = (Y_{t\wedge\sigma} - Z_{t\wedge\sigma})^2$$

is a martingale. Since it is equal to zero at $t = 0$, we conclude that $Y_{t\wedge\sigma} = Z_{t\wedge\sigma}$ a.s. Continuity of paths of Y, Z yields

$$P(Y_{t\wedge\sigma} = Z_{t\wedge\sigma} \quad \forall t) = 1. \tag{1.36}$$

We can use this property to define the stochastic integrals $\int f dW$ for $f \in \mathcal{L}^0(W)$ defined below. Let

$$\mathcal{L}^0(W) = \left\{ f : f \quad \text{predictable}, \quad \int_0^T f_s^2 ds < \infty \quad a.s. \quad \forall T \right\}.$$

For $f \in \mathcal{L}^0(W)$ and $n \geq 1$, let $\sigma_n = \inf\{t \geq 0 : \int_0^t f_s^2 ds \geq n\}$, and let $f_t^n = f_t \cdot 1_{[0,\sigma_n]}(t)$. Then it follows that

$$\int_0^T (f_t^n)^2 dt \leq n \quad \forall T,$$

and hence $f^n \in \mathcal{L}^2(W)$. Let $Z_t^n = \int_0^t f^n dW$. Now (1.35) holds with $f = f^n, g = f^{n+1}, \sigma = \sigma^n$, and thus we have

$$P(Z_t^n = Z_{t\wedge\sigma^n}^{n+1}) = 1. \tag{1.37}$$

Let

$$\Omega' = \{w : Z_t^n(\omega) = Z_{t\wedge\sigma^n(\omega)}^{n+1}(\omega) \quad \forall n\}.$$

Then $P(\Omega') = 1$. Define $Z_t(\omega) = Z_t^n(\omega)$ for $\sigma^{n-1}(\omega) < t \leq \sigma^n(\omega), \omega \in \Omega'$. For $\omega \notin \Omega'$, define $Z_t(\omega) = 0$ for all t. Then $Z_{t\wedge\sigma^n} = Z_t^n$ a.s., and as a consequence, we have that Z_t is a local martingale. By construction, Z_t is a continuous process.

Also $X_t = Z_t^2 - \int_0^t f_s^2 ds$ is a local martingale, since $X_{t\wedge\sigma_n} = (Z_t^n)^2 - \int_0^t (f_s^n)^2 ds$ is a martingale.

If σ is any stopping time such that $f \cdot 1_{[0,\sigma]} \in \mathcal{L}^2(W)$, then as noted (see (1.31))

$$Z_{t\wedge\sigma}^n = \int_0^t f 1_{[0,\sigma]} 1_{[0,\sigma^n]} dW,$$

and hence the process Z_t satisfies

$$P(Z_{t\wedge\sigma} = \int_0^t f 1_{[0,\sigma]} dW \quad \forall t) = 1. \tag{1.38}$$

Definition For $f \in \mathcal{L}^0(W)$, the process (Z_t) satisfying (1.38) for all stopping times σ such that $f \cdot 1_{[0,\sigma]} \in \mathcal{L}^2(W)$ is defined to be the stochastic integral $\int_0^t f dW$.

We record the properties of $\int f dW$ for $f \in \mathcal{L}^0(W)$ below.

Theorem 1.10 *(a) Let $f, g \in \mathcal{L}^0(W)$. Then*

$$\int_0^t (f+g) dW = \int_0^t f dW + \int_0^t g dW.$$

(b) Let $Y_t = \int_0^t f dW$ for $f \in \mathcal{L}^0(W)$. Then Y_t and $X_t = Y_t^2 - \int_0^t f_s^2 ds$ are local martingales.

(c) Let $Y_t = \int_0^t f dW$ for $f \in \mathcal{L}^0(W)$ and let σ be a stopping time. Then

$$\int_0^t f 1_{[0,\sigma]} dW = Y_{t\wedge\sigma}.$$

(d) Let $f \in \mathcal{L}^0(W)$. For all $\lambda > 0$, $\varepsilon > 0$, one has

$$P\{\sup_{0\le t\le T} |\int_0^t f dW| > \lambda\} \le 4\frac{\varepsilon}{\lambda^2} + P\{\int_0^T f_s^2 ds > \varepsilon\}. \tag{1.39}$$

(e) Suppose $f^n, f \in \mathcal{L}^0(W)$ are such that for all $t < \infty$,

$$\int_0^t |f_s^n - f_s|^2 ds \longrightarrow 0 \text{ in probability.}$$

Then for $T < \infty$, one has

$$\sup_{0\le t\le T} |\int_0^t f^n dW - \int_0^t f dW| \longrightarrow 0 \text{ in probability.}$$

Proof We have already observed (b) while constructing the integral and (a) and (c) are obvious. For (d), let

$$\sigma = \inf\{t \geq 0 : \int_0^t f_s^2 ds \geq \varepsilon\}.$$

Now

$$\begin{aligned} P\{\sup_{0\leq t\leq T} |\int_0^t f dW| > \lambda\} &\leq P\{\sigma < T\} + P\{\sup_{0\leq t\leq T\wedge\sigma} |\int_0^t f dW| > \lambda\} \\ &= P\{\sigma < T\} + P\{\sup_{0\leq t\leq T} |\int_0^t f \cdot 1_{[0,\sigma]} dW| > \lambda\} \\ &\leq P\{\sigma < T\} + \frac{1}{\lambda^2}\mathbb{E}[\sup_{0\leq t\leq T} |\int_0^t f \cdot 1_{[0,\sigma]} dW|^2] \\ &\leq P\{\sigma < T\} + 4\frac{1}{\lambda^2}\mathbb{E}[\int_0^{T\wedge\sigma} f_s^2 ds] \\ &\leq P\{\sigma > T\} + 4\frac{\varepsilon}{\lambda^2}. \end{aligned}$$

The required estimate follows since

$$\{\sigma < T\} = \left\{\int_0^T f_s^2 ds > \varepsilon\right\}.$$

It is easy to see that (e) follows from (d). □

Remark The integral $\int_0^t f dW$ for $f \in \mathcal{L}^0(W)$ can be directly defined by first observing that the estimate (1.39) is valid for $f \in \mathcal{S}$, and then proving that given $f \in \mathcal{L}^0(W)$, we can get simple functions $f_n \in \mathcal{S}$ such that

$$\int_0^T (f_s^n - f_s)^2 ds \longrightarrow 0 \quad \forall T < \infty.$$

Then (1.39) would imply that

$$\sup_{0\leq t\leq T} |\int_0^t f^n dW - \int_0^t f^m dW| \longrightarrow 0 \text{ in probability}$$

as $n, m \longrightarrow \infty$. Now $\int_0^t f dW$ can be defined as the limit (in probability) of $\int_0^t f^n dW$.

Remark Sometimes one has to deal with stochastic integration over a fixed interval $[0, T]$; for example, in the problem of pricing options (considered in later chapters), one is dealing with a finite horizon and all entities, the stock price, the investment strategies are defined on $[0, T]$ for fixed $T < \infty$. In this case, the integral is defined similarly, except that T is the given fixed number.

We will now show that the stochastic integral can be obtained as a limit of Riemann sums appropriately defined. First we need the following observation.

Lemma 1.11 *Let U be a $\mathcal{F}_u$-measurable random variable and let*

$$h_t(\omega) = U(\omega)1_{(u,r]}(t).$$

Then $h \in \mathcal{L}^0(W)$ and

$$\int_0^t h dW = U(W_{r\wedge t} - W_{u\wedge t}).$$

Proof For $n \geq 1$, let

$$h_t^n(\omega) = U(\omega)1_{\{U(\omega)\leq n\}}1_{(u,r]}(t).$$

Then it follows that $h^n \in \mathcal{S}$ and

$$\int_0^t h^n dW = U1_{\{U\leq n\}}(W_{r\wedge t} - W_{u\wedge t}).$$

The result follows from part (e) in the previous theorem upon noting that

$$\int_0^t |h_s^n - h_s|^2 ds \longrightarrow 0.$$

□

Theorem 1.12 *Let f be a continuous adapted process and let W be Brownian motion. Then*

$$\int_0^t f dW = \lim_{|\pi|\to 0} \sum_{\pi} f_{s_i}\left(W_{s_{i+1}} - W_{s_i}\right),$$

where $\pi = \{0 = s_0 < s_1 < \ldots < s_k = t\}$ is a partition of $[0, t]$ and $|\pi| = \max_i(s_{i+1} - s_i)$.

Proof For a partition π as in the statement of the theorem, let

$$f_t^{\pi} = \sum_{\pi} f_{s_i} 1_{(s_i, s_{i+1}]}.$$

Then, by the linearity of the stochastic integral and Lemma 1.11, we have

$$\int_0^t f^{\pi} dW = \sum_{\pi} f_{s_i}\left(W_{s_{i+1}} - W_{s_i}\right).$$

The result follows from part (e) in Theorem 1.10, upon noting that here

$$\lim_{|\pi|\to 0} \int_0^t |f_s^{\pi} - f_s|^2 ds \longrightarrow 0 \text{ in probability.}$$

□

We will now briefly consider integrals over an infinite horizon as well as integrals up to a (possibly unbounded) stopping time. To define these, we need the following result.

Theorem 1.13 *Let f be a predictable process such that*

$$\int_0^\infty f_s^2 ds < \infty \ a.s. \tag{1.40}$$

and let $Z_t = \int_0^t f dW$. Then Z_t converges in probability as t tends to infinity.

Proof Note that for $g = f 1_{[u,\infty)}$, $\int_0^t g dW = Z_t - Z_u$ for $t \geq u$. Using (1.39) for g, one has

$$P(|Z_t - Z_u| \geq \delta) \leq 4\frac{\varepsilon}{\delta^2} + P\left\{\int_u^t f_s^2 ds > \varepsilon\right\}.$$

Given $\eta > 0$, $\delta > 0$, choose ε such that

$$\frac{\varepsilon}{\delta^2} < \frac{\eta}{8},$$

and then choose (using assumption (1.40)) t_0, such that for $t, u > t_0$,

$$P\left\{\int_u^t f_s^2 ds > \varepsilon\right\} \leq \frac{\eta}{2}.$$

It now follows that for $t, u > t_0$,

$$P(|Z_t - Z_u| \geq \delta) \leq \eta.$$

□

In view of this we make the following definition.

Definition Let f be a predictable process satisfying (1.40). Then the limit in probability of $\int_0^t f dW$ is defined to be $\int_0^\infty f dW$.

Let $f \in \mathcal{L}^0(W)$ and let τ be a finite stopping time. Let $g = f 1_{[0,\tau]}$. It is easy to check that (1.40) holds for g. We have seen earlier that if $Z_t = \int_0^t f dW$, then

$$\int_0^t g dW = Z_{t \wedge \tau}.$$

Hence it follows that

$$\int_0^\infty g dW = Z_\tau.$$

We define $\int_0^\tau f dW$ to be $\int_0^\infty f 1_{[0,\tau]} dW$.

We will prove an important result on integrals upto stopping times.

Theorem 1.14 *Let $f \in \mathcal{L}^0(W)$ and let τ be a finite stopping time. Suppose*

$$\mathbb{E} \int_0^\tau f_s^2 ds < \infty. \tag{1.41}$$

Then

$$\mathbb{E}\int_0^\tau f dW = 0$$

and

$$\mathbb{E}[|\int_0^\tau f dW|^2] = \mathbb{E}\int_0^\tau f_s^2 ds.$$

In particular, if $\mathbb{E}[\tau] < \infty$, *then* $\mathbb{E}[W_\tau] = 0$ *and* $\mathbb{E}[W_\tau^2] = \mathbb{E}[\tau]$.

Proof Let $Z_t = \int_0^t f 1_{[0,\tau]} dW$. Since

$$\int_0^\tau f dW = \lim_{t\to\infty} Z_t,$$

where the limit is in probability, it suffices to prove that Z_t converges in $L^2(P)$ since $\mathbb{E}[Z_t] = 0$ and

$$\mathbb{E}[Z_t^2] = \int_0^{t\wedge\tau} f_s^2 ds.$$

Now

$$\begin{aligned}\mathbb{E}[(Z_t - Z_s)^2] &= \mathbb{E}\int_s^t f_u^2 1_{[0,\tau]}(u) du \\ &= \mathbb{E}\int_{s\wedge\tau}^{t\wedge\tau} f_u^2 du.\end{aligned}$$

In view of the assumption (1.41), it follows that Z_t converges in $L^2(P)$, completing the proof of the first part. The last assertion follows by taking $f = 1$. □

We now compute explicitly $\int_0^T W dW$. We will use the notation used in Theorem 1.6. Consider the sequence of partitions $\pi_n = \{0 = t_0^n < t_1^n < \dots, t_k^n\}$, $k = 2^n$ (with $t_i^n = Ti2^{-n}$). By Theorem 1.12, the stochastic integral $\int W dW$ is the limit in probability of Riemann sums over π_n, i.e.,

$$\sum_{i=0}^{2^n-1} W_{t_i^n}(W_{t_{i+1}^n} - W_{t_i^n}) \longrightarrow \int_0^t W dW.$$

Using the identity $a(b-a) = \frac{1}{2}\{b^2 - a^2 - (b-a)^2\}$ with $a = W_{t_i^n}$ and $b = W_{t_{i+1}^n}$ and summing over i from 0 to $2^n - 1$, we get

$$\sum_{i=0}^{2^n-1} W_{t_i^n}(W_{t_{i+1}^n} - W_{t_i^n}) = \frac{1}{2}\{W_T^2 - Q^n\}$$

where $Q^n = \sum_{i=0}^{2^n-1}(Wt_{i+1}^n - W_{t_i^n})^2$. We have seen in Theorem 1.6 that $Q^n \to T$. We conclude that

$$\int_0^T W^n dW = \frac{1}{2}\{W_T^2 - T\}. \tag{1.42}$$

Remark Let us note that to obtain the estimate (1.14) we did not need to use the distributional properties of Brownian motion, but only needed to use the fact that if (X_t) is Brownian motion, then (X_t) and $(X_t^2 - t)$ are martingales. We will see later that these two properties characterize Brownian motion in the class of continuous processes. Doob had observed that if (M_t) is a martingale such that $M_t^2 - A_t$ is also martingale, where A_t is a process whose paths are increasing in t, then $\int f dM$ can be defined proceeding as above, first for simple functions and then by extension. He had also conjectured the existence of such an increasing process for every square integrable martingale. This was later proved by Meyer and the result is known as the Doob–Meyer decomposition theorem. We will prove a special case of this result for continuous (local) martingales in the next section.

1.4 Quadratic variation of a continuous martingale

The main objective in this section is to show that for a continuous local martingale M, there exists a unique continuous increasing process A such that $M_t^2 - A_t$ is a local martingale.

This is a special case of the Doob–Meyer decomposition theorem. An additional feature of the result proved here is that it yields a pathwise formula for A_t, i.e., we can write

$$A_t(\omega) = \Phi_t(M.(\omega)),$$

where Φ_t is a family of mappings from $C([0, \infty), \mathbb{R})$ into $\mathbb{R}$, which do not depend upon the law of M or the underlying σ-field. This result is taken from (Karandikar, 1983b).

For a function $f : [0, \infty) \longrightarrow \mathbb{R}$, let $\mathrm{Var}(f)_t$ denote the variation of f on $[0, t]$, i.e.,

$$\mathrm{Var}\ (f)_t := \sup \left\{ \sum_{i=0}^{k-1} |f(s_{i+1}) - f(s_i)| : 0 =\leq s_0 < \cdots < s_k = t, k \geq 1 \right\}.$$

A function f is said to have locally bounded variation if $\mathrm{Var}(f)_t < \infty$ for all $t < \infty$. It is easy to see that if a function f can be written as the difference of two increasing functions, $f = f_1 - f_2$, then f has locally bounded variation. Conversely, if f is a function with locally bounded variation, then taking $f_1(t) = \frac{1}{2}(\mathrm{Var}\ (f)(t) + f(t))$; $f_2(t) = \frac{1}{2}(\mathrm{Var}\ (f)(t) - f(t))$, it can be easily checked that f_1, f_2 are increasing functions.

We now define the *quadratic variation* $\Phi(q)$ of a function $q \in C([0, \infty), \mathbb{R})$.

For $n \geq 1$, let $\{t_i^n(q) : i \geq 1\}$ be defined inductively as follows: $t_0^n(q) = 0$ and having defined $\{t_i^n(q) : i \geq 1\}$, let

$$t_{i+1}^n(q) = \begin{cases} \inf\{t \geq t_i^n : |q(t) - q(t_i^n(q))| \geq 2^{-n}\} & \text{if } t_i^n(q) < \infty \\ \infty & \text{if } t_i^n(q) = \infty. \end{cases}$$

Let

$$\psi_n(q)(t) = \sum_{i=0}^{\infty}\Big(q(t_{i+1}^n(q) \wedge t) - q(t_i^n(q) \wedge t)\Big)^2$$

$$\psi(q)(t) = \limsup_n \psi_n(q)(t).$$

Let us note that for $s \le t$, $\psi_n(q)(s) \le \psi_n(q)(t) + 2^{-n}$ and hence $\psi(q)$ is an increasing function.

Let $\psi_+(q)(t) = \lim_{s \searrow t} \psi(q)(s)$ and $\psi_-(q)(t) = \lim_{s \nearrow t} \psi(q)(s)$ for $t > 0$. Also, $\psi_-(q)(0) = 0$. Finally, let

$$t^* = \inf\{t \ge 0 : \psi_+(q)(t) \ne \psi_-(q)(t)\}$$

$$\Phi_t(q) = \psi_-(q)(t \wedge t^*).$$

Then it follows that $\Phi_t(q)$ is a continuous increasing function of t. It is easy to see that if for some t_0, $q_1(s) = q_2(s)$ for $s \le t_0$, then $\Phi_s(q_1) = \Phi_s(q_2)$ for $s \le t_0$. Also, if $\psi_n(q)(s)$ converges uniformly over $[0, t_0]$, then the limit is $\Phi_s(q)$, $s \le t_0$.

Suppose q is such that $\mathrm{Var}(q)_{t_0} < \infty$. Then

$$\begin{aligned}\psi_n(q)(t_0) &\le \sum_{i=0}^{\infty} 2^{-n}|q(t_{i+1}^n \wedge t_0) - q(t_i^n \wedge t_0)| \\ &\le 2^{-n}\,\mathrm{Var}\,(q)_{t_0},\end{aligned}$$

and hence $\psi(q)(t_0) = 0$. As $\psi(q)$ is an increasing function, it follows that $\psi_-(q)(t) = 0$, and hence that $\Phi_t(q) = 0$. For later reference we record this fact here:

$$\mathrm{Var}\,(q)_t < \infty \Rightarrow \Phi(q)(t) = 0. \tag{1.43}$$

The next lemma connects the quadratic variation map Φ and continuous martingales.

Lemma 1.15 *Let $(M_t, \mathcal{F}_t)$ be a continuous martingale with $|M_t| \le C < \infty$, on a probability space $(\Omega, \mathcal{F}, P)$. Suppose $M_0 = 0$. Let*

$$A_t(\omega) = \Phi_t(M.(\omega)).$$

Then (A_t) is an $(\mathcal{F}_t)$ adapted continuous increasing process such that $X_t := M_t^2 - A_t$ is also a martingale.

Proof Replacing q by $M.(\omega)$, let

$$\begin{aligned}A_t^n(\omega) &= \psi_n(M.(\omega))(t) \\ \tau_i^n(\omega) &= t_i^n(M.(\omega)) \\ Y_t^n(\omega) &= M_t^2(\omega) - A_t^n(\omega).\end{aligned}$$

It is easy to see that $\{\tau_i^n : i \geq 1\}$ are stopping times. Observe that

$$\begin{aligned} Y_t^n &= M_t^2 - \sum_{i=0}^{\infty}\Big(M_{\tau_{i+1}^n\wedge t} - M_{\tau_i^n\wedge t}\Big)^2 \\ &= \sum_{i=0}^{\infty}\Big(M^2_{\tau_{i+1}^n\wedge t} - M^2_{\tau_i^n\wedge t}\Big) \\ &\quad -\sum_{i=0}^{\infty}\Big(M_{\tau_{i+1}^n\wedge t} - M_{\tau_i^n\wedge t}\Big)^2 \\ &= 2\sum_{i=0}^{\infty} M_{\tau_i^n\wedge t}\Big(M_{\tau_{i+1}^n\wedge t} - M_{\tau_i^n\wedge t}\Big) \\ &= 2\sum_{i=0}^{\infty} Z_t^{ni}. \end{aligned}$$

As M is assumed to be a bounded martingale, it follows that Z_t^{ni} is a martingale. Using the fact that Z_t^{ni} is $\mathcal{F}_{t\wedge\tau_{i+1}^n}$-measurable and that $\mathbb{E}(Z_t^{ni}|\mathcal{F}_{t\wedge\tau_{i+1}^n}) = 0$, it follows that for $i \neq j$,

$$\mathbb{E}Z_t^{ni}Z_t^{nj} = 0. \tag{1.44}$$

Also

$$\begin{aligned} \mathbb{E}(Z_t^{ni})^2 &\leq C^2\mathbb{E}\Big\{M_{\tau_{i+1}^n\wedge t} - M_{\tau_i^n\wedge t}\Big\}^2 \\ &= C^2\mathbb{E}\Big\{M^2_{\tau_{i+1}^n\wedge t} - M^2_{\tau_i^n\wedge t}\Big\}^2 \end{aligned} \tag{1.45}$$

since M is a martingale. Using (1.44) and (1.45) , it follows that for $s \leq r$,

$$\mathbb{E}\Big(\sum_{i=s}^{r} Z_t^{ni}\Big)^2 \leq C^2\mathbb{E}\Big\{M^2_{\tau_{r+1}^n\wedge t} - M^2_{\tau_s^n\wedge t}\Big\}$$

and as a consequence, one has

$$\sum_{i=1}^{r} Z_t^{ni} \longrightarrow Y_t^n \text{ in } L^2(P) \text{ as } r \longrightarrow \infty.$$

Thus (Y_t^n) is a $(\mathcal{G}_t)$-martingale. For $n \geq 1$, define a process (M^n) by

$$M_t^n = M(\tau_i^n) \text{ if } \tau_i^n \leq t < \tau_{i+1}^n.$$

Observe that by the choice of $\{\tau_i^n : i \geq 1\}$, one has

$$|M_t - M_t^n| \leq 2^{-n} \quad \text{for all } t. \tag{1.46}$$

It is not difficult to verify that for all ω, n,

$$\{\tau_i^n(\omega) : i \geq 1\} \subseteq \{\tau_i^{n-1}(\omega) : i \geq 1\},$$

and thus we can write Y^{n-1} as

$$Y_t^{n-1} = \sum_{j=0}^{\infty} 2M_{t\wedge\tau_j^n}^{n-1}\Big\{M_{t\wedge\tau_{j+1}^n} - M_{t\wedge\tau_j^n}\Big\}.$$

Hence

$$Y_t^n - Y_t^{n-1} = 2\sum_{j=0}^{\infty} V_t^{nj} \tag{1.47}$$

where

$$V_t^{nj} = \Big(M_{t\wedge\tau_j^n} - M_{t\wedge\tau_j^n}^{n-1}\Big)\Big(M_{t\wedge\tau_{j+1}^n} - M_{t\wedge\tau_j^n}\Big).$$

Since $\mathbb{E}(V_t^{nj}|\mathcal{F}_{t\wedge\tau_j^n}) = 0$ and V_t^{nj} is $\mathcal{F}_{t\wedge\tau_{j+1}^n}$-measurable, it follows that for $i \neq j$,

$$\mathbb{E}V_t^{nj}.V_t^{ni} = 0.$$

Thus, recalling that $Y_t^n - Y_t^{n-1}$ is a martingale, invoking Doob's maximal inequality, Theorem 1.1, one has that

$$\begin{aligned}
\mathbb{E}\ \sup_{s\leq t}|Y_s^n - Y_s^{n-1}|^2 &\leq 4\mathbb{E}(Y_s^n - Y_s^{n-1})^2 \\
&= 16\sum_{j=0}^{\infty}\mathbb{E}(V_t^{nj})^2 \\
&\leq \frac{16}{2^{2(n-1)}}\sum_{j=0}^{\infty}\mathbb{E}\{M_{t\wedge\tau_{j+1}^n} - M_{t\wedge\tau_j^n}\}^2 \\
&= \frac{64}{2^{2n}}\sum_{j=0}^{\infty}\mathbb{E}\{M_{t\wedge\tau_{j+1}^n}^2 - M_{t\wedge\tau_j^n}^2\} \\
&= \frac{64}{2^{2n}}\mathbb{E}M_t^2.
\end{aligned} \tag{1.48}$$

It follows that

$$\sum_{n=1}^{\infty}\sup_{s\leq t}|Y_s^n - Y_s^{n-1}| < \infty \ \ a.s.$$

(as its L^2-norm is finite by (1.48)). Hence Y_s^n (and as a consequence A_s^n) converges uniformly in $s \in [0, T]$ for every T a.s. to a continuous process, say Y_s (respectively to $M_s^2 - Y_s$). Further, (1.48) also implies that convergence of Y_s^n to Y_s is also in L^2 and thus Y_s is a martingale. Using the continuity of (Y_s), it can be checked that

$$A_t = M_t^2 - Y_t \ \ a.s.,$$

and it follows that $M_t^2 - A_t$ is a martingale.

We are now in a position to prove the Doob–Meyer decomposition theorem for the square of a continuous local martingale.

Theorem 1.16 *Let (M_t) be a continuous local martingale w.r.t. $(\mathcal{F}_t)$ on a probability space $(\Omega, \mathcal{F}, P)$ with $M_0 = 0$. Let $A_t(\omega) = \Phi_t(M.(\omega))$. Then*

(i) $X_t := M_t^2 - A_t$ is a local martingale w.r.t. $(\mathcal{F}_t)$ and $A_t(\omega) < \infty$ a.s. for all t.

(ii) If M is a martingale with $\mathbb{E}M_t^2 < \infty$ for all t, then $\mathbb{E}A_t < \infty$ for all t and X_t is a martingale.

(iii) If $\mathbb{E}A_t < \infty$ for all t, then $\mathbb{E}M_t^2 < \infty$ for all t and X_t is a martingale.

(iv) If (B_t) is an $(\mathcal{F}_t)$ adapted continuous increasing process such that $B_0 = 0$ and $M_t^2 - B_t$ is a local martingale, then $P(A_t = B_t \;\; \forall t) = 1$.

Proof (i) Let σ_k be the stopping time defined by

$$\sigma_k = \inf\{t > 0 : |M_t| \geq k\}.$$

Then $M_t^k = M_{t\wedge\sigma_k}$ is a bounded martingale and hence by Lemma (1.15), $(M_t^k)^2 - A_t^k$ is a martingale, where $A_t^k = \Phi(M_\cdot^k(\omega))_t$.

It is easy to verify that (using the definition of Φ)

$$A_t^k = A_{t\wedge\sigma_k}$$

and it follows that $M_{t\wedge\sigma_k}^2 - A_{t\wedge\sigma_k}$ is a martingale for all k and $\sigma_k \uparrow \infty$. Thus X_t is a local martingale.

For part (ii), note that (for σ_k as in (i) above),

$$\mathbb{E}A_{t\wedge\sigma_k} = \mathbb{E}M_{t\wedge\sigma_k}^2 \leq \mathbb{E}M_t^2$$

since (M_t^2) is a submartingale and so $\mathbb{E}A_t \leq \mathbb{E}M_t^2$. Now by Doob's maximal inequality,

$$\mathbb{E} \sup_{s\leq t\wedge\sigma_k} |M_s^2| \leq 4\mathbb{E}M_{t\wedge\sigma_k}^2 \leq 4\mathbb{E}M_t^2$$

so that $\mathbb{E}\sup_{s\leq t}|M_s^2| \leq 4\mathbb{E}M_t^2$. Recalling that (A_s) is an increasing process, we conclude that

$$\mathbb{E}\sup_{s\leq t}|X_s| \leq \mathbb{E}\sup_{s\leq t}|M_s^2| + \mathbb{E}A_t < \infty.$$

Thus the martingales $X_s^k := X_{s\wedge\sigma_k}$ converge in L^1 to X_s so that (X_s) is a martingale. Part (iii) is proved similarly.

For (iv), let $N_t = A_t - B_t$. Then N_t is a local martingale with $N_0 = 0$. Let $D_t = \Phi(N)_t$. Then by part (i), $N_t^2 - D_t$ is a local martingale. But N_t is a difference of two increasing processes, and thus

$$\text{Var}\,(N.(\omega))_t < \infty \quad \forall\, \omega, t.$$

Thus as noted in (1.43), $D_t(\omega) = 0$ a.s. Hence by part (iii) above, N_t^2 is a martingale with $N_0 = 0$. Thus $\mathbb{E}N_t^2 = \mathbb{E}N_0^2 = 0$ for all t. This and the continuity of paths of N yields the desired conclusion, namely,

$$P(A_t = B_t \text{ for all } t) = 0. \qquad \square$$

The process $A_t(\omega) := \Phi_t(M.(\omega))$ is called the quadratic variation process of the local martingale M and is denoted by $\langle M, M\rangle_t$. Note that

$$\langle M, M\rangle_t = \lim_n \sum_{i=0}^{\infty} \left(M_{\tau_{i+1}^n \wedge t} - M_{\tau_i^n \wedge t}\right)^2,$$

where τ_i^n are stopping times constructed in the proof and the convergence (as $n \longrightarrow \infty$) is *a.s.* w.r.t. the probability measure P under which M is a local martingale. Thus $\langle M, M\rangle$ does not depend on the underlying filtration. In fact, the version of $\langle M, M\rangle$ given by $\Phi_t(M.)$ does not depend upon the underlying probability measure either. A simple consequence of this is the following observation.

Theorem 1.17 *Suppose (M_t) and $(M_t^2 - B_t)$ are continuous local martingales w.r.t. $(\mathcal{F}_t)$ on $(\Omega, \mathcal{F}, P)$, where B is a continuous increasing process.*
(a) If $(\mathcal{G}_t)$ is a filtration such that M is a local martingale w.r.t. $(\mathcal{G}_t)$, then $(M_t^2 - B_t)$ is also a local martingale w.r.t. $(\mathcal{G}_t)$.
(b) If Q is a probability measure on $(\Omega, \mathcal{F})$ such that Q is absolutely continuous w.r.t. P and (M_t) is a local martingale on $(\Omega, \mathcal{F}, Q)$, then $(M_t^2 - B_t)$ is also a local martingale on $(\Omega, \mathcal{F}, Q)$.

Proof Note that by part (iv) of Theorem 1.16,

$$B_t = \Phi_t(M.) \;\; a.s. \; P, \tag{1.49}$$

and hence part (a) follows from Theorem 1.16, part (i). For (b), note that absolute continuity of Q w.r.t. P implies that equality (1.49) holds *a.s.* Q as well, and hence once again, the conclusion follows from Theorem 1.16. $\square$

For later reference we state a lemma whose proof is essentially given in the proof of part (iv) of the previous theorem.

Lemma 1.18 *Let (N_t) be a continuous local martingale with*

$$Var\,(N.(\omega))_t < \infty \; a.s. \quad \forall t.$$

Then $P(N_t = N_0 \quad \forall t) = 1$.

Using localization via stopping times, one can indeed deduce a local version of this result: If (N_t) is a continuous local martingale, then almost surely, the variation of $\{N_s(\omega) : s \in [a, b]\} = 0$ or ∞ for all a, b (see Karandikar, 1983a)).

For continuous local martingales M, N with $M_0 = 0 = N_0$, we define the cross-quadratic variation $\langle M, N\rangle_t$ by

$$\langle M, N\rangle_t := \frac{1}{4}\Big\{\langle M+N, M+N\rangle_t - \langle M-N, M-N\rangle_t\Big\}. \tag{1.50}$$

Clearly $\langle M, N\rangle_t = \langle N, M\rangle_t$. It is easily seen from the definition of $\langle M, N\rangle_t$ that

$$M_t N_t - \langle M, N\rangle_t$$

is a continuous local martingale.

Indeed, using Lemma 1.18, it follows that $\langle M, N\rangle_t$ is the unique process (B_t) with continuous paths such that $B_0 = 0$, $\mathrm{Var}\,(B.(\omega))_t < \infty$ a.s. and

$$M_t N_t - B_t \text{ is a local martingale.}$$

It follows from this observation that $\langle M, N\rangle$ is linear in M for fixed N, i.e., for continuous local martingales M^1, M^2, N with $M_0^1 = 0$, $M_0^2 = 0$, $N_0 = 0$, and real numbers a, b,

$$\langle aM^1 + bM^2, N\rangle_t = a\langle M^1, N\rangle_t + b\langle M^2, N\rangle_t \quad \forall t, \quad a.s. \tag{1.51}$$

It is easy to check that for $M = aM^1 + bM^2$ where a, b, M^1, M^2 are as above,

$$\langle M, M\rangle_t = a^2\langle M^1, M^1\rangle_t + 2ab\langle M^1, M^2\rangle_t + b^2\langle M^2, M^2\rangle_t \quad \forall t \; a.s.$$

Thus we get

$$P(a^2\langle M^1, M^1\rangle_t + 2ab\langle M^1, M^2\rangle_t + b^2\langle M^2, M^2\rangle_t \geq 0 \quad \forall t, \\ \text{for all } a, b \text{ rationals}) = 1.$$

From elementary properties of quadratic forms, it follows that

$$|\langle M^1, M^2\rangle_t| \leq (\langle M^1, M^1\rangle_t \cdot \langle M^2, M^2\rangle_t)^{\frac{1}{2}} \quad \forall t \; a.s. \tag{1.52}$$

Remark The following is a direct proof of Lemma 1.18 when (N_t) is a continuous L^2-martingale.

Let $M_t = N_t - N_0$ be a continuous L^2-martingale. Then it can be shown that

$$\langle M, M\rangle_t = \underset{|\Pi_n|\to\infty}{L^1\text{-lim}} \sum_{\Pi_n}\Big(M_{t_{j+1}^n} - M_{t_j^n}\Big)^2$$

where $\Pi_n = \{t_j^n\}$ is a finite partition of $[0, 1]$ and $|\Pi_n| = \max_j\big(t_{j+1}^n - t_j^n\big)$. From the assumption of Lemma 1.18 and the continuity of M_t, we have

$$\begin{aligned}\sum_{\Pi_n}\Big(M_{t_{j+1}^n} - M_{t_j^n}\Big)^2 &\leq \mathrm{Var}\,(M(\cdot, \omega))_t \cdot \max_{\Pi_n} |M_{t_{j+1}^n}(\omega) - M_{t_j^n}(\omega)| \\ &\to 0 \quad \text{a.s.}\end{aligned}$$

It follows that $\langle M, M\rangle_t = 0$ a.s. for all t, i.e., $\langle N. - N_0, N. - N_0\rangle = 0$ a.s. for all t, which yields the conclusion of Lemma 1.18.

1.5 The stochastic integral w.r.t. continuous local martingales

Let $\mathcal{M}_c^2$ be the class of all continuous martingales with $M_0 = 0$ and $\mathbb{E}M_t^2 < \infty$ for all t. We begin by defining $\int f dM$ in a natural fashion for $f \in \mathcal{S}$; note some of its properties and then extend the integral to a larger class by continuity.

For $f \in \mathcal{S}$ given by

$$f_t(\omega) = U_0(\omega)1_{\{0\}}(t) + \sum_{j=1}^{m-1} U_j(\omega)1_{(s_j,s_{j+1}]}(t) \tag{1.53}$$

where $0 = s_0 < s_1 < \cdots < s_m$, U_j is a $\mathcal{F}_{s_j}$-measurable, bounded random variable for $0 \leq j < m$ and $m \geq 1$; and for $M \in \mathcal{M}_c^2$, we define

$$\Big(\int_0^t fdM\Big)(\omega) = \sum_{j=1}^{m-1} U_j(\omega)\Big\{M_{s_{j+1}\wedge t}(\omega) - M_{s_j\wedge t}(\omega)\Big\}. \tag{1.54}$$

We list below some of its properties.

Theorem 1.19 *Let $f, g \in \mathcal{S}, M, N \in \mathcal{M}_c^2$. Then*

(i) $\int (f+g)dM = \int fdM + \int gdM$.

(ii) $\int fd(M+N) = \int fdM + \int fdN$.

(iii) If $X_t = \int_0^t fdM$, then $X \in \mathcal{M}_c^2$ and

$$\langle X, N\rangle_t = \int_0^t f_s d\langle M, N\rangle_s \tag{1.55}$$

and

$$\langle X, X\rangle_t = \int_0^t f_s^2 d\langle M, M\rangle_s. \tag{1.56}$$

(iv) For all $T < \infty$,

$$\mathbb{E}\Big\{\sup_{t\leq T}\Big|\int_0^t fdM\Big|^2\Big\} \leq 4\mathbb{E}\int_0^T f^2 d\langle M, M\rangle. \tag{1.57}$$

Proof (i) and (ii) are easily verified. Since $\mathcal{M}_c^2$ is clearly a linear space and $X \longrightarrow \langle X, N\rangle$ is also linear, it suffices to check (iii) when

$$f_t = U1_{(s,u]}(t)$$

where $s < u$ and U is an $\mathcal{F}_s$-measurable bounded random variable. In this case it is easily checked that $X_t = \int_0^t fdM$ is a martingale and also that $X_tN_t -$

$\int_0^t fd\langle M,N\rangle_s$ is a martingale. Clearly $\int_0^t f_s d\langle M,N\rangle_s$ is continuous and thus (1.55) follows. Taking $N = X$ in (1.55), we can deduce that

$$\langle X,X\rangle_t = \int_0^t f_s d\langle M,X\rangle_s.$$

and now using $\langle M,X\rangle_s = \langle X,M\rangle_s$ and once again using (1.55) with $M = N$, we get (1.56). The last part follows from (1.56) and Doob's maximal inequality, Theorem 1.1. □

The estimate (1.57) suggests the following way of extending the integral $\int fdM$ to a larger class $\mathcal{L}^2(M)$ defined as follows. For $M \in \mathcal{M}_c^2$, let

$$\mathcal{L}^2(M) = \{f : f \text{ predictable and } \int_0^T f^2 d\langle M,M\rangle < \infty \quad \forall T < \infty\}$$

For $n \geq 1$, define a countably additive measure μ_n on $\mathcal{P}$ by, for $C \in \mathcal{P}$,

$$\mu_n(C) = \int_\Omega \int_0^n 1_C((t,\omega))d\langle M,M\rangle_t(\omega)dP(\omega).$$

Then $\mu_n(\bar{\Omega}) = \mathbb{E}M_n^2 < \infty$ and $\mathcal{S}$ is dense in $L^2(\bar{\Omega},\mathcal{P},\mu_n)$ (see Theorem 1.3) for every $n \geq 1$. Thus given $g \in \mathcal{L}^2(M)$, we can get $g^n \in \mathcal{S}$ such that

$$\int |g^n - g|^2 d\mu_n = \int_\Omega \int_0^n |g_t^n(\omega) - g_t(\omega)|^2 d\langle M,M\rangle_t dP(\omega) \leq 2^{-2n}.$$

Let $Y_t^n = \int_0^t g^n dM$. Then using (1.57) for $f = g_{n+1} - g_n$, we get

$$\begin{aligned}
\mathbb{E}\sup_{t\leq n}|Y_t^{n+1} - Y_t^n|^2 &\leq 4\int |g^{n+1} - g^n|^2 d\mu_n \\
&\leq 4.2.\Big\{\int |g^{n+1} - g|^2 d\mu_n + \int |g^n - g|^2 d\mu_n\Big\} \\
&\leq 4.2.\Big\{2^{-(2n+2)} + 2^{-2n}\Big\} \\
&\leq 16.2^{-2n}.
\end{aligned}$$

It follows that for all T, the $L^2(P)$ norm of $\sum_{n=T}^\infty \sup_{t\leq T}|Y_t^{n+1} - Y_t^n|$ is finite and thus

$$\sum_{n=1}^\infty \sup_{t\leq T}|Y_t^{n+1}(\omega) - Y_t^n(\omega)| < \infty \quad \text{a.s.} \tag{1.58}$$

Let Ω_T^0 be the null set where the series in (1.58) does not converge and let $\Omega^0 = \cup_{T=1}^\infty \Omega_T^0$. Then for $\omega \notin \Omega^0$, $Y_t^n(\omega)$ converges uniformly to $Y_t(\omega)$.

For $\omega \in \Omega^0$, let us define $Y_t(\omega) = 0$ for all t. Then Y_t is a continuous process by definition, and further, one also has

$$\mathbb{E}\Big(\sup_{t\leq T}|Y_t^n - Y_t|^2\Big) \longrightarrow 0. \tag{1.59}$$

Further, for any $f \in \mathcal{S}$

$$\mathbb{E}\Big(\sup_{t\le T}|Y_t^n - \int_0^t f dM|^2\Big) \le 4.\mathbb{E}\int_0^T |g_s^n - f_s|^2 d\langle M, M\rangle_s,$$

and hence taking the limit as $n \longrightarrow \infty$, one has

$$\mathbb{E}\Big(\sup_{t\le T}|Y_t - \int_0^t f dM|^2\Big) \le 4.\mathbb{E}\int_0^T |g_s - f_s|^2 d\langle M, M\rangle_s. \tag{1.60}$$

This estimate implies that the process (Y_t) does not depend upon the choice of $\{g_n\}$, as it is easy to see that there is a unique process satisfying this inequality, namely the process Y constructed above. Thus we are justified in making the following definition.

Definition Let $M \in \mathcal{M}_c^2$ and $f \in \mathcal{L}^2(M)$. The unique process Y satisfying (1.60) is defined as the stochastic integral of f with respect to M and is denoted as

$$Y_t := \int_0^t g dM.$$

Theorem 1.20 *Let $M, N \in \mathcal{M}_c^2$ and $f, g \in \mathcal{L}^2(M) \cap \mathcal{L}^2(N)$. Then all the assertions in Theorem (1.19) are valid.*

Proof It is easy to check the linearity properties. For the other two, first note that $Y^n \in \mathcal{M}_c^2$ and in view of (1.59), Y_t^n converges to Y_t in L^2 and (Y_t) is a martingale. (1.59) now yields $\mathbb{E}\sup_{t\le T}|Y_t|^2 < \infty$. Thus $Y \in \mathcal{M}_c^2$.

Also, $Y_t^n \longrightarrow Y_t$ in $L^2(P)$ implies that $(Y_t^n)^2 \longrightarrow Y_t^2$ and $Y_t^n N_t \longrightarrow Y_t N_t$ in $L^1(P)$ where $N \in \mathcal{M}_c^2$ is fixed. Similarly,

$$\int_0^T |g_s^n - g_s|^2 d\langle M, M\rangle_s \longrightarrow 0$$

implies

$$\int_0^T (g_s^n)^2 d\langle M, M\rangle_s \longrightarrow \int_0^T (g_s)^2 d\langle M, M\rangle_s \text{ in } L^1(P)$$

and in view of (1.52), also

$$\mathbb{E}\int_0^T g_n d\langle M, N\rangle_s \longrightarrow \int_0^T g d\langle M, N\rangle_s \text{ in } L^1(P).$$

Thus the martingales $Y_t^n N_t - \int_0^t g_s^n d\langle M, N\rangle_s$ and $(Y_t^n)^2 - \int_0^t (g_s^n)^2 d\langle M, N\rangle_s$ converge to $X_t^1 = Y_t N_t - \int_0^t g_s d\langle M, N\rangle_s$ and $Z_t^2 = (Y_t)^2 - \int_0^t (g_s)^2 d\langle M, M\rangle_s$ in $L^1(P)$ for every t. Thus (Z_t^1), (Z_t^2) are martingales. Thus (1.55) and (1.56) hold for $g \in \mathcal{L}^2(M)$. □

We will now extend $\int f dM$ to a large class of f's and M's. First, let $M \in \mathcal{M}_c^2$ and let us consider a predictable f such that for a sequence of stopping times τ_n,

$$\mathbb{E}\int_0^{\tau_n} f_s^2 d\langle M, M\rangle_s < \infty.$$

Let $f_s^n = f_s 1_{[0,\tau_n]}(s)$. Then it follows that $f^n \in \mathcal{L}^2(M)$. Let $X_t^n = \int_0^t f^n dM$. Then it is easy to check (first for simple functions, then via approximation) that

$$X_t^n = X_t^{n+1} \text{ for } t \leq \tau_n \text{ a.s.} \tag{1.61}$$

Then we can piece together $\{X_t^n \ : \ n \geq 1\}$ and define

$$\int_0^t f dM = X_t^n \text{ for } \tau_{n-1} \leq t < \tau_n.$$

We can carry out this *localization* on the martingale M as well as on f. We need an analogue of (1.61) for this, which is given in the next lemma.

Lemma 1.21 *Let $M, N \in \mathcal{M}_c^2$, $f \in \mathcal{L}^2(M)$, $g \in \mathcal{L}^2(N)$. Suppose σ is a stopping time such that $M = N$ and $f = g$ on $[0, \sigma]$ a.s. i.e.,*

$$P(M_{t\wedge\sigma} = N_{t\wedge\sigma} \quad \forall t) = P(f_{t\wedge\sigma} = g_{t\wedge\sigma} \quad \forall t) = 1.$$

Let $X_t = \int_0^t f dM$ and $Y_t = \int_0^t g dN$. Then

$$P(X_{t\wedge\sigma} = Y_{t\wedge\sigma} \quad \forall t) = 1.$$

Proof One has that $U_t = M_{t\wedge\sigma}^2 - \langle M, M\rangle_{t\wedge\sigma}$ is a martingale and hence so are $V_t = M_{t\wedge\sigma}N_{t\wedge\sigma} - \langle M, M\rangle_{t\wedge\sigma}$ and $R_t = N_{t\wedge\sigma}^2 - \langle M, M\rangle_{t\wedge\sigma}$. Then Lemma 1.18 implies that

$$P(\langle M, M\rangle_{t\wedge\sigma} = \langle M, N\rangle_{t\wedge\sigma} = \langle N, N\rangle_{t\wedge\sigma} \quad \forall t) = 1.$$

Now let $Z_t = X_t - Y_t$. Then (Z_t) is a martingale and

$$\langle Z, Z\rangle_t = \int_0^t f_s^2 d\langle M, M\rangle_s - 2\int_0^t f_s g_s d\langle M, N\rangle_s + \int_0^t g_s^2 d\langle N, N\rangle_s;$$

(we need to use (1.55) twice to get the middle term above) and thus it follows that $\langle Z, Z\rangle_{t\wedge\sigma} = 0$ a.s. Hence, $Z_{t\wedge\sigma}^2$ is a martingale which implies that $\mathbb{E}Z_{t\wedge\sigma}^2 = \mathbb{E}Z_{0\wedge\sigma}^2 = 0$, and thus $Z_{t\wedge\sigma} = 0$ a.s. Now the continuity of paths of X yields

$$P(Z_{t\wedge\sigma} = 0 \quad \forall t) = 1.$$

This completes the proof. □

Now let M be a continuous local martingale with $M_0 = 0$ and let $\mathcal{L}^0(M)$ denote the class of predictable processes f such that

$$\int_0^t f_s^2 d\langle M, M\rangle_s < \infty \text{ a.s.} \quad \forall t.$$

For f, M as above, let

$$\tau_n = \inf\{f \geq 0 : \langle M, M\rangle_t \geq n \text{ or } \int_0^t f_s^2 d\langle M, M\rangle_s \geq n\}. \tag{1.62}$$

Let $f_t^n = f_t 1_{[0,\tau_n]}(t)$, $M_t^n = M_{t\wedge\tau_n}$. Then $M^n \in \mathcal{M}_c^2$ and $f^n \in L^2(\mu_{M^n})$. (the predictability of f^n follows from the fact that $1_{[0,\tau_n]}(t)$ is a left continuous process and hence is predictable).

Let $X_t^n = \int_0^t f^n dM^n$. Then Lemma 1.21 implies that

$$P(X_{t\wedge\tau_n}^n = X_{t\wedge\tau_n}^{n+1} \quad \forall t) = 1.$$

Let $\Omega^0 = \{\omega : X_{t\wedge\tau_n(\omega)}^n(\omega) = X_{t\wedge\tau_n(\omega)}^{n+1}(\omega) \quad \forall t \ \forall n\}$. Then $P(\Omega^0) = 1$. Let us define $X_t(\omega) = X_t^n(\omega)$ for $\tau_{n-1}(\omega) \leq t < \tau_n(\omega)$, if $\omega \in \Omega^0$. For $\omega \notin \Omega^0$, let $X_t(\omega) = 0$. Then X_t is a continuous process, $X_{t\wedge\tau_n} = X_{t\wedge\tau_n}^n$ a.s. and hence X_t is a local martingale.

We define X_t to be the stochastic integral $\int_0^t f dM$.

Theorem 1.22 *Let M, N be continuous local martingales with $M_0 = 0 = N_0$ and let $f, g \in \mathcal{L}^0(M) \cap \mathcal{L}^0(N)$. Let $X_t = \int_0^t f dM$ (defined above). Then X is a continuous local martingale. Further,*

$$\langle X, X\rangle_t = \int_0^t (f_s)^2 d\langle M, M\rangle_s. \tag{1.63}$$

$$\langle X, N\rangle_t = \int_0^t f_s d\langle M, N\rangle_s. \tag{1.64}$$

$$\int_0^t (f+g) dM = \int_0^t f dM + \int_0^t g dM \tag{1.65}$$

and

$$\int_0^t f d(M+N) = \int_0^t f dM + \int_0^t f dN. \tag{1.66}$$

Proof We have already noted that X is a continuous local martingale. Since

$$(X_{t\wedge\tau_n}^n)^2 - \int_0^{t\wedge\tau_n} (f_s^n)^2 d\langle M^n, M^n\rangle_s$$

is a martingale and is equal to $(X_{t\wedge\tau_n})^2 - \int_0^{t\wedge\tau_n} (f_s)^2 d\langle M, M\rangle_s$ (by definition of τ_n, f^n, M^n, X^n, X) it follows that $X_t^2 - \int_0^t (f_s)^2 d\langle M, M\rangle_s$ is a local martingale and thus (1.63) is true. The validity of the remaining three assertions can be verified easily. □

We now record a consequence of the identification of the quadratic variation of a stochastic integral given above and (1.52). The inequality given below is known as the Kunita–Watanabe inequality. It was originally used in defining the integral.

Theorem 1.23 *Let U, V be continuous local martingales and let f, g be predictable processes such that $\int_0^T f_s^2 d\langle U, U\rangle_s < \infty$ and $\int_0^T g_s^2 d\langle V, V\rangle_s < \infty$ for all $T < \infty$. Then*

$$\left|\int_0^T f_s g_s d\langle U, V\rangle_s\right| \leq \int_0^T f_s^2 d\langle U, U\rangle_s \int_0^T g_s^2 d\langle V, V\rangle_s. \tag{1.67}$$

Proof Let $M_t = \int_0^t f dU$ and $N = \int_0^t g dV$. Then as noted in the previous theorem,

$$\langle M, M\rangle_t = \int_0^T f_s^2 d\langle U, U\rangle_s \tag{1.68}$$

$$\langle N, N\rangle_t = \int_0^T g_s^2 d\langle V, V\rangle_s \tag{1.69}$$

$$\langle M, N\rangle_t = \int_0^T f_s d\langle U, N\rangle_s \tag{1.70}$$

$$\langle U, N\rangle_t = \int_0^T g_s d\langle U, V\rangle_s. \tag{1.71}$$

Putting together (1.70) and (1.71) we can deduce that

$$\langle M, N\rangle_t = \int_0^T f_s g_s d\langle U, V\rangle_s. \tag{1.72}$$

The required inequality (1.67) is a consequence of (1.52), (1.68), (1.69) and (1.72). □

For later use, let us note here an estimate on the growth of the integral $\int f dM$.

Lemma 1.24 *Let M be a continuous local martingale with $M_0 = 0$ and $f \in \mathcal{L}^0(M)$. Then for all finite stopping times σ,*

$$\mathbb{E}\sup_{t\leq\sigma}\left|\int_0^t f dM\right|^2 \leq 4\mathbb{E}\int_0^\sigma f_s^2 d\langle M, M\rangle_s. \tag{1.73}$$

Proof Let $\sigma_k = \inf\{t : \int_0^t f_s^2 d\langle M, M\rangle_s \geq k\}$. Let $f_t^k = f_t 1_{[0,\sigma_k]}(t)$. Let $X_t = \int_0^t f dM$ and $X_t^k = \int_0^t f dM$. By Lemma 1.21, $X_{t\wedge\sigma_k} = X_{t\wedge\sigma_k}^k$. Also $\mathbb{E}\int_0^\infty (f_s^k)^2 d\langle M, M\rangle_s \leq k$ and hence $X^k \in \mathcal{M}_c^2$. Thus by Doob's maximal inequality

$$\begin{aligned}
\mathbb{E}\sup_{t\leq\sigma\wedge\sigma_k}(X_t^k)^2 &\leq 4.\mathbb{E}(X_{\sigma\wedge\sigma_k}^k)^2 \\
&= 4\mathbb{E}\int_0^{\sigma\wedge\sigma_k}(f_s^k)^2 d\langle M, M\rangle_s \\
&= 4\mathbb{E}\int_0^{\sigma\wedge\sigma_k}(f_s)^2 d\langle M, M\rangle_s \\
&\leq 4\mathbb{E}\int_0^{\sigma}(f_s)^2 d\langle M, M\rangle_s.
\end{aligned}$$

Hence

$$\mathbb{E} \sup_{t \leq \sigma \wedge \sigma_k} (X_t)^2 \leq 4\mathbb{E} \int_0^{\sigma} (f_s)^2 d\langle M, M \rangle_s.$$

The required estimate (1.73) now follows by using Fatou's lemma. □

We end this section with a version of the dominated convergence theorem.

Theorem 1.25 *Let (M_t) be a continuous local martingale and let $f^n, f \in \mathcal{L}^0(M)$ be such that $f^n \longrightarrow f$ pointwise. Suppose $|f^n| \leq g$ where $g \in \mathcal{L}^0(M)$. Then*

$$\sup_{t \leq T} \left| \int_0^t f^n dM - \int_0^t f dM \right| \longrightarrow 0 \textit{ in probability } \quad \forall T. \tag{1.74}$$

Proof Let us define $\tau_k = \inf\{t : \int_0^t (g_s)^2 d\langle M, M \rangle_s \geq k\}$. Then

$$\mathbb{E} \int_0^{\tau_k} (g_s^2) d\langle M, M \rangle_s \leq k. \tag{1.75}$$

The estimate (1.73) gives

$$\begin{aligned} b_{k,n} &:= \mathbb{E} \sup_{t \leq \tau_k} \left| \int_0^t f^n dM - \int_0^t f dM \right|^2 \\ &\leq 4\mathbb{E} \int_0^{\tau_k} |f_s^n - f_s|^2 d\langle M, M \rangle_s = d_{k,n}. \end{aligned}$$

The dominated convergence theorem for the Lebesgue–Stieltjes integrals along with $|f_n - f| \leq 2g$ and (1.75) imply that $d_{k,n} \longrightarrow 0$ as $n \longrightarrow 0$ for each k.

For $\varepsilon > 0$ and $\nu > 0$ given, first choose k such that $P(\tau_k < T) < \nu/2$ (can be done as $\tau_k \uparrow \infty$ a.s.) and n_0 such that for $n \geq n_0$, $d_{k,n} < \frac{1}{2}\nu\varepsilon^2$. Then

$$\begin{aligned} P\left(\sup_{t \leq T} \left| \int_0^t f^n dM - \int_0^t f dM \right|^2 > \varepsilon \right) &\leq \varepsilon^{\frac{1}{2}} b_{k,n} + P(\tau_k < T) \\ &\leq \nu, \end{aligned}$$

for $n \geq n_0$. This completes the proof. □

1.6 Stochastic integral w.r.t. continuous semimartingales

A continuous process (X_t) (on a probability space $(\Omega, \mathcal{F}, P)$ with filtration $(\mathcal{F}_t)$ satisfying the usual hypotheses) is said to be a continuous semimartingale if (X_t) can be written as $X_t = X_0 + M_t + A_t$, where (M_t) is a continuous local martingale with $M_0 = 0$ and (A_t) is a (necessarily continuous) process such that

$\text{Var}(A(\omega))_t < \infty$ for all $t < \infty$. As seen in the previous section, this decomposition is unique and is also called the canonical decomposition of the semimartingale (X_t).

More generally, the unique decomposition of a semimartingale $X_t = X_0 + M_t + A_t$, where M is a local martingale with $M_0 = 0$ and A is predictable with $\text{Var}(A(\omega))_t < \infty$ for all t, is called the canonical decomposition of X.

For a continuous semimartingale (X_t), let $\mathcal{L}^0(X)$ be the class of all predictable processes f such that

$$\int_0^t f_s^2 d\langle M, M\rangle_s < \infty, \quad \int_0^t |f_s| d|A|_s < \infty \text{ a.s.} \quad \forall t, \tag{1.76}$$

where $X_t = X_0 + M_t + A_t$ is the canonical decomposition of (X_t) and $|A|_t(\omega) := \text{Var}\,(A.(\omega))_t$. For $f \in \mathcal{L}^0(X)$, the stochastic integral $\int_0^t f dM$ has been defined in the previous section and the integral $\int_0^t f dA$ denotes the Lebesgue–Stieltjes integral. We define

$$\int_0^t f dX = \int_0^t f dM + \int_0^t f dA. \tag{1.77}$$

Before proceeding let us note that if Y is an r.c.l.l. adapted process, the process Y_-, defined by

$$(Y_-)_0 := 0, \quad (Y_-)_t = \lim_{s \nearrow t} Y_s,$$

is left continuous and hence predictable. Further, since

$$\int_0^t (Y_-)_s^2 d\langle M, M\rangle_s < \infty \quad \int_0^t |(Y_-)_s| d|A|_s < \infty \quad \text{for all } \omega,$$

$Y_- \in \mathcal{L}^0(X)$ for all semimartingales X.

The following properties of $\int f dX$ are easily deduced by decomposing X into $M + A$ and using the corresponding properties for the dM integral and the dA integral. Here, X, Y are continuous semimartingales.

(i) For $f, g \in \mathcal{L}^0(X)$

$$\int_0^t (f+g) dX = \int_0^t f dX + \int_0^t g dX.$$

(ii) For $f \in \mathcal{L}^0(X) \cap \mathcal{L}^0(Y)$

$$\int_0^t f d(X+Y) = \int_0^t f dX + \int_0^t f dY.$$

(iii) For $f \in \mathcal{S}$ given by

$$f_t(\omega) = U_0(\omega) 1_{\{0\}}(t) + \sum_{j=1}^{m-1} U_j(\omega) 1_{(s_j, s_{j+1}]}$$

where $0 = s_0 < s_1 < \cdots < s_m$, U_j is an $\mathcal{F}_{s_j}$-measurable, bounded random variable

$$\int_0^t f dX = U_0 X_0 + \sum_{j=1}^{m-1} U_j (X_{s_{j+1}\wedge t} - X_{s_j \wedge t}).$$

(iv) If $f \in \mathcal{L}^0(X)$, $g \in \mathcal{L}^0(Y)$ and σ is a stopping time such that

$$P(f_{t\wedge\sigma} = g_{t\wedge\sigma} \quad \forall t) = P(X_{t\wedge\sigma} = Y_{t\wedge\sigma} \quad \forall t) = 1,$$

then

$$P\Big(\int_0^{t\wedge\sigma} f dX = \int_0^{t\wedge\sigma} g dY \quad \forall t\Big) = 1. \tag{1.78}$$

We will now prove an analogue of the dominated convergence theorem (DCT) for the stochastic integral and then use it to deduce properties of the integral.

Theorem 1.26 *Let (X_t) be a continuous semimartingale and let $f^n, f \in \mathcal{L}^0(X)$ be such that $f^n \longrightarrow f$ pointwise. Suppose $|f^n| \leq g$ where $g \in \mathcal{L}^0(X)$. Then*

$$\sup_{t\leq T} \Big| \int_0^t f^n dX - \int_0^t f dX \Big| \longrightarrow 0 \textit{ in probability} \quad \forall T. \tag{1.79}$$

Proof Let $X_t = M_t + A_t$ be the canonical decomposition of X. It suffices to prove that (1.79) holds for $X = M$ and $X = A$. Now $g \in \mathcal{L}^0(X)$ implies that $\int_0^t |g_s| d|A|_s < \infty$ for all t, a.s. and hence the usual dominated convergence theorem implies (1.79) with A in place of X (and a.s. convergence in place of convergence in probability). By definition of $\mathcal{L}^0(X)$, we have that $g \in \mathcal{L}^0(M)$ as well and hence (1.79) with $X = M$ follows from Theorem 1.25. □

The dominated convergence theorem (DCT) allows one to use usual measure theoretic tools for dealing with stochastic integrals. We will illustrate this by two examples.

Theorem 1.27 *Let X be a continuous semimartingale, $f \in \mathcal{L}^0(X)$, g be a predictable process and $Y_t = \int_0^t f dX$. Then $g \in \mathcal{L}^0(Y)$ iff $fg \in \mathcal{L}^0(X)$ and*

$$\int_0^t g dY = \int_0^t f g dX \textit{ a.s.}. \tag{1.80}$$

Proof For the first part, let $X = M + A$ be the canonical decomposition of X. Then $Y = N + B$ is the canonical decomposition of Y where $N_t = \int_0^t f dM$, $B_t = \int_0^t f_s dA_s$. We note that $\langle N, N\rangle_t = \int_0^t (f_s)^2 d\langle M, M\rangle_s$ and further that $\mathrm{Var}(B.)_t = \int_0^t |f_s|\, d\, |A|_s$. We can easily verify the first part.

When $f, g \in \mathcal{S}$, (1.80) can be directly verified. Now, fix $g \in \mathcal{S}$ and let $\mathcal{H}$ be the class of bounded predictable processes f such that (1.80) holds. Then in view of the DCT, Theorem 1.26, $\mathcal{H}$ is closed under bounded pointwise convergence, and

$\mathcal{S} \subseteq \mathcal{H}$. Thus by the monotone class theorem (Theorem 1.2), we get that $\mathcal{H}$ is the class of bounded predictable processes, i.e., (1.80) holds for bounded predictable processes. By approximating f by f^n, defined by $f_t^n := (f_t \wedge n) \vee (-n)$, and once again using Theorem 1.26, we conclude that (1.80) holds for all $f \in \mathcal{L}^0(X)$, $g \in \mathcal{S}$.

Now fixing $f \in \mathcal{L}^0(X)$, we can similarly prove that (1.80) holds for all $g \in \mathcal{L}^0(Y)$. □

Let us note that the argument given above is the same as the corresponding result in measure theory. Our second application of the DCT is

Theorem 1.28 *Let X be a continuous semimartingale, let Y be an r.c.l.l. adapted process, let σ, τ be stopping times with $\sigma \le \tau$. Let g be defined by*

$$g_t(\omega) = Y_{\sigma(\omega)}(\omega) 1_{(\sigma(\omega), \tau(\omega)]}(t).$$

Then $g \in \mathcal{L}^0(X)$ and

$$\int_0^t g dX = Y_\sigma (X_{\tau \wedge t} - X_{\sigma \wedge t}).$$

Further, let $\{\tau_i\}$ be a sequence of stopping times such that $\tau_i \le \tau_{i+1}$ for all i. Let

$$f_t(\omega) = Y_0(\omega) 1_{\{0\}}(t) + \sum_{j=1}^{\infty} Y_{\tau_i(\omega)} 1_{(\tau_i(\omega), \tau_{i+1}(\omega)]}(t).$$

Then $f \in \mathcal{L}^0(X)$ and

$$\int_0^t f dX = Y_0 X_0 + \sum_{j=1}^{\infty} Y_{\tau_i} \Big(X_{\tau_{i+1} \wedge t} - X_{\tau_i \wedge t} \Big).$$

Proof It suffices to consider σ, τ bounded. We have noted earlier that g is predictable since it is adapted and left continuous. Let stopping times σ^m and τ^m be defined by

$$\sigma^m(\omega) = ([2^m \sigma(\omega)] + 1) 2^{-m}; \ \tau^m(\omega) = ([2^m \tau(\omega)] + 1) 2^{-m}.$$

Then $\sigma^m \downarrow \sigma$; $\tau^m \downarrow \tau$, $\sigma < \sigma^m$, $\tau < \tau^m$. Right continuity of Y implies that $g^m := Y_{\sigma^m} 1_{(\sigma^m, \tau^m]}$ converges pointwise to g. It is easy to see that $g^m \in \mathcal{S}$ and that

$$\int_0^t g^m dX = Y_{\sigma^m} (X_{\tau^m \wedge t} - X_{\sigma^m \wedge t}).$$

Thus

$$\int_0^t g^m dX \longrightarrow Y_\sigma (X_{\tau \wedge t} - X_{\sigma \wedge t}).$$

The first part of the result will follow from DCT (Theorem 1.26) once we show that $|g^m| \leq h$ for $h \in \mathcal{L}^0(X)$. Take $h_t(\omega) := \sup_{s<t} Y_s(\omega)$. Clearly, it is predictable since it is left continuous and adapted. It is easy to see that $h \in \mathcal{L}^0(X)$. Of course $|g^m| \leq h$. This completes the proof of the first part.

The second part with finitely many stopping times would follow from the linearity of the integral and the first part. From here we can use the dominated convergence Theorem (1.26) to complete the proof. The same function h defined above works as a dominating function. □

Let Y be an r.c.l.l. adapted process. We have already seen that $Y_- \in \mathcal{L}^0(X)$ for every semimartingale X. The next result shows that the stochastic integral $\int Y_- dX$ can be approximated via the usual Riemann sums over deterministic partitions or over partitions using stopping times.

Theorem 1.29 *Let X be a continuous semimartingale and let Y be an r.c.l.l. process. Let $\{\tau_i^n : i \geq 1\}_{n \geq 1}$ be a sequence of partitions of $(0, \infty)$ by stopping times (i.e., each τ_i^n is a stopping time) such that*

$$0 = \tau_0^n(\omega) < \tau_1^n(\omega) < \cdots < \tau_i^n(\omega) < \cdots ; \tau_i^n(\omega) \uparrow \infty \text{ as } i \longrightarrow \infty \quad \forall \omega \tag{1.81}$$

and

$$\lim_n \Big[\sup_{i \geq 0} (\tau_{i+1}^n - \tau_i^n) \Big] = 0. \tag{1.82}$$

Let

$$Z_t^n = \sum_{i=0}^{\infty} Y_{\tau_i^n} \Big(X_{\tau_{i+1}^n \wedge t} - X_{\tau_i^n \wedge t} \Big). \tag{1.83}$$

Then $\sup_{t \leq T} |Z_t^n - \int_0^t Y_- dX| \longrightarrow 0$ *in probability for every* $T < \infty$.

Proof Define Y^n by

$$Y_t^n = \sum_{i=0}^{\infty} Y_{\tau_i^n} 1_{(\tau_i^n, \tau_{i+1}^n]}(t). \tag{1.84}$$

As noted in (1.28), $Z_t^n = \int_0^t Y^n dX$. Clearly $Y^n \longrightarrow Y_-$ pointwise and if $h_t := \sup_{s<t} |Y_s|$, then h is also locally bounded and thus $h \in \mathcal{L}^0(X)$. Now $|Y_t^n| \leq \sup_{0 \leq s < t} |Y_s| = h_t$. Thus, in view of DCT, Theorem 1.26

$$\sup_{t \leq T} \Big| \int_0^t Y^n dX - \int_0^t Y_- dX \Big| \longrightarrow 0$$

in probability for every $T < \infty$. This completes the proof. □

As a consequence of the preceding result we have:

Theorem 1.30 *Let X be a continuous semimartingale and let Y be an r.c.l.l. process. Then*

$$\sum_{i=0}^{n-1} Y_{\frac{iT}{n}}(X_{\frac{(i+1)T}{n}} - X_{\frac{iT}{n}}) \longrightarrow \int_0^T Y_- dX \tag{1.85}$$

in probability for every $T < \infty$.

Our next result is on the quadratic variation of a semimartingale.

Theorem 1.31 *Let X be a continuous semimartingale and let $\{\tau_i^n : i \geq 1\}$ be a sequence of partitions of $[0, \infty)$ by stopping times satisfying (1.81) and (1.82). Let*

$$Q_t^n = \sum_{i=1}^{\infty} \left(X_{\tau_{i+1}^n \wedge t} - X_{\tau_i^n \wedge t} \right)^2.$$

Then Q_t^n converges in probability to a continuous increasing process, denoted by $[X, X]_t$. The process $[X, X]_t$ satisfies

$$X_t^2 = X_0^2 + 2 \int_0^t X_- dX + [X, X]_t \tag{1.86}$$

and hence does not depend on the choice of the partitions $\{\tau_i^n\}$. Further

$$\sup_{t \leq T} |Q_t^n - [X, X]_t| \longrightarrow 0 \textit{ in probability.}$$

Proof Let

$$W_t^n := \sum_{i=1}^{\infty} X_{\tau_i^n \wedge t} \left(X_{\tau_{i+1}^n \wedge t} - X_{\tau_i^n \wedge t} \right).$$

By Theorem 1.29, $\sup_{t \leq T} |W_t^n - \int_0^t X_- dX| \longrightarrow 0$ in probability for every $T < \infty$. Using the identity

$$b^2 - a^2 = 2a(b - a) + (b - a)^2$$

for $b = X_{\tau_{i+1}^n \wedge t}$, $a = X_{\tau_i^n \wedge t}$ and summing over i, one gets

$$X_t^2 - X_0^2 = 2W_t^n + Q_t^n.$$

It follows that Q_t^n converges in probability to a process defined to be $[X, X]_t$ and also that (1.86) holds. Since $Q_{\tau_i^n}^n \leq Q_t^n$ for $t \geq \tau_i^n$, it follows that the limit is an increasing process and is continuous by (1.86). □

Let M be a continuous local martingale with $M_0 = 0$. It follows that $[M, M]_t$ is an increasing process with $M_0 = 0$ and that

$$M_t^2 - [M, M]_t = 2 \int_0^t M_- dM$$

is a local martingale. Hence by Lemma 1.15

$$[M, M]_t = \langle M, M\rangle_t. \tag{1.87}$$

For continuous semimartingales X, Y, let us define

$$[X, Y]_t = \frac{1}{4}\Big\{[X+Y,\ X+Y]_t - [X-Y,\ X-Y]_t\Big\}. \tag{1.88}$$

Using the identity $ab = \frac{1}{4}\Big\{(a+b)^2 - (a-b)^2\Big\}$, and the bilinearity of the integral $\int f dZ$ and the relation (1.86) for $X, Y, X+Y, X-Y$ one can deduce that

$$X_t Y_t = X_0 Y_0 + \int_0^t X_- dY + \int_0^t Y_- dX + [X, Y]_t. \tag{1.89}$$

This formula is called the integration by parts formula. This in turn gives the symmetry of $[X, Y]$ as well as bilinearity of $[X, Y]$.

1.7 Integration w.r.t. semimartingales

In this section, we will briefly discuss the construction and properties of the stochastic integral $\int f dX$ when X is an r.c.l.l. semimartingale. A complete treatment of this requires a large number of results from the so-called general theory of processes. We will state below without proof a few of these results that are necessary for our treatment. These results can be found in (Dellacherie and Meyer, 1978; Jacod, 1979; Metivier, 1982; Protter, 1990).

Theorem 1.32 [Doob–Meyer decomposition theorem]
(a) Let M be a local L^2-martingale with $M_0 = 0$. There exists a predictable increasing process with $A_0 = 0$ such that $M_t^2 - A_t$ is a local martingale.
(b) If N is a predictable local martingale such that $N_0 = 0$ and $N \in \mathcal{V}$, then

$$P(N_t = 0 \quad \forall t) = 1.$$

As a consequence, the process (A_t) in part (a) above is unique in the class of predictable increasing processes (B_t) satisfying $B_0 = 0$.

The process A_t is denoted by $\langle M, M\rangle_t$. For local L^2-martingales M, N, $\langle M, N\rangle_t$ is defined by (1.50). All the observations made about $\langle M, N\rangle_t$ at the end of section 1.4, including (1.51), (1.52) are valid. We need to use part (b) in Theorem 1.32 in place of Lemma 1.18 and remember that $\langle M, N\rangle_t$ is predictable but may not be continuous. We record this observation here: Let M, N be r.c.l.l. local L^2-martingales. Then

$$|\langle M^1, M^2\rangle_t| \le (\langle M^1, M^1\rangle_t \cdot \langle M^2, M^2\rangle_t)^{\frac{1}{2}} \quad \forall t \ a.s. \tag{1.90}$$

The other two results that we need are stated below.

Theorem 1.33 *Let A be a predictable process with $A_0 = 0$ and $A \in \mathcal{V}$. Then $(|A|_t)$ is locally integrable, i.e., there exist stopping times (τ_n) increasing to ∞ such that*

$$\mathbb{E}(Var(A_{\tau_n})) < \infty \quad \forall n.$$

Theorem 1.34 *Let X be a semimartingale. Then X admits a decomposition $X = M + A$ with M being a local L^2-martingale with $M_0 = 0$ and $A \in \mathcal{V}$.*

Note that in the result given above, no assertion about the uniqueness of the decomposition is made. Also, this result implies in particular that every local martingale M admits a decomposition $M + A$ as in Theorem 1.34.

We need one more observation, which is given below along with its proof.

Lemma 1.35 *Let f be predictable and let A be a predictable increasing process such that*

$$\int_0^t |f_s(\omega)| dA_s(\omega) < \infty \quad \text{for all } (t, \omega). \tag{1.91}$$

Then $B_t = \int_0^t f_s(\omega) dA_s(\omega)$ is a predictable locally integrable process.

Proof Fix $f \in \mathcal{S}$ given by (1.53). $\int_0^t f dX$ can be defined for all r.c.l.l. processes X by (1.54) (with X replacing M). The class $\mathcal{H}$ of the process X such that $\int_0^t f dX$ is predictable contains all continuous processes, since then $\int_0^t f dX$ is continuous. Using Theorem 1.2, it follows that it includes all bounded predictable processes. By localizing, it follows that the result is true if $f \in \mathcal{S}$. Another application of Theorem 1.2 yields that the result is true for all bounded predictable processes f. A truncation argument (using $f_k = (f \wedge k) \vee (-k)$) now completes the proof. □

We can now proceed exactly as in section 1.5 and define $X_t = \int_0^t f dM$ when M is a local L^2-martingale, f is predictable and

$$\int_0^t f_s^2 d\langle M, M\rangle_s < \infty \quad a.s. \quad \forall t. \tag{1.92}$$

Let $\mathcal{L}^0(M)$ denote the class of predictable f satisfying (1.92). Most of the arguments in section 1.5 hold verbatim. The exceptions are noted here. The integral is no longer a continuous process but an r.c.l.l. process that is a local L^2-martingale. Also we need to use Lemma 1.35 and part (b) in Theorem 1.32 to identify $\langle X, N\rangle$ (in place of Lemma 1.18). Also for an f satisfying (1.92), note that

$$B_t = \int_0^t f_s^2 d\langle M, M\rangle_s$$

is a predictable increasing process with $B_0 = 0$. In view of Theorem 1.34, we can get a sequence $\{\sigma_n\}$ of stopping times increasing to ∞ such that $\mathbb{E}B_{\sigma_n} < \infty$. Let

$\{\sigma_n'\}$ be a localizing sequence for the local L^2-martingale M. Instead of defining τ_n by (1.62) take $\tau_n = \sigma_n \wedge \sigma_n'$.

We note here that Theorem 1.22, Theorem 1.23, Lemma 1.24 and Theorem 1.25 continue to be true if M, N are local L^2-martingales.

Let us now come to integration w.r.t. a semimartingale X. For a decomposition $X = M + A$ as in Theorem 1.34, we could define $\int f dX$ via (1.77). However, we need to ensure that the integral so defined does not depend upon the decomposition.

Let $X = M + A = N + B$ where M, N are local L^2-martingales and A, B are processes such that $Var(A(\omega))_t < \infty$, $Var(B(\omega))_t < \infty$ for all t, ω. Clearly if $f \in \mathcal{S}$, then

$$\int_0^t f dM + \int_0^t f dA = \int_0^t f dN + \int_0^t f dB. \tag{1.93}$$

Using the functional version of the monotone class theorem (Theorem 1.2), Theorem 1.25 (DCT for stochastic integral) and the DCT for the Riemann–Stieltjes integral, we conclude that (1.93) holds for all bounded predictable processes. Now via truncation we can check that (1.93) holds for predictable processes f, satisfying

$$\int_0^t f_s^2 d\langle M, M\rangle_s < \infty \quad \text{a.s.,} \quad \int_0^t |f_s| d|A|_s < \infty \quad \text{a.s.} \quad \forall t \tag{1.94}$$

and

$$\int_0^t f_s^2 d\langle N, N\rangle_s < \infty \quad \text{a.s.,} \quad \int_0^t |f_s| d|B|_s < \infty \quad \text{a.s.} \quad \forall t, \tag{1.95}$$

where $|A|_t = Var(A)_t$, $|B|_t = Var(B)_t$.

Let $\mathcal{L}^0(X)$ denote the class of predictable processes f satisfying (1.94) for some decomposition $X = M + A$, as in Theorem 1.34. In view of the observation made earlier, we can define $\int f dX$ for $f \in \mathcal{L}^0(X)$ unambiguously. All the results in section 1.6 carry over verbatim for an (r.c.l.l.) semimartingale X, since the proofs of these results only used the dominated convergence theorem. We record these observations here for future reference.

Theorem 1.36 *Let X be an r.c.l.l. semimartingale, $f \in \mathcal{L}^0(X)$, let g be a predictable process and $Y_t = \int_0^t f dX$. Then $g \in \mathcal{L}^0(Y)$ iff $fg \in \mathcal{L}^0(X)$ and*

$$\int_0^t g dY = \int_0^t fg dX \text{ a.s.}. \tag{1.96}$$

Theorem 1.37 *Let X be an r.c.l.l. semimartingale, let Y be an r.c.l.l. adapted process, let σ, τ be stopping times with $\sigma \le \tau$. Let g be defined by*

$$g_t(\omega) = Y_{\sigma(\omega)}(\omega) 1_{(\sigma(\omega), \tau(\omega)]}(t).$$

Then $g \in \mathcal{L}^0(X)$ *and*

$$\int_0^t g dX = Y_\sigma (X_{\tau\wedge t} - X_{\sigma\wedge t}).$$

Further let $\{\tau_i\}$ *be a sequence of stopping times such that* $\tau_i \le \tau_{i+1}$ *for all* i.
Let

$$f_t(\omega) = Y_0(\omega) 1_{\{0\}}(t) + \sum_{j=1}^{\infty} Y_{\tau_i(\omega)} 1_{(\tau_i(\omega),\tau_{i+1}(\omega)]}(t).$$

Then $f \in \mathcal{L}^0(X)$ *and*

$$\int_0^t f dX = Y_0 X_0 + \sum_{j=1}^{\infty} Y_{\tau_i}\Big(X_{\tau_{i+1}\wedge t} - X_{\tau_i \wedge t}\Big).$$

Theorem 1.38 *Let* X *be an r.c.l.l. semimartingale and let* Y *be an r.c.l.l. process. Let* $\{\tau_i^n : i \ge 1\}_{n\ge 1}$ *be a sequence of partitions of* $(0, \infty)$ *by stopping times (i.e., each* τ_i^n *is a stopping time) such that*

$$0 = \tau_0^n(\omega) < \tau_1^n(\omega) < \cdots < \tau_i^n(\omega) < \cdots ; \tau_i^n(\omega) \uparrow \infty \text{ as } i \longrightarrow \infty \quad \forall\omega \tag{1.97}$$

and

$$\lim_n \Big[\sup_{i\ge 0}(\tau_{i+1}^n - \tau_i^n)\Big] = 0. \tag{1.98}$$

Let

$$Z_t^n = \sum_{i=0}^{\infty} Y_{\tau_i^n}\Big(X_{\tau_{i+1}^n\wedge t} - X_{\tau_i^n \wedge t}\Big) \tag{1.99}$$

Then $\sup_{t\le T}|Z_t^n - \int_0^t Y_- dX| \longrightarrow 0$ *in probability for every* $T < \infty$.

Theorem 1.39 *Let* X *be semimartingale and let* $\{\tau_i^n : i \ge 1\}$ *be a sequence of partitions of* $[0, \infty)$ *by stopping times satisfying (1.97) and (1.98).*
Let

$$Q_t^n = \sum_{i=1}^{\infty}\Big(X_{\tau_{i+1}^n\wedge t} - X_{\tau_i^n\wedge t}\Big)^2.$$

Then Q_t^n *converges in probability to an increasing process, denoted by* $[X, X]_t$.
The process $[X, X]_t$ *satisfies*

$$X_t^2 = X_0^2 + 2\int_0^t X_- dX + [X, X]_t \tag{1.100}$$

and hence does not depend on the choice of the partitions $\{\tau_i^n\}$. *Further*

$$\sup_{t\le T}|Q_t^n - [X, X]_t| \longrightarrow 0 \text{ in probability.}$$

Proof Let

$$W_t^n := \sum_{i=1}^{\infty} X_{\tau_i^n \wedge t}\left(X_{\tau_{i+1}^n \wedge t} - X_{\tau_i^n \wedge t}\right).$$

By Theorem 1.38, $\sup_{t \le T} |W_t^n - \int_0^t X_- dX| \longrightarrow 0$ in probability for every $T < \infty$. Again using the identity

$$b^2 - a^2 = 2a(b-a) + (b-a)^2$$

for $b = X_{\tau_{i+1}^n \wedge t}$, $a = X_{\tau_i^n \wedge t}$, and summing over i, one gets

$$X_t^2 - X_0^2 = 2W_t^n + Q_t^n.$$

It follows that Q_t^n converges in probability to a process defined to be $[X,X]_t$ and also that (1.100) holds. Since $Q_{\tau_i^n}^n \le Q_t^n$ for $t \ge \tau_i^n$, it follows that the limit is an increasing process. □

For semimartingales X, Y, we define the cross-quadratic variation $[X,Y]$ by (1.88) as for continuous semimartingales and can deduce that the integration by parts formula (1.89) is valid here as well. In particular, for local L^2-martingales M, N,

$$M_t N_t = \int_0^t M_- dN + \int_0^t N_- dM + [M,N]_t$$

and in particular, $M_t N_t - [M,N]_t$ is a local martingale. Here, $M_t N_t - \langle M,N\rangle_t$ is also a local martingale. However, $[M,N]$ need not be predictable and hence here we cannot deduce that $[M,N] = \langle M,N\rangle$.

Let M, N be local martingales. We have noted above that if they are local L^2-martingales, then $M_t N_t - [M,N]_t$ is a local martingale. The next result shows that this is valid for all local martingales M, N.

Theorem 1.40 *Let M, N be local martingales and let f be a predictable locally bounded process, i.e., $f 1_{[0,\sigma_n]}$ is bounded for a sequence of stopping times σ_n increasing to infinity. Then $Y_t = \int_0^t f dM$ is a local martingale. As a consequence,*

$$M_t N_t - [M,N]_t$$

is a local martingale.

Proof We need to use Theorem 1.34 for the local martingale M and write it as $M = U + A$ where U is a local L^2 martingale and $A \in \mathcal{V}$. It follows that A is also a local martingale. Since $\int f dU$ is a local martingale, we need to prove that $\int f dA$ is a local martingale.

Let τ_n^1 be defined by

$$\tau_n^1 = \inf\{t \ge 0 : |A|_t \ge n\}$$

and let τ_n^2 be a sequence such that $A_{t\wedge\tau_n^2}$ is a martingale for every n and τ_n^2 increasing to infinity. Let $\tau_n = \sigma_n \wedge \tau_n^1 \wedge \tau_n^2$. Then τ_n increases to infinity. Note that for $t < \tau_n$, $|A|_t \le n$ and hence $|A|_{\tau_n} \le 2n + |A_{\tau_n}|$. Thus $|A|_{t\wedge\tau_n}$ is integrable.

Fix n and let us write $g = f1_{[0,\tau_n]}$ and $B_t = A_{t\wedge\tau_n}$. Since $Y_{t\wedge\tau_n} = \int_0^t gdB$, if we show that $\int_0^t gdB$ is a martingale, then it would follow that Y is a local martingale. Note that g is uniformly bounded and $|B|_t$ is integrable. Let $\mathcal{H}$ be the class of bounded predictable processes h such that $\int_0^t hdB$ is a martingale. Then $\mathcal{H}$ includes all simple predictable processes since B is a martingale and $\mathcal{H}$ is closed under bounded pointwise convergence since $|B|_t$ is integrable. Hence $\mathcal{H}$ contains all bounded predictable processes. As noted above, this proves that Y is a local martingale.

For the other part, note that M_-, N_- are locally bounded and from what has just been shown, it follows that $\int_0^t M_-dN$ and $\int_0^t N_-dM$ are local martingales. The result now follows from the identity

$$M_tN_t - [M,N]_t = \int_0^t M_-dN + \int_0^t N_-dM. \qquad \square$$

For an r.c.l.l. process X, let $\Delta X = X - X_-$. Thus $\Delta X_t \neq 0$ if and only if the process has a jump at time t. Since X is assumed to be r.c.l.l., it follows that for any fixed ω, $\Delta X_t(\omega) \neq 0$ for at most countably many t's. We can deduce the following results on jumps of semimartingales and stochastic integrals.

Theorem 1.41 *Let X be a semimartingale and let Y be an r.c.l.l. process. Let $Z = \int Y_-dX$. Then*
(i) $\Delta Z = Y_-(\Delta X)$
(ii) $\Delta[X,X] = (\Delta X)^2$
(iii) $\sum_{0<s\le t}(\Delta X)^2 \le [X,X]_t$.

Proof Let Y^n, Z^n be defined by (1.84) and (1.83), respectively. Then $Y^n \to Y_-$ and $\sup_{t\le T}\left|Z_t^n - \int_0^t Y^ndX\right| \to 0$ in probability (see Theorem 1.38). Since

$$\Delta Z^n = Y^n\Delta X$$

part (i) of the theorem follows. For (ii) note that

$$X_t^2 = X_0^2 + 2\int_0^t X - dX + [X,X]_t$$

and hence

$$\Delta(X^2) = 2X_-(\Delta X) + \Delta[X,X]_t.$$

Note that $X_t^2 - X_{t-}^2 - 2X_{t-}(X_t - X_{t-}) = (X_t - X_{t-})^2$. Hence we get

$$(\Delta X_t)^2 = \Delta[X,X]_t.$$

(iii) follows from (i) since $[X, X]$ is an increasing process. □

Remark It follows as a consequence of (iii) above that for a semimartingale X,

$$\sum_{0 \langle s \leq t} (\Delta X_s)^2 < \infty \quad \text{a.s.} \quad \forall t. \tag{1.101}$$

The following result on quadratic variation of semimartingales is very useful in applications of Itô's formula.

Lemma 1.42 *Suppose X is a semimartingale and $A \in \mathcal{V}$.*

(a) *If A is continuous, then* $[X, A] = 0$.

(b) *If X is continuous, then* $[X, A] = 0$.

(c) $[X, A] = \sum_{0<s\leq t} (\Delta X)_s (\Delta A)_s$.

Proof Fix t and let $t_i^n = it2^{-n}$, $0 \leq i \leq 2^n$, $n \geq 1$. From Lemma (2.3), it follows that

$$[X, A]_t = \lim_n \sum_{i=0}^{2^n-1} (X_{t_{i+1}^n} - X_{t_i^n})(A_{t_{i+1}^n} - A_{t_i^n}), \tag{1.102}$$

where the limit above is in probability. On the other hand, $A \in \mathcal{V}$ implies that

$$\lim_n \sum_{i=0}^{2^n-1} X_{t_{i+1}^n} (A_{t_{i+1}^n} - A_{t_i^n}) = \int_0^t X_s dA_s$$

and

$$\lim_n \sum_{i=0}^{2^n-1} X_{t_i^n} (A_{t_{i+1}^n} - A_{t_i^n}) = \int_0^t X_{s-} dA_s.$$

These identities give us (c) above and (a), (b) are special cases of (c). □

As a consequence of this lemma, it follows that if $X = M + A$ where M is a continuous local martingale and $A \in \mathcal{V}$ is a continuous process, then

$$[X, X] = \langle M, M \rangle \tag{1.103}$$

since $[X, X] = [M, M] + [M, A] + [A, M] + [A, A] = [M, M] = \langle M, M \rangle$.

We will end this section with a result connecting the cross-quadratic variation $[X, Y]$ between two semimartingales and the stochastic integral.

Theorem 1.43 *Let X, Y be semimartingales and let $f \in \mathcal{L}^0(X)$. Then*

$$[\int_0^\bullet f dX, Y]_t = \int_0^t f d[X, Y]. \tag{1.104}$$

Proof Let $Z_t = \int_0^t f dX$ and let $\xi_t = Z_t Y_t$ and $\eta_t = X_t Y_t$. Using the integration by parts formula (1.89), the equation (1.104) can be equivalently written as

$$Z_t Y_t = f_0 X_0 Y_0 + \int_0^t Z_- dY + \int_0^t f Y_- dX + \int_0^t f d[X, Y]. \tag{1.105}$$

Here we have also used the fact that $\int_0^t Y_- dZ = \int_0^t f X_- dX$. Now the equation (1.105) is easily checked for simple predictable functions f by first taking f to be $U 1_{(r,s]}$ with U $\mathcal{F}_r$-measurable, and then using the linearity of the integrals. Now the class of bounded predictable processes f such that (1.105) holds is easily seen to be closed under bounded pointwise convergence and thus (1.105) is true for bounded predictable processes (in view of Theorem 1.2). The general case now follows by truncation and the dominated convergence theorem, Theorem 1.25. □

2

Itô's Formula and its Applications

2.1 Preliminaries

Itô's formula is the change of variable formula for the stochastic integral. We will see that the usual change of variable formula does not hold for the stochastic integral. In fact, we have seen that

$$\int_0^t W dW = \frac{1}{2}(W_t^2 - t)$$

where as the usual change of variable formula would have given $\frac{1}{2}W_t^2$. The additional term $-\frac{1}{2}t$ is often called the Itô correction.

Itô's formula is one of the the most useful tools in stochastic calculus. It gives a representation for $f(X_t^1, X_t^2, \ldots, X_t^d)$, where the leading term corresponds to the usual change of variable formula along with a correction term.

We have already seen two special cases: (1.86) for $f(x) = x^2$ and (1.89) for $f(x^1, x^2) = x^1 x^2$. Here is the formula for the case of continuous semimartingales.

Let $X^1, X^2, \ldots, X^d$ be continuous semimartingales and let $f \in C^{1,2}([0, \infty) \times \mathbb{R}^d)$. Then

$$\begin{aligned} f(t, X_t) &= f(0, X_0) + \int_0^t f_t(s, X_s)ds + \textstyle\sum_{j=1}^d \displaystyle\int_0^t f_j(s, X_s)dX_s^j \\ &\quad + \frac{1}{2}\textstyle\sum_{j=1}^d \sum_{k=1}^d \displaystyle\int_0^t f_{jk}(s, X_s)d[X^j, X^k]_s. \end{aligned}$$

Here,

$$f_t = \frac{\partial f}{\partial t}, \quad f_j = \frac{\partial f}{\partial x^j} \quad \text{and} \quad f_{jk} = \frac{\partial^2 f}{\partial z^j \partial x^k}.$$

We begin with the proof of the Itô formula in a special case. For better understanding, we will take a one-dimensional case and assume that f is thrice continuously differentiable.

Theorem 2.1 *Let X be a continuous semimartingle and let f be a thrice continuously differentiable function on $\mathbb{R}$. Then*

$$f(X_t) = f(X_0) + \int_0^t f'(X_s)dX_s + \frac{1}{2}\int_0^t f''(X_s)d[X,X]_s. \tag{2.1}$$

Let us recall here the following result on approximation of a stochastic integral via Riemann sums. This is a simple consequence of the dominated convergence theorem for stochastic integrals. (See Theorem 1.29).

Let $\{t_i^n : 0 \le i \le k_n\}$ be any sequence of partitions of $[0, T]$ with

$$\max_{i \le k_n} |t_i^n - t_{i-1}^n| \longrightarrow 0.$$

Let X be a continuous semimartingale and let Y be a continuous (adapted) process. Then

$$\sum_{i=1}^{k_n} Y_{t_{i-1}^n}(X_{t_i^n} - X_{t_{i-1}^n}) \longrightarrow \int_0^T Y dX \text{ in probability.} \tag{2.2}$$

Proof (of Theorem 2.1) It suffices to prove (2.1) for a bounded semimartingale. The general case can be deduced by considering $X_t^k = X_{t\wedge\tau_k}$ where $\tau_k = \inf\{t : |X_t| \ge k\}$.

So let us assume that $|X_s| \le M,\ 0 \le s \le T$. Let $t_i^n = \frac{i}{n} \cdot t, 0 \le i \le n$. By Taylor's theorem, we have

$$\begin{aligned} f(X_{t_{i+1}^n}) - f(X_{t_i^n}) &= f'(X_{t_i^n})(X_{t_{i+1}^n} - X_{t_i^n}) + \frac{1}{2} f''(X_{t_i^n})(X_{t_{i+1}^n} - X_{t_{i-1}^n})^2 \\ &\quad + \frac{1}{3!} f'''(\xi_i^n)(X_{t_{i+1}^n} - X_{t_i^n})^3. \end{aligned}$$

Summing over $i = 0$ to $n-1$, we write (using $(b-a)^2 = b^2 - a^2 - 2a(b-a)$),

$$f(X_t) - f(X_0) = H_t^n + G_t^n + I_t^n + R_t^n \tag{2.3}$$

where

$$
\begin{aligned}
H_t^n &= \sum_{i=0}^{n-1} f'(X_{t_i^n})(X_{t_{i+1}^n} - X_{t_i^n}) \\
G_t^n &= \sum_{i=0}^{n-1} \frac{1}{2} f''(X_{t_i^n})(X^2_{t_{i+1}^n} - X^2_{t_i^n}) \\
I_t^n &= -\sum_{i=0}^{n-1} f''(X_{t_i^n}) X_{t_i^n}(X_{t_{i+1}^n} - X_{t_i^n}) \\
R_t^n &= \sum_{i=0}^{n-1} \frac{1}{3!} f'''(\xi_i^n)(X_{t_{i+1}^n} - X_{t_i^n})^3.
\end{aligned}
$$

Recalling that X_t^2 is also a semimartingale (in view of the fact that $X_t^2 = X_0^2 + 2\int_0^t X_s dX_s + [X,X]_t$), we get using (2.2)

$$
\begin{aligned}
H_t^n &\longrightarrow \int_0^t f'(X_s) dX_s \\
G_t^n &\longrightarrow \frac{1}{2}\int_0^t f''(X_s) d(X^2)_s. \\
I_t^n &\longrightarrow -\int_0^t f''(X_s) X_s dX_s.
\end{aligned}
$$

Since $X_t^2 = X_0^2 + 2\int_0^t X_s dX_s + [X,X]_t$, we have that

$$
\frac{1}{2}\int_0^t f''(X_s) d(X^2)_s = \int_0^t f''(X_s) X_s dX_s + \frac{1}{2}\int_0^t f''(X_s) d[X,X]_s,
$$

and the required result (2.1) follows from (2.3) once we prove that $R_t^n \longrightarrow 0$ in probability. Let C be the upper bound of f''' on $[-M, M]$. C is finite as f''' is continuous. Then

$$
\begin{aligned}
|R_t^n| &\leq C.\sum_{i=0}^{n-1} |X_{t_{i+1}^n} - X_{t_i^n}|^3 \\
&\leq C.\max_{o\leq i<n} |X_{t_{i+1}^n} - X_{t_i^n}| \sum_{i=0}^{n-1} (X_{t_{i+1}^n} - X_{t_i^n})^2 \\
&\longrightarrow 0 \text{ in probability}
\end{aligned}
$$

since $\max_{0\leq i<n} |X_{t_{i+1}^n} - X_{t_i^n}| \longrightarrow 0$ in probability and $\sum_{i=0}^{n-1}(X_{t_{i+1}^n} - X_{t_i^n})^2 \longrightarrow [X,X]_t$ in probability. This completes the proof. □

Remark Let $X_t = X_0 + M_t + A_t$ be the canonical decomposition of X, where M_t is a (continuous) local martingale and A_t is a process with bounded variation paths. Then

$$[X, X]_t = [M, M]_t = \langle M, M \rangle_t.$$

The Ito formula thus takes the form

$$f(X_t) = f(X_0) + \int_0^t f'(X_s) dX_s + \frac{1}{2} \int_0^t f''(X_s) d\langle M, M \rangle_s.$$

We now come to the proof of the formula in the multidimensional case and remove the requirement that f is thrice differentiable, twice differentiable is enough.

The proof given here, like the one given above in the special case, is based on the Taylor series expansion of a function. This idea is classical and the proof given here is a simplification of the proof presented in (Metivier, 1982). We state the required version of Taylor's theorem, with a proof. Here, $|\cdot|$ denotes the Euclidian norm on $\mathbb{R}^d$, and $C^{1,2}([0,\infty) \times \mathbb{R}^d)$ denotes the class of functions $f(t,x)$ that are once continuously differentiable in t and twice continuously differentiable in x. Also, for $f \in C^{1,2}([0,\infty) \times \mathbb{R}^d)$, f_t denotes the partial derivative of f in the t variable, f_j denotes the partial derivative of f w.r.t. the jth coordinate of x and f_{jk} denotes the partial derivative of f_j w.r.t. the kth coordinate of x.

Lemma 2.2 *Let $f \in C^{1,2}([0,\infty) \times \mathbb{R}^d)$. Define $h : [0,T] \times \mathbb{R}^d \times \mathbb{R}^d \to \mathbb{R}^d$ by*

$$\begin{aligned} h(t,y,x) &= f(t,y) - f(t,x) - \textstyle\sum_j (y^j - x^j) f_j(t,x) \\ &\quad - \frac{1}{2} \textstyle\sum_{jk} (y^j - x^j)(y^k - x^k) f_{jk}(t,x), \end{aligned} \tag{2.4}$$

where $y = (y^1, \ldots, y^d), x = (x^1, \ldots, x^d) \in \mathbb{R}^d$. (Here and in the proof, the sum over j and k is from 1 to d.)

For $K < \infty$ and $\delta > 0$, let

$$\Gamma(K,\delta) = \sup\{|h(t,y,y)||y-x|^{-2} : t \le K, |x| \le K, |y| \le K, |x-y| \le \delta\}$$

$$\text{and } \Lambda(K) = \sup\{|h(t,x,y)||y-x|^{-2} : t \le K, |x| \le K, |y| \le K\}.$$

Then $\Lambda(K) < \infty$ and $\lim_{\delta \to 0} \Gamma(K,\delta) = 0$ for all $K < \infty$.

Proof Fix $x, y \in \mathbb{R}^d$ and $t \in [0,\infty)$ and let $z = y - x$. For $0 \le s \le 1$, define

$$g(s) = f(t, x+sz) - f(t,x) - s \textstyle\sum_j z^j f_j(t,x) - \frac{s^2}{2} \sum_{jk} z^j z^k f_{jk}(t,x).$$

Then $g(0) = 0 = g'(0)$ and hence by Taylor's theorem

$$\begin{aligned} h(t,x,y) &= g(1) \\ &= \int_0^1 (1-s) g''(s) dt \\ &= \int_0^1 (1-s) \big(\textstyle\sum_{jk} (f_{jk}(t, x+sz) - f_{jk}(t,x)) z^j z^k\big) ds. \end{aligned}$$

The desired estimates follow from this. □

We will also need the following result.

Lemma 2.3 *Let $\{\tau_i^n : i \geq 1\}, n \geq 1$ be a sequence of partitions of $[0, \infty)$ via stopping times satisfying (1.81), (1.82). Let X^1, X^2 be continuous semimartingales and let Y be an r.c.l.l. process. Let*

$$U_t^n = \textstyle\sum_{i=0}^{\infty} Y_{\tau_i^n}(X^1_{\tau_{i+1}^n \wedge t} - X^1_{\tau_i^n \wedge t})(X^2_{\tau_{i+1}^n \wedge t} - X^2_{\tau_j^n \wedge t}).$$

Then for all $T < \infty$

$$\sup_{t \leq T} \Big| U_t^n - \int_0^t Y_- d[X^1, X^2] \Big| \to 0 \quad \textit{in probability},$$

where $[X^1, X^2]$ is the cross quadratic variation between the semimartingales X^1, X^2.

Proof Let us write $U_t^n = R_t^n - V_t^n - W_t^n$ where

$$\begin{aligned} V_t^n &= \textstyle\sum_{i=1}^{\infty} Y_{\tau_i^n} X^1_{\tau_i^n}(X^2_{\tau_{i+1}^n \wedge t} - X^2_{\tau_i^n \wedge t}) \\ W_t^n &= \textstyle\sum_{i=0}^{\infty} Y_{\tau_i^n} X^2_{\tau_i^n}(X^1_{\tau_{i+1}^n \wedge t} - X^1_{\tau_i^n \wedge t}) \\ R_t^n &= \textstyle\sum_{i=0}^{\infty} Y_{\tau_i^n}(X^1_{\tau_{i+1}^n \wedge t} X^2_{\tau_{i+1}^n \wedge t} - X^1_{\tau_i^n \wedge t} X^2_{\tau_i^n \wedge t}). \end{aligned}$$

By Theorem 1.29 it follows that $\sup_{t \leq T} |U_t^n - U_t| \to 0$ in probability where

$$U_t = \int_0^t Y_- d(X^1 X^2) - \int_0^t Y_- X^1 dX^2 - \int_0^t Y_- X^2 dX^1.$$

Using Theorem 1.28 it follows that $U_t = \int_0^t Y_- dZ$ where

$$Z_t = X_t^1 X_t^2 - X_0^1 X_0^2 - \int_0^t X_-^1 dX^2 - \int_0^t X_-^2 dX^1.$$

The integration by parts formula (1.89) yields $Z_t = [X^1, X^2]_t$ completing the proof. □

2.2 Itô's formula for continuous semimartingales

Theorem 2.4 *Let $X^1, X^2, \ldots, X^d$ be continuous semimartingales and let $f \in C^{1,2}([0, \infty) \times \mathbb{R}^d)$. Then*

$$\begin{aligned} f(t, X_t) &= f(0, X_0) + \int_0^t f_t(s, X_s) ds + \textstyle\sum_{j=1}^{d} \int_0^t f_j(s, X_s) dX_s^j \\ &\quad + \frac{1}{2} \textstyle\sum_{j=1}^{d} \sum_{k=1}^{d} \int_0^t f_{jk}(s, X_s) d[X^j, X^k]_s. \end{aligned} \tag{2.5}$$

Note that $f_j(X_s)$ is a continuous adapted process, and thus the stochastic integral $\int_0^t f_j(X_s)dX_s^j$ is defined and is itself a continuous semimartingale.

Proof For each $n \geq 1$, define a sequence $\{\tau_i^n : i \geq 1\}$ of stopping times inductively as follows: $\tau_0^n = 0$ and for $i \geq 0$,

$$\tau_{i+1}^n = \inf\{t \geq \tau_i^n : \textstyle\sum_{j=1}^d |X_t^j - X_{\tau_i^n}^j|^2 \geq 2^{-2n}\} \wedge (\tau_i^n + 2^{-n}) \tag{2.6}$$

It is easy to see that $\{\tau_i^n : i \geq 1\}_{n \geq 1}$ satisfies (1.81), (1.82). Fix t. Let

$$U_i^n = \textstyle\sum_{j=1}^d (X_{\tau_{i+1}^n \wedge t}^j - X_{\tau_i^n \wedge t}^j) f_j(\tau_i^n \wedge t, X_{\tau_i^n \wedge t}) \tag{2.7}$$

$$V_i^n = \frac{1}{2} \textstyle\sum_{j=1}^d \sum_{k=1}^d (X_{\tau_{i+1}^n \wedge t}^j - X_{\tau_i^n \wedge t}^j)(X_{\tau_{i+1}^n \wedge t}^k - X_{\tau_i^n \wedge t}^k) f_{jk}(\tau_i^n \wedge t, X_{\tau_i^n \wedge t}) \tag{2.8}$$

$$R_i^n = h(\tau_i^n \wedge t, X_{\tau_{i+1}^n \wedge t}, X_{\tau_i^n \wedge t}). \tag{2.9}$$

Recalling the definition of h in (2.4), note that

$$f(\tau_i^n \wedge t, X_{\tau_{i+1}^n \wedge t}) - f(\tau_i^n \wedge t, X_{\tau_i^n \wedge t}) = U_i^n + V_i^n + R_i^n. \tag{2.10}$$

Hence

$$f(\tau_{i+1}^n \wedge t, X_{\tau_{i+1}^n \wedge t}) - f(\tau_i^n \wedge t, X_{\tau_i^n \wedge t}) = S_i^n + U_i^n + V_i^n + R_i^n \tag{2.11}$$

where

$$S_i^n = f(\tau_{i+1}^n \wedge t, X_{\tau_{i+1}^n \wedge t}) - f(\tau_i^n \wedge t, X_{\tau_{i+1}^n \wedge t}). \tag{2.12}$$

Thus

$$f(t, X_t) - f(0, X_0) = \textstyle\sum_{i=0}^\infty S_i^n + \sum_{i=0}^\infty U_i^n + \sum_{i=0}^\infty V_i^n + \sum_{i=0}^\infty R_i^n. \tag{2.13}$$

From Theorem 1.29 and Lemma 2.3 it follows that

$$\textstyle\sum_{i=0}^\infty U_i^n \longrightarrow \sum_{j=1}^d \displaystyle\int_0^t f_j(X_-)dX^j \tag{2.14}$$

and

$$\textstyle\sum_{i=0}^\infty V_i^n \longrightarrow \frac{1}{2} \sum_{j=1}^d \sum_{k=1}^d \displaystyle\int_0^t f_j(X_-)d[X^j, X^k] \tag{2.15}$$

in probability. Next we will prove that for all ω

$$\textstyle\sum_{i=0}^\infty S_i^n(\omega) \longrightarrow \displaystyle\int_0^t f_t(s, X_s(\omega))ds. \tag{2.16}$$

Fix ω and let $t_i^n = \tau_i^n$. Then it follows that for any continuous function $g : \mathbb{R}^d \longrightarrow \mathbb{R}$,

$$\begin{aligned} \textstyle\sum_i [f(t_{i+1}^n, g(t_{i+1}^n)) - f(t_i^n, g(t_{i+1}^n))] &= \textstyle\sum_i [f(r_i^n, g(t_{i+1}^n))] (t_{i+1}^n - t_i^n) \\ &\longrightarrow \int_0^t f_t(s, g(s)) ds. \end{aligned} \tag{2.17}$$

Here the first step follows from Taylor's theorem with $t_i^n \le r_i^n \le t_{i+1}^n$ and the second step follows from the dominated convergence theorem for Lebesgue integrals. Taking $g(s) = X_s(\omega)$ above, we conclude that (2.16) holds.

To complete the proof, it suffices to show that

$$\textstyle\sum_{i=0}^{\infty} R_i^n \longrightarrow 0 \quad \text{in probability.} \tag{2.18}$$

The result follows from (2.13)–(2.18).

Let $K(t, \omega) = \sup_{t \le T} |X_t(\omega)|$. Note that $K(t, \omega) < \infty$ for all (t, ω) since X is a continuous process. Continuity of X also implies that

$$\textstyle\sum_{j=1}^{d} |X^j_{\tau_{i+1}^n \wedge t} - X^j_{\tau_i^n \wedge t}|^2 \le 2^{-2n}$$

and hence $|X_{\tau_{i+1}^n \wedge t} - X_{\tau_i^n \wedge t}| \le 2^{-n}$. Thus

$$|R_i^n(\omega)| \le \Gamma(K(t, \omega), 2^{-n}) \textstyle\sum_{j=1}^{d} \left(X^j_{\tau_{i+1}^n \wedge t}(\omega) - X^j_{\tau_i^n \wedge t}(\omega)\right)^2. \tag{2.19}$$

Since $\sum_{i=1}^{\infty} \left(X^j_{\tau_{i+1}^n \wedge t} - X^j_{\tau_i^n \wedge t}\right)^2 \to \left[X^j, X^j\right]_t$ in probability (see Theorem 1.31) and $\lim_n \Gamma(K(t, \omega), 2^{-n}) = 0$ (see Lemma 2.2) the required conclusion (2.18) follows from (2.19). □

Remark It is clear from the proof that the formula is valid for $0 \le t \le T$, if $f \in C^{1,2}([0, T] \times G)$ where G is an open convex set in $\mathbb{R}^d$ and

$$P(X_t \in G,\ 0 \le t \le T) = 1.$$

An important case is $d = 1,\ G = (0, \infty)$ and $f(t, x) = x^{-1}$.

For applications it is useful to recast Itô's formula as follows:

Theorem 2.5 *Let $X^1, \dots, X^d$ be continuous semimartingales with $X^i = M^i + A^i$ where M^i are continuous local martingales and $A^i \in \mathcal{V}$.*

Let $f \in C^{1,2}([0, \infty) \times \mathbb{R}^d)$. Then

$$f(t, X_t) = N_t + B_t \tag{2.20}$$

where

$$N_t = \int_0^t f_j(s, X_s) dM_s^j \tag{2.21}$$

and

$$\begin{aligned} B_t &= f(0, X_0) + \int_0^t f_t(s, X_s)ds + \textstyle\sum_{j=1}^d \displaystyle\int_0^t f_j(s, X_s)dA_s^j \\ &\quad + \frac{1}{2}\textstyle\sum_{j=1}^d \sum_{k=1}^d \displaystyle\int_0^t f_{jk}(s, X_s)d\langle M^j, M^k\rangle_s. \end{aligned} \tag{2.22}$$

To see that this is the same as (2.5), note that as a consequence of Lemma 1.42,

$$\begin{aligned} [X^j, X^k] &= [M^j, M^k] && (2.23) \\ &= \langle M^j, M^k\rangle. && (2.24) \end{aligned}$$

In (2.21) above, N_t is a local martingale since it is a sum of stochastic integrals w.r.t. a local martingale and the process $B \in \mathcal{V}$.

2.3 Itô's formula for r.c.l.l. semimartingales

Theorem 2.6 *Let $X^1, \ldots, X^d$ be semimartingales, $X_t := (X_t^1, \ldots, X_t^d)$. Let $f \in C^{1,2}([0, \infty) \times \mathbb{R}^d)$. Then*

$$\begin{aligned} f(t, X_t) &= f(0, X_0) + \int_0^t f_t(s, X_s)ds \\ &\quad + \textstyle\sum_{j=1}^d \displaystyle\int_0^t f_j(s, X_{s-})dX_s^j \\ &\quad + \frac{1}{2}\textstyle\sum_{j=1}^d \sum_{k=1}^d \displaystyle\int_0^t f_{jk}(s, X_{s-})d[X^j, X^k] \\ &\quad + \textstyle\sum_{0<s\le t}\Big\{ f(s, X_s) - f(s, X_{s-}) - \sum_{j=1}^d f_j(s, X_{s-})\Delta X_s^j \\ &\qquad - \textstyle\sum_{j=1}^d \sum_{k=1}^d \frac{1}{2} f_{jk}(s, X_{s-})(\Delta X_s^j)(\Delta X_s^k)\Big\}. \end{aligned} \tag{2.25}$$

Proof Fix t. Let us begin by noting that the last term is simply

$$\textstyle\sum_{0<s\le t} h(s, X_s, X_{s-}) \tag{2.26}$$

By Lemma 2.2 $|h(s, X_s, X_{s-})| \le \Lambda(K_t)(\Delta X_s)^2$, where

$$K_t(\omega) = \sup_{t\le T} |X_t(\omega)|.$$

As a consequence, we have

$$\textstyle\sum_{0<s\le t} |h(s, X_s, X_{s-})| \le \Lambda(K_t) \sum_{0<s\le t}(\Delta X_s)^2,$$

and hence the series in (2.26) converges uniformly, a.s.

Coming to the proof of (2.25), we proceed exactly as in the proof of the formula for continuous semimartingales. Define U_i^n, V_i^n, R_i^n, S_i^n by (2.7), (2.8), (2.9)and (2.12) respectively. Observe that (2.13) holds. By Theorem 1.38, (2.14) is true and so is (2.15) once we see that Lemma 2.3 holds for all semimartingales. Also, (2.16) continues to be valid, as (2.17) is true for an r.c.l.l. function g as well.

To complete the proof, we need to show

$$\sum_{i=0}^{\infty} R_i^n \to \sum_{0<s\leq t} h(s, X_s, X_{s-}) \quad \text{in probability.} \tag{2.27}$$

Let us write $\sigma_i^n = \tau_i^n \wedge t$. Define

$$\begin{aligned} G(\omega) &= \{s > 0 : |\Delta X_s(\omega)| \neq 0 \quad \text{and} \quad s \leq t\} \\ E^n(\omega) &= \{i : |\Delta X_{\sigma_{i+1}^n}(\omega)| \leq 2 \cdot 2^{-n} \quad \text{or} \quad \sigma_{i+1}^n(\omega) = \sigma_i^n(\omega)\} \\ F^n(\omega) &= \{i : |\Delta X_{\sigma_{i+1}^n}(\omega)| > 2 \cdot 2^{-n} \quad \text{and} \quad \sigma_{i+1}^n(\omega) > \sigma_i^n(\omega)\}. \end{aligned}$$

For all ω, $G(\omega)$ is a countable set. Note that

$$S := \{s \in G(\omega) : |\Delta X_s(\omega)| > 2 \cdot 2^{-n}\} = \{\sigma_{i+1}^n(\omega) : i \in F^n(\omega)\}, \tag{2.28}$$

since if $|\Delta X_s(\omega)| > 2 \cdot 2^{-n}$, then s must be equal to one of $\sigma_j^n(\omega)$.

For $i \in E^n$,

$$\begin{aligned} \left|X_{\sigma_{i+1}^n} - X_{\sigma_i^n}\right| &\leq \left|(X_-)_{\sigma_{i+1}^n} - X_{\sigma_i^n}\right| + \left|\Delta X_{\sigma_{i+1}^n}\right| \\ &\leq 3 \cdot 2^{-n} \end{aligned}$$

and hence

$$\sum_{i\in E^n} |R_i^n| \leq \sum_i \Gamma(K_t, 3 \cdot 2^{-n}) \left|X_{\sigma_{i+1}^n} - X_{\sigma_i^n}\right|^2$$

which goes to zero (as seen in the proof of Theorem 2.4). It remains to show that

$$\sum_{i\in F^n} R_i^n \longrightarrow \sum_{0<s\leq t} h(s, X_s, X_{s-}) \quad \text{in probability.} \tag{2.29}$$

Define

$$a_s^n(\omega) := \sum_{i\in F^n(\omega)} h\big(\sigma_i^n, X_{\sigma_{i+1}^n}(\omega), X_{\sigma_i^n}(\omega)\big) 1_{\{\sigma_{i+1}^n(\omega)=s\}}.$$

Then

$$\sum_{i\in F^n} R_i^n = \sum_{s\in S} a_s^n.$$

If $|\Delta X_s| > 2 \cdot 2^{-n}$, then $a_s^n = h(\sigma_i^n, X_s, X_{\sigma_i^n})$ with $s = \sigma_{i+1}^n$ (as seen in (2.28)) and hence

$$a_s^n(\omega) \to h(s, X_s(\omega), X_{s-}(\omega)) \quad \text{for all } \omega. \tag{2.30}$$

For $i \in F^n$,

$$\begin{aligned} |X_{\sigma_{i+1}^n} - X_{\sigma_i^n}| &\leq |X_{\sigma_{i+1}^n} - X_{\sigma_i^n}| + |\Delta X_{\sigma_{i+1}^n}| \\ &\leq 2^{-n} + \left|\Delta X_{\sigma_{i+1}^n}\right| \\ &\leq 2\left|\Delta X_{\sigma_{i+1}^n}\right|. \end{aligned}$$

Thus if $|a_s^n(\omega)| \neq 0$, then $s = \sigma_{i+1}^n(\omega)$ for some $i \in F^n(\omega)$, and then

$$\begin{aligned} |a_s^n| &\leq \Lambda(K_t)\big|X_{\sigma_{i+1}^n} - X_{\sigma_i^n}\big|^2 \\ &\leq 4\Lambda(K_t)\big|\Delta X_{\sigma_{i+1}^n}\big|^2 \\ &= 4\Lambda(K_t)\big|\Delta X_s\big|^2. \end{aligned}$$

Let $a_s(\omega) = 4\Lambda(K_t(\omega))|\Delta X_s(\omega)|^2$. We know that $\sum_s a_s(\omega) < \infty$ a.s. Fix ω such that

$$\textstyle\sum_s a_s(\omega) < \infty.$$

Using a series version of the dominated convergence theorem along with $|a_s^n(\omega)| \leq a_s(\omega)$ (recall (2.30)), we get

$$\textstyle\sum_{s\in G(\omega)} a_s^n(\omega) \to \sum_{s\in G(\omega)} h\big(s, X_s(\omega), X_{s-}(\omega)\big).$$

We have proved that the convergence in (2.29) holds almost surely and hence in probability. This completes the proof. □

2.4 Applications

An immediate consequence of the Itô formula is that if the X^i are semimartingales, $1 \leq i \leq d$, and f is a smooth function on $\mathbb{R}^d$, then $f(X_t^1, \ldots, X_t^d)$ is a semimartingale. Further, it also gives us a decomposition of this semimartingale into a martingale (or local martingale) and a process with locally bounded variation paths. The stochastic integral w.r.t. the local martingale part of the semimartingale is a local martingale and the remaining terms including the correction terms, constitute a process in $\mathcal{V}$.

As an example, take W to be the Wiener process and $f(x) = x^m$, $m \geq 2$, to be a positive integer. Then $f_1(x) = m\,x^{m-1}$, $f_{11}(x) = m(m-1)x^{m-2}$. The Itô formula gives

$$W_t^m = m\int_0^t W_s^{m-1}\,dW_s + \frac{m(m-1)}{2}\int_0^t W_s^{m-2}\,ds.$$

Taking $m = 2$, one gets

$$W_t^2 = \frac{1}{2}\{W_t^2 - t\} \tag{2.31}$$

Taking $m = 3$ and using (2.31), we have

$$\begin{aligned} W_t^3 &= 3\textstyle\int_0^t W_s^2\,dW_s + 3\int_0^t W_s\,ds \\ &= 3\textstyle\int_0^t\big\{\big(2\int_0^s W_{s_1}\,dW_{s_1}\big) + s\big\}dW_s + 3\int_0^t W_s\,ds \\ &= 6\textstyle\int_0^t\big(\int_0^s W_{s_1}\,dW_{s_1}\big) + 3\int_0^t s\,dW_s + 3\int_0^t W_s\,ds \\ &= 6\textstyle\int_0^t\big(\int_0^s W_{s_1}\,dW_{s_1}\big) + 3\big\{tW_t - \int_0^t W_s\,ds\big\} + 3\int_0^t W_s\,ds, \end{aligned}$$

where the quantity in {..} is obtained by integration by parts. Hence

$$W_t^3 = 6\int_0^t \Big(\int_0^s W_{s_1} dW_{s_1}\Big) dW_s + 3t W_t.$$

From these calculations, it should be clear that $\int e^{W_s} dWs$ is not going to be e^{W_t}. We can use Itô's formula to compute this integral as follows. Take $f(x) = e^x$. Then $f_1(x) = f_{11}(x) = e^x$ and one has

$$e^{W_t} = 1 + \int_0^t e^{W_s} dW_s + \frac{1}{2}\int_0^t e^{W_s} ds;$$

thus rearranging terms we get

$$\int_0^t e^{W_s} dW_s = e^{W_t} - 1 - \frac{1}{2}\int_0^t e^{W_s} ds. \tag{2.32}$$

So, unlike in the case of the Riemann–Stieljes integral, e^{X_t} is not a solution to the integral equation

$$Y_t = 1 + \int_0^t Y_s dY_s. \tag{2.33}$$

To find a solution to this equation, we can proceed as follows. Let us try a solution Y_t given by $\exp\{W_t + A_t\}$ for some continuous process $A \in \mathcal{V}$. We will prove below that for such a process A, $[W + A, W + A] = 0$. For now, using this, and applying Itô's formula for $f(x) = e^x$, $X_t = W_t + A_t$, one gets

$$Y_t = 1 + \int_0^t Y_s dW_s + \int_0^t Y_s dA_s + \frac{1}{2}\int_0^t Y_s ds.$$

We thus get

$$A_t = -\frac{t}{2}.$$

Now it can be verified that $Y_t = \exp\{W_t - \frac{1}{2}t\}$ is a solution to the equation (2.33). We can similarly solve the equation (2.33) by replacing W with any continuous semimartingale X and we need to replace t by $[X, X]_t$.

We are now ready to prove the result on the solution to the exponential equation.

Theorem 2.7 *Let X be a continuous semimartingale. Consider the equation*

$$Z_t = 1 + \int_0^t Z_- dX. \tag{2.34}$$

The equation (2.34) admits a unique solution given by

$$Y_t = \exp\{X_t - \frac{1}{2}[X, X]_t\}. \tag{2.35}$$

Proof Since X is assumed to be continuous, any solution to (2.34) is continuous, and thus $Z_- = Z$. Let $U_t = X_t - \frac{1}{2}[X, X]_t$. It follows from Lemma 1.42 that $[X, X] = [U, U]$. Using Itô's formula for the function $f(x) = e^x$ and the semimartingale U, we get that Y satisfies (2.34). Now to prove uniqueness, let Z be any solution to (2.34) and let $V_t = \exp\{-U_t\}$, $\xi_t = Z_t V_t$. By Itô's formula,

$$V_t = 1 - \int_0^t V_s dU_s + \frac{1}{2}\int_0^t V_s d[-U, -U]_s,$$

and thus, noting that $[-U, -U] = [U, U] = [X, X]$, we get

$$V_t = 1 - \int_0^t V_s dX_s + \int_0^t V_s d[X, X]_s. \tag{2.36}$$

By integration by parts formula (or Itô's formula) we get

$$Z_t V_t = 1 + \int_0^t \{Z dV + V dZ + d[V, Z]\}$$

which reduces to

$$\xi_t = 1 + \int_0^t \xi\{-dX + [X, X] + dX\} + [V, Z]_t. \tag{2.37}$$

Now, V, Z, X, Y, U are all continuous processes. So using Theorem 1.43 and Lemma 1.42 it follows that

$$\begin{aligned} [V, Z]_t &= \left[Z, -\int_0^\bullet V\, dX + \int_0^\bullet V\, d[X, X]\right] && (2.38)\\ &= -\int_0^t V d[X, Z] && (2.39)\\ &= -\int_0^t V Z d[X, X]. && (2.40) \end{aligned}$$

Using this in (2.37), we conclude that $\xi_t = 1$. This proves

$$Z_t = \exp\{X_t - \frac{1}{2}[X, X]_t\}.$$

□

Here is an interesting consequence of this result.

Theorem 2.8 *Let M be a continuous local martingale. Then*

$$N_t = \exp\{M_t - \frac{1}{2}\langle M, M\rangle_t\}$$

is a local martingale and a supermartingale with $\mathbb{E}(N_t) \leq 1$. N_t is a martingale if and only if $\mathbb{E}(N_t) = 1 \;\; \forall t \geq 0$. If for some $\delta > 0$,

$$C_T = \mathbb{E}\exp\{\frac{1}{2}(1 + \delta)\langle M, M\rangle_T\} < \infty, \tag{2.41}$$

then N is a martingale on $[0, T]$.

Proof We have seen earlier that $[M, M] = \langle M, M\rangle$ since M is a continuous local martingale, it follows that N is a stochastic integral with respect to a local martingale, and hence is a local martingale.

Let τ_n be a localizing sequence for this local martingale. Then for $s < t$ one has

$$\mathbb{E}[N_{t\wedge\tau_n}|\mathcal{F}_s] = N_{s\wedge\tau_n}$$

Taking the limit as n tends to infinity and using Fatou's lemma for conditional expectations, we conclude

$$\mathbb{E}[N_t|\mathcal{F}_s] \leq N_s.$$

Thus N is a supermartingale. It is clear that N_t is a martingale if and only if $\mathbb{E}(N_t) = 1 \quad \forall t \geq 0$. This will be the case if $\{N_{t\wedge\tau_n} : n \geq 1\}$ is uniformly integrable, then N will be a martingale.

Fix n and let $p > 1$ and let $\alpha > 1$ (to be chosen later). Let us write

$$N^p_{t\wedge\tau_n} = U_t V_t$$

where

$$U_t = \exp\{pM_{t\wedge\tau_n} - \frac{1}{2}\alpha p^2\langle M, M\rangle_{t\wedge\tau_n}\}$$
$$V_t = \exp\{\frac{p}{2}(\alpha p - 1)\langle M, M\rangle_{t\wedge\tau_n}\}.$$

Now $U_t^\alpha = \exp\{J_t - \frac{1}{2}\langle J, J\rangle_t\}$ with $J_t = \alpha p M_{t\wedge\tau_n}$ and since J is a local martingale, it follows that

$$\mathbb{E}(U_t^\alpha) \leq 1.$$

For $t \leq T$ observe that

$$\begin{aligned} V_t^{\frac{\alpha}{\alpha-1}} &= \exp\{\frac{\alpha p(\alpha p - 1)}{2(\alpha - 1)}\langle M, M\rangle_{t\wedge\tau_n}\} \\ &\leq \exp\{\frac{\alpha p(\alpha p - 1)}{2(\alpha - 1)}\langle M, M\rangle_T\} \end{aligned}$$

and hence, choosing $\alpha = 1 + \varepsilon$, $p = 1 + \varepsilon^2$ with ε such that

$$(1 + \varepsilon)(1 + \varepsilon^2)(1 + \varepsilon + \varepsilon^2) < (1 + \delta),$$

it follows that

$$\mathbb{E}(V_t^{\frac{\alpha}{\alpha-1}}) \leq \mathbb{E}[\exp\{\frac{1}{2}(1 + \delta)\langle M, M\rangle_T\}].$$

Hölder's inequality now gives

$$\begin{aligned} \mathbb{E}[N^p_{t\wedge\tau_n}] &\leq [\mathbb{E}(U_t^\alpha)]^{\frac{1}{\alpha}}[\mathbb{E}(V_t^{\frac{\alpha}{\alpha-1}})]^{\frac{(\alpha-1)}{\alpha}} \\ &\leq \{C_T\}^{\frac{(\alpha-1)}{\alpha}}. \end{aligned}$$

It follows that if (2.41) is satisfied, then $\{N_{t\wedge\tau_n} : n \geq 1\}$ is uniformly integrable and as noted earlier, this completes the proof. □

Remark It is known that in the previous theorem, it is sufficient if (2.41) holds with $\delta = 0$. This is known as the Novikov condition. See (Kallianpur, 1980, Karatzas and Shreve, 1988). Another sufficient condition, due to (Kazamaki, 1976) is that

$$\mathbb{E}\exp\{\frac{1}{2}M_t\} < \infty \quad \forall t \geq 0.$$

Our next application of Itô's formula is Levy's characterization of the Brownian motion.

Theorem 2.9 (Martingale characterization of Brownian motion) *Let $M = (M^1, M^2, \ldots, M^d)$ be a d-dimensional continuous local martingale with respect to a filtration $(\mathcal{F}_t)$ which is not assumed to be right continuous. Further suppose that $M_0^j = 0$ and $\langle M^j, M^j \rangle_t = t$ for $1 \leq j \leq d$ and that $\langle M^j, M^k \rangle_t = 0$ for $j \neq k$. Then*

(i) M is a d-dimensional Brownian motion;

(ii) $\sigma\{M_{s_2} - M_{s_1}, s_2 > s_1 \geq s\}$ is independent of $\mathcal{F}_s$ $\forall s$.

Proof First assume that the filtration $(\mathcal{F}_t)$ is right continuous. Fix $t > 0, 0 \leq s < t$ and let $A \in \mathcal{F}_s$. Let $g(t) := \int_A e^{i\theta \cdot (M_t - M_s)} dP$, where $\theta = (\theta^1, \ldots, \theta^d) \in \mathbb{R}^d$ and $\cdot$ denotes the inner product in $\mathbb{R}^d$. Let us note that the given information on M implies that $\theta \cdot M_t$ and $(\theta \cdot M_t)^2 - \| \theta \|^2 t$ are $\mathcal{F}_t$ martingales. Applying Itô's formula to the real and imaginary parts of $f(x) = e^{i\theta \cdot x}$, we have

$$e^{i\theta \cdot M_t} = e^{i\theta \cdot M_s} + i\sum_{j=1}^d \theta^j \int_s^t \theta^j e^{i\theta \cdot M_r} dM_r^j - \frac{1}{2} \| \theta \|^2 \int_s^t e^{i\theta \cdot M_r} dr.$$

Therefore

$$\begin{aligned} e^{i\theta \cdot (M_t - M_s)} &= 1 + i\sum_{j=1}^d \theta^j \int_s^t e^{i\theta \cdot (M_r - M_s)} dM_r^j \\ &\quad - \frac{1}{2} \| \theta \|^2 \int_s^t e^{i\theta \cdot (M_r - M_s)} dr. \end{aligned}$$

The real and imaginary parts of the stochastic integral on the right side are local martingales and indeed belong to $\mathcal{M}^2$. This can be deduced from Theorem 1.16 by computing its quadratic variation and observing that it is a bounded random variable for each t. Hence

$$\begin{aligned} \int_A e^{i\theta \cdot (M_t - M_s)} dP &= P(A) + i\sum_{j=1}^d \theta^j \int_A \int_s^t e^{i\theta \cdot (M_r - M_s)} dM_r^j \, dP \\ &\quad - \frac{1}{2} \| \theta \|^2 \int_s^t \int_A e^{i\theta \cdot (M_r - M_s)} dP \, dr. \end{aligned} \tag{2.42}$$

The middle term on the right hand side above is zero since $A \in \mathcal{F}_s$ and the stochastic integral is a martingale. Hence defining $g(t)$ to be the quantity on the left hand side in equation (2.42), we have

$$g(t) = P(A) - \frac{1}{2} \parallel \theta \parallel^2 \int_s^t g(r)\,dr. \qquad (2.43)$$

Note that here s, θ, A are fixed. Since M is a continuous process, g is continuous and

$$g'(t) = -\frac{\parallel \theta \parallel^2}{2} g(t), \qquad s \leq t \leq T.$$

Hence, $g(t) = K e^{-\frac{1}{2}\|\theta\|^2 t}$ for $t \geq s$. Since $g(s) = P(A)$ by definition, we get

$$g(t) = P(A) e^{-\frac{1}{2}\|\theta\|^2 (t-s)},$$

i.e.,

$$\int_A e^{i\theta \cdot (M_t - M_s)} dP = \int_A e^{-\frac{1}{2}\|\theta\|^2 (t-s)} dP.$$

Therefore,

$$\mathbb{E}\{e^{i\theta \cdot (M_t - M_s)} | \mathcal{F}_s\} = e^{-\frac{1}{2}\|\theta\|^2 (t-s)}, \quad a.s.$$

Let $s = t_1 < t_2 < \cdots < t_n < t_{n+1} = t$.

$$\mathbb{E}\left\{\prod_{k=1}^n e^{i\theta_k \cdot (M_{t_{k+1}} - M_{t_k})} | \mathcal{F}_s\right\} = \mathbb{E}\left\{\mathbb{E}[\prod_{k=1}^n e^{i\theta_k \cdot (M_{t_{k+1}} - M_{t_k})} | \mathcal{F}_{t_n}] | \mathcal{F}_s\right\}$$

$$= \mathbb{E}[\{\mathbb{E} e^{i\theta_n \cdot (M_{t_{n+1}} - M_{t_n})} | \mathcal{F}_{t_n}\} \prod_{k=1}^{n-1} e^{i\theta_k \cdot (M_{t_{k+1}} - M_{t_k})} | \mathcal{F}_s]$$

$$= e^{-\frac{1}{2}\|\theta_n\|^2 (t - t_n)} \mathbb{E}[\prod_{k=1}^{n-1} e^{i\theta_k \cdot (M_{t_{k+1}} - M_{t_k})} | \mathcal{F}_s].$$

Using this successively, we get

$$\mathbb{E}\{\prod_{k=1}^n e^{i\theta_k \cdot (M_{t_{k+1}} - M_{t_k})} | \mathcal{F}_s\} = e^{-\frac{1}{2} \sum_{k=1}^n \parallel \theta_k \parallel^2 (t_{k+1} - t_k)}.$$

Hence M_t is a d-dimensional Wiener process and

$$\sigma\{M_{t_{k+1}} - M_{t_k}, s < t_2 < \cdots < t_n < t\}$$

is independent of $\mathcal{F}_s$ and we have

$$\sigma\{M_{t_2} - M_{t_1},\ s < t_1 < t_2 \leq T\} \text{ is independent of } \mathcal{F}_s.$$

To conclude the proof we only have to observe that if M_t is a continuous local martingale with respect to a filtration $(\mathcal{F}_t)$ which is not right continuous (which is the assumption in the theorem), then M_t is a continuous local martingale with respect to the right continuous filtration $(\mathcal{F}_{t+})$. □

An interesting consequence of this result is the following:

Theorem 2.10 *Let $W_t = (W_t^i)$ be a d-dimensional Brownian motion and let $(\mathcal{G}_t)$ be a filtration such that W_t^i is a martingale w.r.t. $(\mathcal{G}_t)$. Then for s fixed,*

$$\{W_{s+t} - W_s \ : \ t \geq 0\} \text{ is independent of } \mathcal{G}_s.$$

Proof It follows from Theorem 1.17 that

$$\sum_{i=1}^{d}(a_i W_t^i)^2 - (\sum_{i=1}^{d} a_i^2)t \quad \text{is a } (\mathcal{G}_t) \text{ martingale.}$$

Hence, $\langle W^i, W^i \rangle_t = t$ and for $i \neq j$, $\langle W^i, W^j \rangle_t = 0$ and the conclusion now follows from Theorem 2.9. □

2.5 Application to geometric Brownian motion

A simple application of Itô's formula to geometric Brownian motion (GBM) yields the stochastic differential for the stock price S_t with constant rate of return μ and constant volatility σ.

Taking $F(x) = e^x$, $M_t = \int_0^t f dW$, $A_t = -\frac{1}{2}\int_0^t f^2 ds$. Then

$$\begin{aligned} e^{X_t} &= 1 + \int_0^t e^{X_s} dM_s + \frac{1}{2}\int_0^t e^{X_s} f^2\, ds + \int_0^t e^{X_s} dA_s \\ &= 1 + \int_0^t e^{X_s} dM_s. \end{aligned}$$

Let $X_t = \sigma W_t - \frac{1}{2}\sigma^2 t$. Then, letting $\hat{S}_t := e^{\sigma W_t - \frac{1}{2}\sigma^2 t}$, we have

$$\begin{aligned} \hat{S}_t &= 1 + \int_0^t \hat{S}_u dM_u \text{ where } M_t = \sigma W_t. \\ d\hat{S}_t &= \hat{S}_t dM_t. \text{ Hence} \\ S_t &:= e^{\mu t}\, \hat{S}_t = e^{(\mu - \frac{1}{2}\sigma^2)t + \sigma W_t} \\ dS_t &= e^{\mu t}\, d\hat{S}_t + \mu e^{\mu t}\, \hat{S}_t dt \\ &= \mu S_t dt + S_t dM_t = \mu S_t dt + \sigma S_t dW_t. \end{aligned}$$

2.6 Local time and the Tanaka formula

In this section we give, without proof, an extension of the Itô formula due to Tanaka. For this we need the notion of local time, first introduced by P. Lévy.

Fix $t > 0$ and $a \in \mathbb{R}$. The set

$$\{s \in [0, t] : W_s \in (a - \varepsilon, a + \varepsilon)\}$$

is the time spent by the Brownian motion in $(a-\varepsilon, a+\varepsilon)$ between 0 and t. Denoting Lebesgue measure by λ, we have

$$\lambda\{s \in [0,t] : W_s \in (a-\varepsilon, a+\varepsilon)\} = \int_0^t \mathbf{1}_{(a-\varepsilon,a+\varepsilon)}(W_s)ds.$$

Definition $L_W(t,a) := \lim_{\varepsilon\downarrow 0} \frac{1}{2\varepsilon}\int_0^t \mathbf{1}_{(a-\varepsilon,a+\varepsilon)}(W_s)ds$. The limit exists a.s. or in $\mathbb{L}^2$ and is called the *local time of* W *at* a.

A special case of Tanaka's formula is the following: For each (t,a) we have a.s.

$$|W_t - a| - |W_0 - a| = \int_0^t sgn(W_s - a)dW_s + L_W(t,a)$$

where $sgn\, x = 1$ if $x > 0$, -1 if $x < 0$ and 0 if $x = 0$. For $a = 0$, and $W_0 = 0$, we get

$$|W_t| = \int_0^t (sgn\ W_s)\, dW_s + L_W(t,0).$$

Note that the first term on the right hand side is also a standard Wiener process.

If f is a Borel measurable and integrable function, we have the formula

$$\int_0^t f(W_s)ds = \int_{-\infty}^{\infty} L_W(t,a)\, f(a)da.$$

Local time can also be defined for a continuous semimartingale X:

$$L_X(t,a) := |X_t - a| - |X_0 - a| - \int_0^t sgn\,(X_s - a)\, dX_s.$$

$L_X(\cdot, a)$ is adapted, continuous and nondecreasing in t.

We conclude by giving a general form of Tanaka's formula: Let X be a continuous semimartingale and $f : \mathbb{R} \to \mathbb{R}$ a convex function. Then

$$f(X_t) = f(X_0) + \int_0^t f'(X_s)dX_s + \frac{1}{2}\int_{-\infty}^{\infty} L_X(t,a)\, \mu(da)$$

where f' is the left hand derivative of f and μ is the second distributional derivative of f.

2.7 Brownian motion and the heat equation

Let $W_t = (W_t^1 \cdots, W_t^d)$ be a d-dimensional Brownian motion. For a function f on $[0,T] \times \mathbb{R}^d$ for which $\frac{\partial f}{\partial t}$, $\frac{\partial f}{\partial x_i}$ and $\frac{\partial^2 f}{\partial x_i \partial x_j}$ exist, by Ito's formula one has that

$$f(t, W_t) - \int_0^t \left(\frac{\partial f}{\partial s} + \frac{1}{2}\Delta f\right)(s, W_s)ds$$

equals a stochastic integral w.r.t. Brownian motion and hence is a local martingale. Thus if g satisfies

$$\frac{\partial g}{\partial t} + \frac{1}{2}\Delta g = 0,$$

taking $f(t, y) = g(s+t, x+y)$, one gets (s and x fixed and Δ is w.r.t. y)

$$M_t^{g,x,s} = g(s+t, x+W_t)$$

is a local martingale. If we have an appropriate integrability condition on g that implies that this local martingale is indeed a martingale, it would follow that (equality of expectation at $t = 0$ and $t = T - s$)

$$g(s, x) = \mathbb{E}(g(T, x + W_{T-s})),$$

giving us a representation of g in terms of its *boundary* value $g(T, \cdot)$. Similarly, if h satisfies

$$\frac{\partial h}{\partial t} = \frac{1}{2}\Delta h,$$

taking $g(t, y) = h(T - t, x + y)$ and using the heuristics given above, we could conclude (with $u = T - t$) that

$$h(u, x) = \mathbb{E}(h(0, x + W_u)).$$

This representation assumes that we have conditions that ensure that the local martingale $h(u - t, x + W_t)$ is a martingale.

These ideas can be formalized and lead to the following results on existence and uniqueness of a solution to the initial value problem or the Cauchy problem for the heat equation.

Theorem 2.11 *Suppose f is a continuous function on $\mathbb{R}^d$ and*

$$|f(x)| \leq C_1 \exp(a_1 |x|^2), \quad x \in \mathbb{R}^d \tag{2.44}$$

for $C_1 < \infty$ and $a_1 < \infty$. Then the Cauchy problem (for $u \in C^{1,2}((0,T) \times \mathbb{R}^d)$) for the heat equation

$$\frac{\partial u}{\partial t}(t, x) = (\frac{1}{2}\Delta u)(t, x) \;:\; (t, x) \in (0, T) \times \mathbb{R}^d \tag{2.45}$$

$$\lim_{t \downarrow 0} u(t, x) = f(x) \tag{2.46}$$

admits a unique solution in the class of functions g satisfying

$$|g(t, x)| \leq C_2 \exp(a_2 |x|^2), \quad (t, x) \in (0, T) \times \mathbb{R}^d \tag{2.47}$$

for some $C_2 < \infty$, $a_2 < \infty$. The unique solution is given by

$$u(t, x) = E(f(x + W_t)) \tag{2.48}$$

where $W_t = (W_t', \ldots, W_t^d)$ is a d-dimensional Brownian motion.

Proof The function u defined by (2.48) is given by

$$u(t,x) = \int f(y)p_t(x,y)dy \tag{2.49}$$

where, for $t > 0$, $x, y \in \mathbb{R}^d$,

$$p_t(x,y) = (2\pi t)^{-d/2} \exp(-\frac{1}{2t}|x-y|^2).$$

It is very easy to verify that for $0 \le t \le T$

$$\int \exp(a_1|y|^2)p_t(x,y)dy < \infty \tag{2.50}$$

since $a_1 T < \frac{1}{2}$. Indeed, for $0 \le t \le T$

$$\int |y|^2 \exp(a_1|y|^2)p_t(x,y)dy < \infty, \ t \le T. \tag{2.51}$$

Noting that $\Delta = \sum_{i=1}^d \frac{\partial^2}{\partial x_i^2}$, it is easy to see that $p_t(x,y)$ is a fundamental solution of the heat equation. That is

$$\frac{\partial p_t(x,y)}{\partial t} = \frac{1}{2}\,\Delta p_t(x,y), \quad (y \ne x).$$

Next, using (2.51), one can justify interchanging the order of integration and differentiation (w.r.t. t and twice w.r.t. x_i) to conclude that

$$\begin{aligned} \frac{\partial u}{\partial t}(t,x) - \frac{1}{2}\,\Delta u(t,x) &= \int f(y)\Big\{\frac{\partial p_t}{\partial t}(x,y) - \frac{1}{2}\,\Delta p_t(x,y)\Big\}dy \\ &= 0. \end{aligned}$$

Thus u satisfies the heat equation. Finally, from the definition (2.48) it follows that

$$\lim_{t\downarrow 0} u(t,x) = f(x).$$ □

Remark It should be noted that the solution f constructed above belongs to

$$C([0,T]\times\mathbb{R}^d) \cap C^{1,2}((0,T)\times\mathbb{R}^d).$$

The uniqueness part follows from the next result.

Theorem 2.12 *Suppose $u \in C^{1,2}((0,T)\times\mathbb{R}^d)$ satisfies*

$$\frac{\partial u}{\partial t} = \frac{1}{2}\Delta u$$

on $(0, T) \times \mathbb{R}^d$ *and*

$$|u(t,x)| \leq c_3 \exp(a_3|x|^2)$$

for some $c_3 < \infty$, $a_3 < \infty$. *Further, suppose*

$$\lim_{t \downarrow 0} u(t,x) = 0 \quad \forall x \in \mathbb{R}^d. \tag{2.52}$$

Then $u(t,x) = 0$ *for all* $(t,x) \in (0,T) \times \mathbb{R}^d$.

Proof Fix $t_0 \in (0,T)$ such that $a_3 t_0 \leq \frac{1}{4}$ and $\varepsilon > 0$ such that $t_0 + \varepsilon < T$. Since u satisfies the heat equation, M_s^ε defined by

$$M_s^\varepsilon = u(t_0 + \varepsilon - s, x + W_s), \quad 0 \leq s \leq t_0$$

equals a stochastic integral with respect to Brownian motion and hence is a local martingale. Now

$$|M_s^\varepsilon| \leq c_3 \exp(a_3|x + W_s|^2). \tag{2.53}$$

Now $|x + W_s|^2$ is a submartingale and so is $\exp(a_3|x + W_s|^2)$. Then setting $N_s = \exp\left(\frac{1}{2} a_3|x + W_s|^2\right)$ by Doob's maximal inequality, we have

$$\begin{aligned} \mathbb{E} \sup_{0 \leq s \leq t_0} |N_s|^2 &\leq 4\mathbb{E} N_{t_0}^2 \\ &= 4\mathbb{E} \exp(a_3|x + W_{t_0}|^2 \\ &< \infty \end{aligned}$$

as $a_3 t_0 < \frac{1}{2}$. It thus follows from (2.53) that

$$\mathbb{E} \sup_{0 < \varepsilon < T - t_0} \sup_{0 \leq s \leq t_0} |M_s^\varepsilon| < \infty. \tag{2.54}$$

As a consequence, for $\varepsilon > 0$ fixed, $\{M_s^\varepsilon : 0 \leq s \leq t_0\}$ is a martingale and for any sequence τ_k of stopping times, $\{M_{s \wedge \tau_k}^\varepsilon : k \geq 1\}$ is uniformly integrable in view of (2.54). Further,

$$\lim_{\varepsilon \downarrow 0} M_s^\varepsilon = M_s^0$$

where $M_s^0 = u(t_0 - s, x + W_s)$, $s < t_0$ and $M_{t_0}^0 = 0$ (recall (2.52)). Uniform integrability of $\{M_s^\varepsilon : 0 < \varepsilon < T - t\}$ for every s (again in view of (2.54)) implies that $\{M_s^0 : 0 \leq s \leq t_0\}$ is a martingale. Now $M_{t_0}^0 = 0$ implies $M_s^0 \equiv 0$. In particular, $M_0^0 = 0$, or

$$u(t_0, x) = 0.$$

We have thus proved that $u(t_0, x) = 0$ for all $x \in \mathbb{R}^d$, for all $t_0 \leq \frac{1}{2a_3} = \delta$.

Repeating this argument, we can prove

$$u(a, x) = 0 \ \forall x \in \mathbb{R}^d, \quad \forall a \leq 2\delta.$$

Iterating this, we conclude $u \equiv 0$.

We present an analytical proof of the above result since it seems to us to be of independent interest. The proof uses the ideas in [Friedman, 1964].

Theorem 2.13 (Analytical proof) *Let $u \in C^{1,2}[(0,T) \times \mathbb{R}]$ such that*

$$\begin{aligned} \frac{\partial u}{\partial t} &= \frac{1}{2}\frac{\partial^2 u}{\partial x^2}, \quad t > 0,\ x \in \mathbb{R}, \\ \sup_{0 \le t \le T} |u(t,x)| &\le K e^{a|x|^2} \quad \forall x \in \mathbb{R} \end{aligned}$$

for some K and $0 < a < \frac{1}{2T}$. $u(0,x) = 0$. Then $u(t,x) \equiv 0$ in $[0,T] \times \mathbb{R}$.

Proof We introduce the following notation:

$$\begin{aligned} Lu &\equiv \frac{1}{2}\frac{\partial^2 u}{\partial x^2} - \frac{\partial u}{\partial t}, \\ L^* v &\equiv \frac{1}{2}\frac{\partial^2 v}{\partial x^2} + \frac{\partial v}{\partial t}, \\ p(t,x,\zeta) &= \frac{1}{\sqrt{2\pi t}} e^{-\frac{(x-\zeta)^2}{2t}}, \qquad t > 0 \\ B_R &= S(\overline{x}, R) = \{|x - \overline{x}| \le R\}. \end{aligned}$$

Let $h(\zeta)$ be a C^∞ function such that

$$h(\zeta) = \begin{cases} 1 \text{ on } B_R, \\ 0 \text{ on } B_\rho^c \end{cases}$$

where $\rho > R$ and $0 \le h(\zeta) \le 1 \quad \forall \zeta$. We have

$$|h'(\zeta)| + |h''(\zeta)| \le H \quad \forall \zeta \quad \text{(for some constant } H).$$

In order to prove the theorem we have to show that

$$\begin{aligned} (A) \qquad Lu(t,x) &\equiv 0, \\ \sup_{0 \le t \le T} |u(t,x)| &\le K e^{ax^2} \quad \text{for some } K, a > 0 \qquad (2.55) \\ \text{and } u(0,x) &= 0 \end{aligned}$$

for all x implies $u(t,x) \equiv 0$. Fix $(\overline{t}, \overline{x})$ arbitrary $0 < \overline{t} < T$ and $-\infty < \overline{x} < \infty$. We have to show that $u(\overline{t}, \overline{x}) = 0$.

$$\begin{aligned} vLu - uL^*v &= \frac{1}{2}\left(v\frac{\partial^2 u}{\partial x^2} - u\frac{\partial^2 v}{\partial x^2} \right) - \frac{\partial}{\partial t}(uv) \\ &= \frac{1}{2}\frac{\partial}{\partial x}\left(v\frac{\partial u}{\partial x} - u\frac{\partial v}{\partial x} \right) - \frac{\partial}{\partial t}(uv) \qquad (2.56) \end{aligned}$$

$$\begin{aligned} \text{Take} \qquad v(\tau,\zeta) &= h(\zeta) p(\tau, \overline{x}, \zeta) \\ &= h(\zeta) \cdot \frac{1}{\sqrt{2\pi\tau}} e^{-\frac{(\zeta - \overline{x})^2}{2\tau}} \qquad \tau > 0. \end{aligned}$$

Take $u = u(\tau, \zeta)$, $v(\tau, \zeta) = h(\zeta)p(\bar{t} - \tau, \bar{x}, \zeta)$ $\tau < \bar{t}$. Integrate $vLu - uL^*v$ over $-\rho \leq \zeta \leq \rho, 0 < \tau < \bar{t} - \epsilon$. Recalling our assumption about Lu, we get

$$-\int_0^{\bar{t}-\epsilon} \int_{\bar{x}-\rho}^{\bar{x}+\rho} u(\tau, \zeta) L^* v(\tau, \zeta) d\zeta d\tau$$

$$= \frac{1}{2} \int_0^{\bar{t}-\epsilon} \int_{\bar{x}-\rho}^{\bar{x}+\rho} \left\{ \frac{\partial}{\partial \zeta} \left(v \frac{\partial u}{\partial \zeta} - u \frac{\partial v}{\partial \zeta} \right) - \frac{\partial}{\partial \tau}(uv) \right\} d\zeta d\tau.$$

The first term on the right hand side integral

$$\begin{aligned} &= \frac{1}{2} \int_0^{\bar{t}-\epsilon} \left[v(\tau, \zeta) \frac{\partial u}{\partial \zeta} - u(\tau, \zeta) \frac{\partial v}{\partial \zeta} \right]_{\zeta = \bar{x}-\rho}^{\zeta = \bar{x}+\rho} d\tau \\ &= 0 \end{aligned}$$

since $h(\zeta)$ and $h'(\zeta) = 0$ at $\zeta = \bar{x} \pm \rho$. Hence we get

$$\begin{aligned} \int_0^{\bar{t}-\epsilon} \int_{\bar{x}-\rho}^{\bar{x}+\rho} uL^* v d\zeta d\tau &= \int_0^{\bar{t}-\epsilon} \int_{\bar{x}-\rho}^{\bar{x}+\rho} \frac{\partial}{\partial \tau}(uv) d\zeta d\tau \\ &= \int_{\bar{x}-\rho}^{\bar{x}+\rho} u(\bar{t} - \epsilon, \zeta) v(\bar{t} - \epsilon, \zeta) d\zeta \qquad (2.57) \end{aligned}$$

since $u(0, \zeta) = 0$. The integral on the right hand side

$$\begin{aligned} &= \int_{\bar{x}-\rho}^{\bar{x}+\rho} u(\bar{t} - \epsilon, \zeta) h(\zeta) p(\epsilon, \bar{x}, \zeta) d\zeta \\ &= \int_{\bar{x}-\rho}^{\bar{x}+\rho} u(\bar{t} - \epsilon, \zeta) h(\zeta) \frac{1}{\sqrt{2\pi\epsilon}} e^{-\frac{(\bar{x}-\zeta)^2}{2\epsilon}} d\zeta \\ &\to u(\bar{t}, \bar{x}) h(\bar{x}) = u(\bar{t}, \bar{x}) \quad \text{as } \epsilon \searrow 0. \qquad (2.58) \end{aligned}$$

Now,

$$\begin{aligned} L^* v(\tau, \zeta) &= L^* h(\zeta) p(\bar{t} - \tau; \bar{x}, \zeta) \\ &= \left(\frac{1}{2} \frac{\partial^2}{\partial \zeta^2} + \frac{\partial}{\partial \tau} \right) \left[h(\zeta) p(\bar{t} - \tau; \bar{x}, \zeta) \right] \\ &= \frac{1}{2} \left[h''(\zeta) p(\bar{t} - \tau; \bar{x}, \zeta) + 2h'(\zeta) \frac{\partial p}{\partial \zeta} + h(\zeta) \frac{\partial^2 p}{\partial \zeta^2} + h(\zeta) \frac{\partial p}{\partial \tau} \right] \\ &= \frac{1}{2} h''(\zeta) p(\bar{t} - \tau; \bar{x}, \zeta) + h'(\zeta) \frac{\partial p}{\partial \zeta} = 0 \end{aligned}$$

using the fact that $\frac{1}{2} \frac{\partial^2 p(\bar{t}-\tau, \bar{x}, \zeta)}{\partial \zeta^2} + \frac{\partial p(t-\tau, \bar{x}, \zeta)}{\partial \tau} = 0$ and $h(\zeta) = 1$ on B_R. The integral on the left hand side of (2.57) then becomes

$$\int_0^{\bar{t}-\epsilon} \int_{R < |\zeta - \bar{x}| \leq \rho} u(\tau, \zeta) \left\{ h'(\zeta) \frac{\partial p}{\partial \zeta} + \frac{1}{2} h''(\zeta) p \right\} d\zeta d\tau.$$

Using (2.58) we get (with $\epsilon \to 0$),

$$|u(\bar{t},\bar{x})| \leq H \int_0^{\bar{t}} \int_{R<|\zeta-\bar{x}|\leq\rho} |u(\tau,\zeta)| \cdot \left(|\frac{\partial p}{\partial \zeta}| + |p| \right) d\zeta d\tau. \tag{2.59}$$

Next,

$$\frac{\partial p}{\partial \zeta} = \frac{1}{\sqrt{2\pi}(\bar{t}-\tau)^{3/2}} \cdot (\bar{x}-\zeta) e^{-\frac{(\bar{x}-\zeta)^2}{2(\bar{t}-\tau)}},$$

and we get the following bound in the region $R \leq |\zeta - \bar{x}| \leq \rho$:

$$|p| + \left|\frac{\partial p}{\partial \zeta}\right| \leq \frac{1}{\sqrt{2\pi}} \left\{ (\bar{t}-\tau)^{-1/2} + (\bar{t}-\tau)^{-3/2} \right\} (1+\rho) e^{-\frac{R^2}{2T}}.$$

Hence from (2.59) we have

$$|u(\bar{t},\bar{x})| \leq H_1 (1+\rho) e^{-\frac{R^2}{2T}} \int_0^{\bar{t}} \int_{R\leq|\zeta-\bar{x}|\leq\rho} |u(\tau,\zeta)| d\zeta d\tau \tag{2.60}$$

where

$$H_1 = \frac{1}{\sqrt{2\pi}} H \left\{ (\bar{t}-\tau)^{-1/2} + (\bar{t}-\tau)^{-3/2} \right\}.$$

Since by assumption $\sup_{0\leq\tau\leq T} |u(\tau,\zeta)| \leq K e^{a|\zeta|^2}$ we have

$$\int_0^T \int_{-\infty}^{\infty} |u(\tau,\zeta)| e^{-c|\zeta|^2} d\zeta d\tau \quad < \quad \infty \qquad \text{for} \qquad c > a$$

$$\text{and} \quad \int_0^T \int_{-\infty}^{\infty} |u(\tau,\zeta)| e^{-c|\zeta-\bar{x}|^2} d\zeta d\tau \quad < \quad \infty. \tag{2.61}$$

It follows that

$$\int_0^T \int_{R<|\zeta-\bar{x}|<\rho} |u(\tau,\zeta)| e^{-c|\zeta-\bar{x}|^2} d\zeta d\tau \to 0 \quad \text{as} \quad R \to \infty. \tag{2.62}$$

From (2.60), taking $\rho = R + 1$ we get

$$|u(\bar{t},\bar{x})| \leq H_1 (1+\rho) e^{-\frac{R^2}{2T}+c\rho^2} \int_0^T \int_{R\leq|\zeta-\bar{x}|<\rho} |u(\tau,\zeta)| e^{-c|\zeta-\bar{x}|^2} d\zeta d\tau.$$

The integral on the right hand side $\to 0$ as $R \to \infty$ from (2.62) so that the quantity on the right hand side $\to 0$ as $R \to \infty$ if c is chosen such that $a < c < \frac{1}{2T}$. Such a choice is possible since by assumption $0 < a < \frac{1}{2T}$. Thus we finally have $u(\bar{t},\bar{x}) = 0$. $\square$

3

Representation of Square Integrable Martingales

In the last chapter, we saw that if $M \in \mathcal{M}^2$, then the indefinite stochastic integral $Y_t := \int_0^t f dM$ is a square integrable martingale. Can every L^2-martingale on $(\Omega, \mathcal{F}, (\mathcal{F}_t), P)$ be expressed as a stochastic integral with respect to M? We can also pose the question another way. Can any square integrable functional on $(\Omega, \mathcal{F}_T, P)$ be represented as a stochastic integral? The present chapter will be devoted to this question.

We begin with the case when M is the Wiener process. Itô had shown that in this case, all square integrable martingales adapted to the natural filtration of the Wiener process can be represented as a stochastic integral.

3.1 The Itô representation

Let $W_t = (W_t^1, \ldots, W_t^d)$ be a d-dimensional Wiener process (defined on $(\Omega, \mathcal{F}, P)$. Let $\mathcal{G}$ be the completed σ-field generated by $\{W_t : 0 \leq t \leq T\}$. Define $\mathbf{L}^2 = L^2(\Omega, \mathcal{G}, P)$ to be the space of square integrable functionals on the Wiener space. Ω may be taken to be $C([0, T], \mathbb{R}^d)$ and P, the standard d-dimensional Wiener measure on the completed σ-field of Borel sets of $C([0, T], \mathbb{R}^d)$. This is only one possible choice (perhaps a natural one) and the result to be proved below does not depend on any particular choice.

Theorem 3.1 *Let $F \in \mathbf{L}^2$. Then F can be represented as*

$$F = \mathbb{E}(F) + \sum_{j=1}^{d} \int_0^T f_j(s) dW_s^j \tag{3.1}$$

where the $\{f_j : j = 1, \dots, d\}$ *are predictable (w.r.t. the filtration* $(\mathcal{F}_t^W)$*) and*

$$\sum_{j=1}^{d} E\int_0^T f_j^2(s)ds < \infty. \tag{3.2}$$

Proof Let $\mathcal{H}$ be the class of random variables $F \in \mathbf{L}^2$ admitting a representation (3.1) for some (f_j) satisfying (3.2). We wish to show that $\mathcal{H} = \mathbf{L}^2$.

We will first show that $\mathcal{H}$ is a closed subspace of $\mathbf{L}^2$. The linearity of the stochastic integral implies that $\mathcal{H}$ is a linear subspace. Let $F^n \in \mathcal{H}$ be such that $F^n \longrightarrow F$ in $\mathbf{L}^2$. It follows that $\mathbb{E}(F^n) \to \mathbb{E}(F)$ and thus, writing $G^n = F^n - \mathbb{E}(F^n)$ and $G = F - \mathbb{E}(F)$, we can conclude that $G^n \longrightarrow G$ in $\mathbf{L}^2$ and $G^n \in \mathcal{H}$. Let

$$G^n = \sum_{j=1}^{d} \int_0^T f_j^n(s)dW_s^j.$$

Then we have

$$\mathbb{E}[(G^n - G^m)^2] = \mathbb{E}\Bigg[\sum_{j=1}^{d} \int_0^T (f_j^n(s) - f_j^m(s))^2 ds\Bigg].$$

Since G^n converges to G, the expression above goes to zero as $n, m \to \infty$. It follows that there exist predictable processes f_j such that

$$\mathbb{E}\Bigg[\sum_{j=1}^{d} \int_0^T (f_j^n(s) - f_j(s))^2 ds\Bigg] \longrightarrow 0. \tag{3.3}$$

The relation (3.3) implies that

$$\sum_{j=1}^{d} \int_0^T f_j^n(s)dW_s^j \longrightarrow \sum_{j=1}^{d} \int_0^T f_j(s)dW_s^j$$

in $\mathbf{L}^2$ and thus we have

$$G = \sum_{j=1}^{d} \int_0^T f_j(s)dW_s^j.$$

This proves completeness of $\mathcal{H}$.

To complete the proof, it suffices to show that any random variable F, orthogonal to $\mathcal{H}$, is identically equal to 0. So let F be orthogonal to $\mathcal{H}$. Fix $g_1, g_2, \dots, g_d \in L^2[0, T]$ and let

$$Z_t = \sum_{j=1}^{d} \int_0^t g_j(s)dW_s^j.$$

We saw in Chapter 2 that $Y_t = \exp\{Z_t - \frac{1}{2}\langle Z, Z\rangle_t\}$ satisfies

$$Y_t = 1 + \sum_{j=1}^{d} \int_0^t Y_s g_j(s) dW^j(s).$$

Here $\langle Z, Z\rangle_t = \sum_{j=1}^d \int_0^t (g_j(s))^2 ds$ and Z_t has normal distribution with variance $\langle Z, Z\rangle_t$. It follows that $Y_T \in \mathcal{H}$, and then that $\exp\{Z_T\}$ belongs to $\mathcal{H}$ as well. Thus orthogonality of F to $\mathcal{H}$ implies

$$\mathbb{E}\left[F \exp\{\sum_{j=1}^{d} \int_0^T g_j(s) dW_s^j\} \right] = 0. \tag{3.4}$$

Fix k, rational numbers λ_{jl}, $1 \leq j \leq d$, $1 \leq l \leq 2^k$ and let

$$g_j(s) = \lambda_{jl} \quad \text{for} \ \ (l-1)2^{-k}T \leq s \leq l2^{-k}T.$$

Let us write $Z_{jl} = (W^j_{l2^{-k}} - W^j_{(l-1)2^{-k}})$. Using (3.4) for this choice of (g_j), we conclude that

$$\mathbb{E}\big[F \exp\{\sum_{jl} \lambda_{jl} Z_{jl}\}\big] = 0. \tag{3.5}$$

Let $\mathcal{G}_k = \sigma(Z_{jl})$, the σ-field generated by the Z_{jl}, and let f be a function (on $\mathbb{R}^{dm}$ with $m = 2^k$) such that

$$\mathbb{E}\big[F | \mathcal{G}_k\big] = f(Z_{11}, Z_{12}, \dots, Z_{dm}).$$

Then the relation (3.5) gives

$$\int \dots \int f(z_{11}, \dots, z_{dm}) e^{\{-\frac{m}{2T}\sum_{jl} z_{jl}^2 + \sum_{jl}\lambda_{jl} z_{jl}\}} dz_{11} dz_{12} \dots dz_{dm} = 0.$$

Since this holds for all real numbers λ_{jl}, unicity of the bilateral Laplace transform implies that $f = 0$ *a.s.* (Lebesgue). It follows that

$$\mathbb{E}\big[F | \mathcal{G}_k\big] = 0 \tag{3.6}$$

Note that $\mathcal{G}_k$ is an increasing family of sub-σ fields of $\mathcal{F}$ and that

$$\mathcal{G} = \sigma\big\{\cup_k \mathcal{G}_k\big\}.$$

It follows from the martingale convergence theorem and (3.6) that

$$\mathbb{E}\big[F | \mathcal{G}\big] = 0. \tag{3.7}$$

Since F is assumed to be $\mathcal{G}$ measurable, we conclude that $F = 0$. This completes the proof. □

An alternative way to state this result is the following.

Theorem 3.2 *Every square integrable martingale $(M_t, \mathcal{F}_t^W)$ can be represented as*

$$M_t = M_0 + \sum_{j=1}^{d} \int_0^t f_s^j dW_s^j$$

for some $(\mathcal{F}_t^W)$-predictable process f_j satisfying

$$\sum_{j=1}^{d} E \int_0^T f_j^2(s) ds < \infty.$$

This follows from the previous result by taking $F = M_T$ and noting that M_0 must be a constant and must equal $\mathbb{E}M_t$.

3.2 The Kunita–Watanabe representation

Let $(X_t, \mathcal{F}_t)$ be a square integrable martingale on $(\Omega, \mathcal{F}, P)$ where $(\mathcal{F}_t), 0 \leq t \leq T$ is a filtration satisfying the usual conditions. Introduce the notation $\mathbf{L}^2 = L^2(\Omega, \mathcal{F}_T, P)$.

In this section we will obtain a representation of L^2-martingales adapted to the filtration $(\mathcal{F}_t)$. The following result is important in discussing the possible incompleteness of markets. It is of basic theoretical value.

Theorem 3.3 (Kunita–Watanabe) *Let $(X_t, \mathcal{F}_t)$ be a square integrable martingale. Then for any $M_T \in \mathbf{L}^2$ we have*

$$M_T = \mathbb{E}(M_T) + \int_0^T h_s dX_s + Z_T \tag{3.8}$$

where $h \in \mathcal{L}^2(X)$, i.e., h is predictable,

$$\mathbb{E} \int_0^T h^2(s) d\langle X, X\rangle_s < \infty, \tag{3.9}$$

$\mathbb{E}[Z_T] = 0$, $\mathbb{E}[Z_T^2] < \infty$, *and*

$$\mathbb{E}[Z_T \textstyle\int_0^T f dX] = 0 \tag{3.10}$$

for all $f \in \mathcal{L}^2(X)$. Equivalently, if $(M_t, \mathcal{F}_t)$ is any L^2-martingale, then it admits a representation

$$M_t = M_0 + \int_0^t h_s dX_s + Z_t \tag{3.11}$$

where $h \in \mathcal{L}^2(X)$ and Z is an L^2-martingale with $Z_0 = 0$. Further, the martingales Z and X are orthogonal in the sense that

$$\langle X, Z\rangle_t = 0 \quad a.s. \quad \forall t \in [0, T]. \tag{3.12}$$

Remark The processes (Y_t), $Y_t = \int_0^t h dX$ and (Z_t) are r.c.l.l. in general. The orthogonality relation (3.12) is sometimes referred to as *strong orthogonality* since (3.12) implies $X_t \perp Z_t \quad \forall t$:

$$\mathbb{E}(X_t Z_t) = \mathbb{E}\langle X, Z\rangle_t = 0 \quad \forall t.$$

Proof Define $\mathbf{L}_0^2 = \{F \in \mathbf{L}^2 : \mathbb{E}(F) = 0\}$, and let

$$\mathcal{H} = \{F_T \in \mathbf{L}_0^2 : F_T = \int_0^T f dX, \quad \mathbb{E}\int_0^T f^2(s) d\langle X, X\rangle_s < \infty\}.$$

Then $\mathcal{H}$ is a closed linear subspace of the Hilbert space $\mathbf{L}^2$. Linearity is obvious. Let $F_T^n \to F_T^*$ in $\mathbf{L}_0^2$, where $F_T^n \in \mathcal{H}$. Then writing $F_T^n = \int_0^T f^n dX$ we have

$$\begin{aligned} \mathbb{E}[F_T^n - F_T^m]^2 &= \mathbb{E}\Big[\int_0^T (f^n - f^m) dX\Big]^2 = \mathbb{E}\int_0^T (f^n - f^m)^2 d\langle X, X\rangle_s \\ &= \| f^n - f^m \|_{L^2(\mu)}^2 \to 0 \end{aligned}$$

as $m, n \to \infty$, where μ is the measure on $(\Omega \times [0, T], \mathcal{P})$ defined by

$$\mu(C) = \mathbb{E}\Big[\int_0^T 1_C(\omega, t) d\langle X, X\rangle_t\Big].$$

Since $L^2(\mu)$ is a Hilbert space, the Cauchy sequence $\{f^n\}$ converges to $f^* \in L^2(\mu)$ (and hence f^* is $\mathcal{P}$-measurable and $(\mathcal{F}_t)$ adapted) such that $\| f^n - f^* \|_{L^2(\mu)} \to 0$. Clearly, $\mathbb{E}\int_0^T (f^*)^2 d\langle X, X\rangle_s < \infty$ so that $\tilde{F}_T^* := \int_0^T f^* dX$ is defined. Further, $\mathbb{E}[F_T^n - \tilde{F}_T]^2 \to 0$ and hence $F_T^* = \tilde{F}_T$ *a.s.*, $\mathbb{E}(\tilde{F}_T) = 0$, i.e., $F_T^* \in \mathcal{H}$ and thus $\mathcal{H}$ is closed.

Let $\mathcal{H}^\perp$ be the orthogonal complement of $\mathcal{H}$ in $\mathbf{L}_0^2$.

We now show that $\mathcal{H}$ has the following stability property. Let $F_T \in \mathcal{H}$ and let $\tau \le T$ be any stopping time for $(\mathcal{F}_t)$. Let $F_t = \mathbb{E}[F_T | \mathcal{F}_t]$ be the martingale associated with F_T and let $F_T^\tau = \mathbb{E}[F_T | \mathcal{F}_\tau]$. Then F_T^τ also belongs to $\mathcal{H}$. To see this, note that if $F_T = \int_0^T f dX$, then $F_T^\tau = \int_0^T f 1_{[0,\tau]} dX$, and clearly

$$E\int_0^T [1_{[0,\tau]} f]^2 d\langle X, X\rangle \le E\int_0^T f^2 d\langle X, X\rangle < \infty.$$

Hence $F_T^\tau \in \mathcal{H}$, i.e., $\mathcal{H}$ is a *stable* subspace.

If $\mathcal{H} = \mathbf{L}_0^2$, the proof is complete. Otherwise, let $\mathcal{H}^\perp$ denote the orthogonal complement of $\mathcal{H}$ in $\mathbf{L}_0^2$. We now show that $\mathcal{H}^\perp$ is also stable.

Let $N_T \in \mathcal{H}^{\perp}$. Let $\tau \leq T$ be a stopping time and let $N_T^{\tau} = \mathbb{E}[N_T|\mathcal{F}_{\tau}]$. To show that $N_T^{\tau} \in \mathcal{H}^{\perp}$, take $F_T \in \mathcal{H}$. Then

$$\begin{aligned}
\mathbb{E}(F_T N_T^{\tau}) &= \mathbb{E}[\mathbb{E}(F_T N_T^{\tau}|\mathcal{F}_{\tau})] \\
&= \mathbb{E}[N_T^{\tau}\mathbb{E}(F_T|\mathcal{F}_{\tau})] \\
&= \mathbb{E}[N_T^{\tau} F_T^{\tau}] \\
&= \mathbb{E}[\mathbb{E}(N_T|\mathcal{F}_{\tau})F_T^{\tau}] \\
&= \mathbb{E}[N_T F_T^{\tau}] \\
&= 0
\end{aligned}$$

where the last step follows from the fact that $\mathcal{H}$ is stable and so $F_T^{\tau} \in \mathcal{H}$. Thus $\mathcal{H}^{\perp}$ is a stable subspace.

Let us now consider the given L^2-martingale M. Without loss of generality we assume $M_0 = 0$; otherwise we work with $M_t - M_0$. Then $M_T \in \mathbf{L}_0^2$. Let Y_T be the orthogonal projection of M_T onto $\mathcal{H}$. Setting $Z_T = M_T - Y_T$, $M_t = \mathbb{E}[M_T|\mathcal{F}_t]$, $Y_t = \mathbb{E}[Y_T|\mathcal{F}_t]$, $Z_t = \mathbb{E}[Z_T|\mathcal{F}_t]$, we have the following relations:

$$M_t = Y_t + Z_t,$$

where $Y_t = \int_0^t h dX$ and Z_t is an r.c.l.l. L^2-martingale with $Z_0 = 0$.

Next, since $\mathcal{H}$ and $\mathcal{H}^{\perp}$ are stable, for *any* $L_T \in \mathcal{H}$, we have $E(L_T^{\tau} Z_T^{\tau}) = 0$ for every finite stopping time τ. Hence $L_t Z_t$ is a martingale with respect to $(\mathcal{F}_t)$. We separate the proof of this statement in the form of a lemma following this proof. In particular $X_t Z_t$ is a martingale and thus $\langle X, Z\rangle = 0$. (Recall that $\langle X, Z\rangle$ is the unique increasing predictable process A such that $X_t Z_t - A_t$ is a martingale). □

Lemma 3.4 *Let $\zeta_t, 0 \leq t \leq T$ be a right continuous adapted process with $E|\zeta_{\tau}| < \infty$ and $E(\zeta_{\tau}) = 0$ for every finite stopping time τ. Then $(\zeta_t, \mathcal{F}_t)$ is a uniformly integrable martingale.*

Proof Fix $t \in [0, T]$ and let $A \in \mathcal{F}_t$. Define $\tau(\omega) = t$ if $\omega \in A$ and $\tau(\omega) = T$, if $\omega \in A^c$. The right continuity of (ζ_t) ensures that ζ_{τ} is a r.v. By our assumptions

$$0 = E(\zeta_{\tau}) = \int_A \zeta_t dP + \int_{A^c} \zeta_T dP.$$

Also,

$$\int_A \zeta_T dP + \int_{A^c} \zeta_T dP = E(\zeta_T) = 0.$$

Hence

$$\int_A \zeta_T dP = \int_A \zeta_t dP.$$

Since this equality holds for all $A \in \mathcal{F}_t$, we have $\zeta_t = E(\zeta_T|\mathcal{F}_t)$. The uniform integrability is obvious). □

Let us note here that the Z_T in the representation (3.8) is orthogonal to every element in $\mathcal{H}$. This follows from the proof or by noting that for $L_T \in \mathcal{H}$, $L_t = \mathbb{E}[L_T|\mathcal{F}_t]$, $L_t = \int_0^t g dX$

$$\langle L, Z\rangle_t = \int_0^t g\, d\langle X, Z\rangle = 0.$$

A vector-valued version of Theorem (3.3) is often useful and can be proved following the arguments in the proof of the 1-dimensional case.

Theorem 3.5 *Let $X = (X^1, \ldots, X^d)$ be a d-dimensional L^2-martingale. If $M_T \in \mathbf{L}^2$, then*

$$M_t = M_0 + \sum_{i=1}^{d} \int_0^t H_s^i dX_s^i + Z_t,$$

where

$$\mathbb{E}\Big[\sum_{i=1}^{d} \int_0^T (H_s^i)^2 d\langle X^i, X^i\rangle_s\Big] < \infty$$

and Z is an L^2-martingale with $Z_0 = 0$ and

$$\langle X^i, Z\rangle_t = 0 \quad \forall t,\ 1 \le i \le d. \tag{3.13}$$

It may be noted that the conclusion (3.13) can be rephrased as

$$Z_t X_t^i \text{ is a martingale for } 1 \le i \le d. \tag{3.14}$$

When $X = (X^1, \ldots, X^d)$ is a continuous local martingale, we can get a result analogous to the decomposition theorem given above for any local martingale M.

Theorem 3.6 *Let $X = (X^1, \ldots, X^d)$ be a d-dimensional continuous local martingale and let M be any local martingale. Then (M_t) can be decomposed as*

$$M_t = M_0 + \sum_{i=1}^{d} \int_0^t h_s^i dX_s^i + Z_t, \tag{3.15}$$

where h is a predictable process with

$$\int_0^t \left[\sum_{i=1}^{d} |h_s^i|^2 d\langle X^i, X^i\rangle_s\right] < \infty \ \ a.s. \quad \forall t < \infty$$

and Z is a local martingale that is orthogonal to X^i for $1 \le i \le d$, i.e.,

$$Z_t X_t^i \textit{ is a local martingale for } 1 \le i \le d.$$

Proof First, using Theorem 1.34, let us write M as

$$M_t = N_t + A_t$$

where N is a local square integrable martingale and $A \in \mathcal{V}$. Here A is also a local martingale as M, N are local martingales. Hence $[X^i, A] = 0$ by Theorem 1.42 since X^i is continuous. Hence,

$$X_t^i A_t \text{ is a local martingale for } 1 \leq i \leq d.$$

Let τ_n be a sequence of bounded stopping times increasing to infinity such that $X^{n,i} = X^i_{t\wedge\tau_n}$, $N_{t\wedge\tau_n}$ are square integrable martingales. Using Theorem 3.5, we can get $H^{n,i}$ predictable processes and Z^n such that

$$N_{t\wedge\tau_n} = N_0 + \int_0^t H_s^{n,i} dX_s^{n,i} + Y_t^n,$$

where Y^n and $X^{n,i}$ are martingales. Since $H^{n,i}$ and Y^n are uniquely determined by these conditions, it follows that they determine predictable processes h^i and Z such that

$$\int_0^{t\wedge\tau_n} |h^i|^2 d\langle X^i, X^i\rangle = \int_0^{t\wedge\tau_n} |h^{n,i}|^2 d\langle X^{n,i}, X^{n,i}\rangle \quad \forall t \ \ a.s.$$

$$\int_0^{t\wedge\tau_n} h^i dX^i = \int_0^{t\wedge\tau_n} h^{n,i} dX^{n,i} \quad \forall t \ \ a.s.$$

and

$$Y_{t\wedge\tau_n} = Y^n_{t\wedge\tau_n} \quad \forall t \ \ a.s.$$

Hence it follows that

$$N_t = N_0 + \sum_{i=1}^d \int_0^t h_s^i dX_s^i + Y_t$$

where Y and YX^i are local martingales. The result now follows by taking $Z = Y + A$. □

4

Stochastic Differential Equations

In this chapter, we consider the stochastic differential equations of diffusion type and present a result on the existence and uniqueness of solution. We also prove a version of the Feynman–Kac formula.

4.1 Preliminaries

We saw earlier that if (W_t) is a Brownian motion, then $X_t = \exp(W_t - \frac{1}{2}t)$ satisfies

$$X_t = 1 + \int_0^t X_s dW_s. \tag{4.1}$$

This is also expressed as

$$dX_t = X_t dW_t, \; X_0 = 1. \tag{4.2}$$

The equation (4.1) is called a stochastic differential equation (SDE) which is to be interpreted via the integral equation (4.2). We will consider, more generally,

$$dX_t = \sigma(t, X_t) dW_t + b(t, X_t) dt, \; X_0 = 1, \tag{4.3}$$

where σ, b satisfy Lipschitz conditions. We will also consider the case when X, W, b are vector-valued and σ is matrix-valued (of appropriate dimensions). The solution to (4.3) is called a diffusion process.

We begin with some preliminaries.

Lemma 4.1 *Let $\phi(t)$ be a bounded measurable function on $[0, T]$ satisfying, for some $0 < \alpha < \infty$, $0 < \beta < \infty$,*

$$\phi(t) \leq \alpha + \beta \int_0^t \phi(s)ds, \quad 0 \leq t \leq T. \tag{4.4}$$

Then

$$\phi(t) \leq \alpha \exp(\beta t).$$

Proof Writing

$$g(t) = e^{-\beta t} \int_0^t \phi(s)\, ds$$

we have

$$g'(t) = e^{-\beta t}\phi(t) - \beta e^{-\beta t} \int_0^t \phi(s)\, ds.$$

Substituting for $\phi(t)$ from (4.4),

$$g'(t) \leq \alpha e^{-\beta t} \quad \text{a.e. in } [0, T].$$

Hence $g(t) = \frac{\alpha}{\beta}(1 - e^{-\beta t})$ from which we get

$$\int_0^t \phi(s)\, ds \leq \frac{\alpha}{\beta}(e^{\beta t} - 1).$$

The conclusion $\phi(t) \leq \alpha\, e^{\beta t}$ follows immediately from (4.4). □

Let $L(d, m)$ denote the class of $d \times m$ real matrices and let $\|\cdot\|$ denote the matrix norm.

Lemma 4.2 *Let $f = (f_{ij})$ be an $L(d, m)$ valued $(\mathcal{F}_t)$-predictable process. Let (W_t) be an $\mathbb{R}^m$-valued Wiener process s.t. $(W_t^i, \mathcal{F}_t)$ is a martingale, $1 \leq i \leq m$. Suppose*

$$\mathbb{E} \int_0^T \|f_s\|^2 ds < \infty.$$

Then

$$\mathbb{E}[\sup_{t \leq T} | \int_0^T f_s dW_s|^2] \leq 4\mathbb{E} \int_0^t \|f_s\|^2 ds$$

Proof

$$\begin{aligned}
\mathbb{E}[\sup_{t\le T} |\int_0^t f_s dW_s|^2] &\le \sum_{i=1}^d \mathbb{E}|\sup_{t\le T} |\sum_{j=1}^m \int_0^t f_s^{ij} dW_s^j|^2 \\
&\le 4\sum_{i=1}^d \mathbb{E}|\sum_{j=1}^m \int_0^T f_s^{ij} dW_s^i|^2 \\
&= 4\sum_{i=1}^d \mathbb{E}(\int_0^T \sum_{j=1}^m (f_s^{ij})^2 ds) \\
&= 4\mathbb{E}\int_0^T \| f_s\|^2 ds. \qquad \square
\end{aligned}$$

4.2 Existence and uniqueness of solutions

In preparation for our SDE, we introduce the following quantities:

(1) $b(t,x) : [0,T] \times \mathbb{R}^d \to \mathbb{R}^d$ is a Borel measurable function of (t,x);

(2) $\sigma(t,x) = \big(\sigma_{ij}(t,x) \quad i \in \{1,\dots,d\}$ and $j = \{1,\dots,m\}\big)$ is a $d \times m$ matrix function such that each $\sigma_{ij}(t,x) : [0,T] \times \mathbb{R}^d \to \mathbb{R}$ is Borel measurable. Also write $|b|^2 = \sum_{i=1}^d (b^i)^2$ and $\| \sigma \|^2 = \sum_{i=1}^d \sum_{j=1}^m \sigma_{ij}^2$, the square of the norm of the matrix σ.

(3) An SDE

$$dX_t = b(t,X_t)dt + \sigma(t,X_t)\,dW_t, \qquad 0 < t \le T,\ X_0 = \xi, \qquad (4.5)$$

is really the integral equation:

(4)

$$X_t = X_0 + \int_0^t b(s,X_s)\,ds + \int_0^t \sigma(s,X_s)\,dW_s, \qquad 0 \le t \le T. \quad (4.6)$$

In (4.6), $W = (W^1,\dots,W^m)$ is an m-dimensional standard Wiener process defined on a probability basis $(\Omega, \mathcal{F}, (\mathcal{F}_t), P)$ such that $(W_t, \mathcal{F}_t)$ is a martingale and ξ is a d-dimensional random variable on $(\Omega, \mathcal{F}, P)$, independent of W. Under these assumptions, the problem is to find a process (X_t) defined on the same probability space, satisfying (4.6).

The stochastic integral in (4.6) is a vector-valued Itô integral whose i^{th} component is $\sum_{j=1}^m \int_0^t \sigma_{ij}(s,X_s)\,dW_s^j$. Equation (4.6), written in terms of the components of X, takes the form:

$$(5)\quad X_t^i = \xi^i + \int_0^t b^i(s,X_s)ds + \sum_{j=1}^m \int_0^t \sigma_{ij}(s,X_s)dW_s^j, \quad i = 1,...,d. \quad (4.7)$$

Theorem 4.3 *Let the coefficient functions b and σ satisfy the following conditions: There exists a positive constant K_T such that for $0 \le t \le T$, and $x, y \in \mathbb{R}^d$,*

$$|b(t,x) - b(t,y)|^2 + \|\sigma(t,x) - \sigma(t,y)\|^2 \le K_T\, |x-y|^2 \text{ (Lipschitz condition)} \tag{4.8}$$

$$|b(t,x)|^2 + \|\sigma(t,x)\|^2 \le K_T \left(1 + |x|^2\right) \text{ (Growth condition)}. \tag{4.9}$$

Also let

$$E\ |\xi|^2 < \infty. \tag{4.10}$$

Then the SDE (4.5) has a unique solution (X_t), $0 \le t \le T$ on the given probability space such that

$$X_0 = \xi \text{ a.s.}; \tag{4.11}$$

X has the following additional properties:

$$X \text{ is adapted to } (\mathcal{F}_t); \tag{4.12}$$

$$E \int_0^T \left(|b(t,X_t)| + \|\sigma(t,X_t)\|^2\right) dt < \infty. \tag{4.13}$$

The process X satisfying (4.6), (4.11), (4.12) and (4.13) is called a *strong solution.*

Proof (Existence) We use the method of successive approximations. Successively define

$$\begin{array}{rcl} X_0(t) & = & \xi \quad \text{for all } t \in [0,T], \\ X_1(t) & = & \xi + \int_0^t b(s, X_0(s))ds + \int_0^t \sigma(s, X_0(s))dW_s, \\ \vdots & & \\ X_{n+1}(t) & = & \xi + \int_0^t b(s, X_n(s))ds + \int_0^t \sigma(s, X_n(s))dW_s \\ \cdots & \text{etc.} & \end{array} \tag{4.14}$$

We will also use the martingale inequality for the vector-valued martingale $\int_0^t \sigma(s, X_0(s))dW_s \equiv M_t$ say. Writing M^i for the ith component of M, we have

$$\begin{aligned} E \sup_{0\le t\le T} |M_t|^2 &= \sum_{i=1}^d E \sup_{0\le t\le T} \left|M_t^i\right|^2 \le 4 \sum_{i=1}^d E\left(M_T^i\right)^2 \\ &= 4 \sum_{i-1}^d E\left(\sum_{j=1}^m \int_0^T \sigma_{ij}(s, X_0(s))dW_s^j\right)^2 \\ &= 4E \int_0^T \|\sigma(s, X_0(s))\|^2 ds. \end{aligned}$$

From

$$X_{n+1}(s) - X_n(s) = \int_0^s \left[b(u, X_n(u)) - b(u, X_{n-1}(u))\right] du \quad (4.15)$$
$$+ \int_0^s \left[\sigma(u, X_n(u)) - \sigma(u, X_{n-1}(u))\right] dW_u,$$

we get

$$E \sup_{0 \le s \le t} |X_{n+1}(s) - X_n(s)|^2 \le \quad (4.16)$$
$$2t \int_0^t E\,|b(u, X_n(u)) - b(u, X_{n-1}(u))|^2\, du$$
$$+ 2E \sup_{0 \le s \le t} \left| \int_0^s \left[\sigma(u, X_n(u)) - \sigma(u, X_{n-1}(u))\right] dW_u \right|^2$$
$$\le 2(T+4) \int_0^t E\left(|b(u, X_n(u)) - b(u, X_{n-1}(u))|^2 \right.$$
$$\left. + \|\sigma(u, X_n(u)) - \sigma(u, X_{n-1}(u))\|^2 \right) du$$
$$\le 2(T+4)K_T \int_0^t E\,|X_n(s) - X_{n-1}(s)|^2\, ds, \quad 0 \le t \le T,$$

from condition (4.8). The right hand side is dominated by

$$\le 2(T+4)K_T \int_0^t E \sup_{0 \le u \le s} |X_n(u) - X_{n-1}(u)|^2\, ds. \quad (4.17)$$

Writing $\Delta_{n+1}(t) = E \sup_{0 \le s \le t} |X_{n+1}(s) - X_n(s)|^2$ and $L = 2(T+4)K_T$, we then have

$$\Delta_{n+1}(t) \le L \int_0^t \Delta_n(s)\, ds. \quad (4.18)$$

Iterating the above inequality and using integration by parts, we have

$$\Delta_{n+1}(t) \le \frac{L^{n+1}T^{n+1}}{n!} \Delta_1(T), \quad (4.19)$$

where $\Delta_1(T) = E \sup_{0 \le s \le T} |X_1(s) - \xi|^2 < \infty$. Hence

$$P\left[\sup_{0 \le t \le T} |X_{n+1}(t) - X_n(t)| > \frac{1}{2^n} \right] \le \frac{(4LT)^{n+1}}{n!} \Delta_1(T). \quad (4.20)$$

The Borel–Cantelli lemma then implies that

$$\sum_{n=1}^{\infty} \sup_{0 \le t \le T} |X_{n+1}(t) - X_n(t)| < \infty \quad \text{a.s.} \quad (4.21)$$

Thus $\lim_{n\to\infty} X_n(t)$ exists a.s. Denoting this limit by $X(t)$ (and setting $X(t,\omega)=0$ for all t and ω in the exceptional P-null set) we see at once that $t \to X(t)$ is continuous and $X(t)$ is adapted.

For $m > n$ we have

$$\begin{aligned} |X_m(t) - X_n(t)|^2 &= \left| \sum_{j=n+1}^{m} \left[X_j(t) - X_{j-1}(t)\right] \right|^2 \\ &\le 2^{-n} \sum_{j=n+1}^{m} 2^j \left|X_j(t) - X_{j-1}(t)\right|^2 . \end{aligned} \tag{4.22}$$

So

$$\sup_{0\le t\le T} |X_m(t) - X_n(t)|^2 \le 2^{-n} \sum_{j=1}^{\infty} 2^j \sup_{0\le t\le T} \left|X_j(t) - X_{j-1}(t)\right|^2 \tag{4.23}$$

and hence

$$\begin{aligned} E \sup_{0\le t\le T} |X_m(t) - X_n(t)|^2 &\le 2^{-n} \sum_{j=1}^{\infty} 2^j \Delta_j(T) \\ &\le 2^{-n} \sum_{j=1}^{\infty} \frac{2^j (LT)^j}{(j-1)!} \Delta_1(T) \quad \text{from (4.23)} \\ &\le 2^{-n+1}(LT) \sum_{j=0}^{\infty} \frac{(2LT)^j}{j!} \Delta_1(T) < \infty . \end{aligned} \tag{4.24}$$

Putting $n = 1$ and changing m to n in (4.24), we have

$$E \sup_{0\le t\le T} |X_n(t)|^2 \le A_T, \tag{4.25}$$

which is a constant independent of N. Next, keeping n fixed in (4.24) and making $m \to \infty$

$$\sup_{0\le t\le T} |X_m(t) - X_n(t)|^2 \to \sup_{0\le t\le T} |X(t) - X_n(t)|^2 \text{ a.s.} \tag{4.26}$$

From (4.23) and the dominated convergence theorem, we have that

$$E \sup_{0\le t\le T} |X(t) - X_n(t)|^2 \le 2^{-n+1}(LT) \sum_{j=0}^{\infty} \frac{(2LT)^j}{j!} \Delta_1(T). \tag{4.27}$$

It immediately follows from (4.25) and (4.27) that

$$E \sup_{0\le t\le T} |X(t)|^2 < \infty. \tag{4.28}$$

From

$$\begin{aligned} E & \left\{ \int_0^T |b(s, X_n(s))|^2 ds + \int_0^T \|\sigma(s, X_n(s))\|^2 ds \right\} \\ & \leq K_T \int_0^T E\left[1 + |X_n(s)|^2\right] ds \qquad (4.29) \\ & \leq K_T \left[T + T\, E \sup_{0 \leq s \leq T} |X_n(s)|^2\right] \end{aligned}$$

and (4.25), it follows (using Fatou's lemma) that

$$E\left[\int_0^T |b(s, X(s))|^2 ds + \int_0^T \|\sigma(s, X(s))\|^2 ds\right] < \infty. \qquad (4.30)$$

which yields assertion (4.13) of the theorem. It remains to verify that $X(t)$ is a solution of our SDE. Temporarily setting

$$\tilde{X}(t) := \xi + \int_0^t b(s, X(s))\, ds + \int_0^t \sigma(s, X(s)) dW_s, \qquad (4.31)$$

we have

$$\begin{aligned} E\left|X_{n+1}(t) - \tilde{X}(t)\right|^2 & \leq 2E\left|\int_0^t [b(s, X_n(s)) - b(s, X(s))]\, ds\right|^2 \\ & \quad + 2E\int_0^t \|\sigma(s, X_n(s)) - \sigma(s, X(s))\|^2 ds \\ & \leq 2(T+1)K_T \int_0^T E\, |X_n(s) - X(s)|^2 ds \\ & \leq 2(T+1)K_T \cdot T \cdot E \sup_{0 \leq s \leq T} |X_n(s) - X(s)|^2 \\ & \to 0 \qquad \text{as } n \to \infty \text{ by (4.27).} \qquad (4.32) \end{aligned}$$

Therefore by Fatou's lemma we have $E\left|X(t) - \tilde{X}(t)\right|^2 = 0$. Since X and $\tilde{X}$ are both continuous we have, recalling the definition of $\tilde{X}$ that $X(t) = \xi + \int_0^t b(s, X(s))ds + \int_0^t \sigma(s, X(s))dW_s$ a.s. for $0 \leq t \leq T$. Hence X is a strong solution. From (4.27) we also see that

$$E \sup_{0 \leq s \leq t} |X(s)|^2 < \infty. \qquad (4.33)$$

A more diligent calculation will yield the inequality

$$E \sup_{0 \leq s \leq t} |X(s)|^2 \leq 2c_1(1 + E|\xi|^2)e^{ct}, \quad 0 \leq t \leq T, \qquad (4.34)$$

where c_1 and c are constants independent of t. □

Proof (Uniqueness) Let Y be another strong solution of (4.6) with $Y_0 = \xi$. Then, for $0 \le t \le T$,

$$X_t - Y_t = \int_0^t [b(s, X_s) - b(s, Y_s)]\,ds + \int_0^t [\sigma(s, X_s) - \sigma(s, Y_s)]\,dW_s \quad (4.35)$$

and

$$\begin{aligned} \sup_{0\le t\le T} |X_t - Y_t|^2 &\le 2T \int_0^T |b(s, X_s) - b(s, Y_s)|^2\,ds \quad (4.36)\\ &\quad + 2 \sup_{0\le t\le T} \left| \int_0^t [\sigma(s, X_s) - \sigma(s, Y_s)]\,dW_s \right|^2. \end{aligned}$$

Using the martingale inequality for the second term and then condition (4.8),

$$\begin{aligned} &E \sup_{0\le t'\le t} |X_{t'} - Y_{t'}|^2 \quad (4.37)\\ &\le 2T \cdot K_T \int_0^t E|X_s - Y_s|^2\,ds + 8K_T \int_0^t E|X_s - Y_s|^2\,ds. \end{aligned}$$

Writing $H(t)$ for the quantity on the left hand side, we have the inequality

$$H(t) \le 2K_T(T+4) \int_0^t H(s)\,ds,$$

from which it follows that $H(t) = 0$ for $0 \le t \le T$. Hence $P(X_t = Y_t$ for all $t \in [0, T]) = 1$ and uniqueness is proved. □

Remark It follows from the approximation procedure that the law of the solution (X_t) is also uniquely determined by the law of $(X_0, , \sigma, b)$. Thus if (ξ_t) is a solution to the SDE

$$d\xi_t = \sigma(t, \xi_t)dW'_t + b(t, \xi_t)dt,$$

where (W'_t) is an m-dimensional Wiener process on some probability space $(\Omega', \mathcal{F}', P)$, and if

$$P' \circ \xi_0^{-1} = P \circ X_0^{-1},$$

then

$$P' \circ (\xi)^{-1} = P \circ (X)^{-1} \text{ (as measures on } C([0, T], \mathbb{R}^d)).$$

Remark The proof given above also shows that if σ, b are functions on $[0, \infty) \times \mathbb{R}^d$ satisfying the assumed conditions on $[0, T] \times \mathbb{R}^d$ for every $T < \infty$, then the SDE admits a unique solution on $[0, T]$ for every $T < \infty$, and hence a solution for $t \in [0, \infty)$.

For $(s, x) \in [0, T] \times \mathbb{R}^d$, consider the SDE

$$\begin{aligned} dX_t &= \sigma(t, X_t)dW_t + b(t, X_t)dt, \quad s \le t \le T\\ X_s &= x. \end{aligned}$$

The proof given above (with obvious modifications) shows that this equation admits a unique solution $(X_t^{s,x})$ and that its law

$$P \circ (X_0^{s,x})^{-1} = P_{s,x} \quad \text{(a measure on } C([s,T],\mathbb{R}^d))$$

is uniquely determined.

Let (X_t) be any solution to (4.5) and let $f \in C^{1,2}([0,T]\times\mathbb{R}^d)$. Then by Ito's formula

$$\begin{aligned} f(t,X_t) &= f(0,X_0) + \int_0^t \frac{\partial f}{\partial f}(s,X_s)ds + \sum_{i=1}^d \int_0^t \frac{\partial f}{\partial x_i}(s,X_s)dX_s^i \\ &= \frac{1}{2}\sum_{i=j=1}^d \int_0^t \frac{\partial^2 f}{\partial x_i \partial x_j}(s,X_s)d\langle X^i,X^j\rangle_s. \end{aligned} \tag{4.38}$$

Here

$$\begin{aligned} \langle X^i,X^j\rangle_t &= \sum_k \int_0^t \sigma^{ik}(s,X_s)d\,\sigma^{jk}(s,X_s)ds \\ &= \int_0^t a^{ij}(s,X_s)ds, \end{aligned}$$

where $a(s,x) = \sigma(s,x)\sigma^*(s,x)$. Define A_t (acting on $f \in C^{1,2}([0,T]\times\mathbb{R}^d)$) by

$$(A_t f)(x) = \frac{\partial f}{\partial t}(t,x) + \sum_{i=1}^d b^i(t,x)\frac{\partial f}{\partial x_i} + \frac{1}{2}\sum_{i,j=1}^d a^{ij}(t,x)\frac{\partial f}{\partial x_i \partial x_j}.$$

Then (4.38) can be rewritten as

$$\begin{aligned} &f(t,X_t) - f(0,X_0) - \int_0^t A_u f(X_u)du \\ &= \sum_{i=1}^d \int_0^t \frac{\partial f}{\partial x_i}(u,X_u)\Big[\sum_{j=1}^d \sigma^{ij}(u,X_u)\Big]dW_u^j. \end{aligned}$$

It follows that

$$M_t^f := f(t,X_t) - f(0,X_0) - \int_0^t A_u f(X_u)du$$

is a local martingale since it is a stochastic integral with respect to a Brownian motion. If in addition, we have suitable integrability conditions, then we can conclude that (M_t^f) is a martingale. This is the case if we have, e.g.,

$$\mathbb{E}[\sup_{0\le t\le T} |M_t^f|] < \infty$$

or

$$\sup_{\tau \leq T} \mathbb{E}[|M_\tau^f|^{1+\delta}] < \infty$$

for some $\delta > 0$ where the supremum is taken over all stopping times τ.

It can be shown that under the conditions on σ, b imposed in this section, (X_t) is a Markov process and for $s < t$

$$E[f(X_t)|\sigma(X_u : u \leq s)] = g(X_s)$$

where

$$g(x) = E_{P_{s,x}}[f(\xi_t)]$$

and where $P_{s,x}$ is the law of the solution $X^{s,x}$ constructed earlier and (ξ_t) is the coordinate process on $C([s, T], \mathbb{R}^d)$. One can associate a 2-parameter semigroup with $\{X^{s,x}\}$ and A_t as the (restriction of the) weak generator of this semigroup. (X_t) is also known as a diffusion process.

The solution of the SDE constructed earlier is known as a strong solution; it is constructed on the given probability space where a Wiener process is fixed.

A process X defined on some probability space is said to be a weak solution to the SDE (4.5) if we can construct a Wiener process (W_t), a filtration $(\mathcal{F}_t)$ such that $(W_t, \mathcal{F}_t)$ is a martingale, X_t is $(\mathcal{F}_t)$-adapted and the (X_t, W_t) satisfy (4.5). The notion of uniqueness of a weak solution requires that the law of any weak solution be uniquely determined. If one replaces the Lipschitz condition (4.8) by requiring that a be continuous and uniformly elliptic, then the existence and uniqueness of weak solution holds. We refer the reader to (Stroock and Varadhan, 1978) for this and related material.

Lemma 4.4 *Let*

$$M_t^i = \int_0^t \sum_{j=1}^m f_s^{ij} dW_s^j,$$

where $(W^1, \cdots, W^d)$ *is* d*-dimensional Brownian motion and* $\{f_s^{ij}\}$ *are predictable processes such that, for some* $p \geq 2$,

$$\mathbb{E}(\int_0^T \|f_s\|^2 ds)^{p/2} < \infty.$$

Then

$$\mathbb{E} \sup_{s \leq T} |M_t|^p \leq c(p)\mathbb{E}(\int_0^T \|f_s\|^2 ds)$$

for a constant $c(p)$ *depending only on* p.

Proof For $n \geq 1$, let

$$\tau_n = \inf\{t \geq 0 : |M_t| \geq n \text{ or } \int_0^t \|f_s\|^2 ds \geq n\} \wedge T.$$

Then $M_{t\wedge\tau_n}$ is a bounded martingale for each $n \geq 1$. Applying Ito's formula to the function $h(x) = (\sum_{i=1}^{d} x_i^2)^{p/2}$, one has

$$
\begin{aligned}
|M_{t\wedge\tau_n}|^p &= \sum_{i=1}^{d} \int_0^{t\wedge\tau_n} pM_u^i |M_u|^{p-2} dM_u^i \\
&+\frac{1}{2}\sum_{i=1}^{d} \int_0^{t\wedge\tau_n} p|M_u|^{p-2} d\langle M_u^i, M^i\rangle_u \qquad (4.39) \\
&+\frac{1}{2}\sum_{i,j=1}^{d} \int^{t\wedge\tau_n} p(p-2)M_u^i M_u^j |M_u|^{p-4} d\langle M^i, M^j\rangle_u.
\end{aligned}
$$

The first term in (4.39) is a martingale since the integrand is bounded and the integrator is also a bounded martingale. Further,

$$
\langle M^i, M^j\rangle_u = \int_0^u g_s^{ij}\, ds
$$

where $g_s^{ij} = \sum_k f_s^{ik} f_s^{jk}$. Note that $|g_s^{ij}| \leq \sqrt{g^{ii}g^{jj}}$ and $\sum_i g_s^{ii} = \|f_s\|^2$. Thus for $p \geq 2$,

$$
\begin{aligned}
E\sup_{t\leq T} |M_{t\wedge\tau_n}|^p &\leq \left(\frac{p}{p-1}\right)^p \mathbb{E}|M_{T\wedge\tau_n}|^p \\
&\leq \left(\frac{p}{p-1}\right)^p \{\mathbb{E}\sum_{i=1}^{d} \int_0^{t\wedge\tau_n} p|M_u|^{p-2} g_u^{ii}\, du \qquad (4.40) \\
&+\mathbb{E}\sum_{i,j=1}^{d} \int_0^{t\wedge\tau_n} p(p-1)|M_u|^{p-4} \cdot |M_u^i|\, |M_u^j|\, |g_u^{ij}| du\}.
\end{aligned}
$$

Here we have also used Doob's maximal inequality.

Note that

$$
\begin{aligned}
\sum_{i,j=1}^{d} |M_u^i|\, |M_u^j|\, |g_u^{ij}| &\leq \sum_{i,j=1}^{d} (|M_u^i|\, |M_u^j| \sqrt{g_u^{ii}}\sqrt{g_u^{jj}}) \\
&= \left(\sum_{i=1}^{d} |M_u^i| \sqrt{g_u^{ii}}\right)^2 \\
&\leq \sum (M_u^i)^2 \cdot (\sum_i g_u^{ii}) \\
&= |M_u|^2 \cdot \|f_u\|^2.
\end{aligned}
$$

Using this in (4.40) we conclude

$$\mathbb{E}\sup_{t\leq T}|M_{T\wedge\tau_n}|^p \leq \left(\frac{p}{p-1}\right)^p \cdot p(p-1)\mathbb{E}\int_0^{t\wedge\tau_n}|M_u|^{p-2}\|f_u\|^2du. \quad (4.41)$$

From this, it follows that

$$\begin{aligned}\mathbb{E}\sup_{t\leq T}|M_{t\wedge\tau_n}|^p &\leq C_1(p)\mathbb{E}[(\sup_{t\leq T}|M_{t\wedge\tau_n}|^p)^{\frac{p-2}{p}}\int_0^{t\wedge\tau_n}\|f_u\|^2du] \\ &\leq C_1(p)(\mathbb{E}\sup_{t\leq T}|M_{t\wedge\tau_n}|^p)^{\frac{p-2}{2}}(\mathbb{E}(\int_0^{t\wedge\tau_n}\|f_u\|^2du)^{\frac{p}{2}})^{\frac{2}{p}}.\end{aligned}$$

Since $\mathbb{E}\sup_{t\leq T}|M_{t\wedge\tau_n}|^p < \infty$, we have

$$\begin{aligned}\mathbb{E}\left(\sup_{t\leq T}|M_{t\wedge\tau_n}|^p\right) &\leq C(p)\mathbb{E}((\int_0^{t\wedge\tau_n}\|f_u\|^2du)^{\frac{p}{2}}) \\ &\leq C(p)\mathbb{E}[(\int_0^T\|f_u\|^2du)^{\frac{p}{2}}].\end{aligned}$$

The result follows by taking the limit as $n \longrightarrow \infty$ and using Fatou's lemma. Note that here $C_1(p) = (\frac{p}{p-1})^p \cdot p(p-1)$ and $C(p) = (C_1(p))^{\frac{p}{2}}$. □

Remark Proceeding exactly as above and using the inequality (1.67), one can show the following: If M is a continuous local martingale such that for $p \geq 2$, $\mathbb{E}[(\langle M, M\rangle_T)^{\frac{p}{2}}] < \infty$, then

$$\mathbb{E}[\sup_{0\leq s\leq T}|M_s|^p] \leq C_p\mathbb{E}[(\langle M, M\rangle_T)^{\frac{p}{2}}] \quad (4.42)$$

for a constant C_p. This inequality is indeed true if $p > 1$ and M is an r.c.l.l. local martingale, in which case $\langle M, M\rangle$ above has to be replaced by $[M, M]$. This result is known as Burkholder's inequality.

Theorem 4.5 *Suppose (X_t) is a solution to the SDE (4.5) with $\mathbb{E}|X_0|^p < \infty, 2 \leq p < \infty$. Then there exist finite constants K_1, K_2 (depending on p, d, T, K_T, appearing in (4.8) and (4.9)) such that*

$$E\sup_{t\leq T}|X_t|^p \leq K_1\mathbb{E}|X_0|^p + K_2.$$

Proof The proof uses the L^p-estimate obtained in the previous lemma. Let $\tau_n = \inf\{t : |X_t| \leq n\} \wedge T$. Then

$$\begin{aligned}\mathbb{E}\sup_{u\leq t}|X_{u\wedge\tau_n}|^p &\leq 3^{p-1}\{\mathbb{E}|X_0|^p + \mathbb{E}(\int_0^{t\wedge\tau_n}\|\sigma(u, X_u)\|^2du)^{p/2} \\ &\quad +\mathbb{E}(\int_0^{t\wedge\tau_n}|b(u, X_u| \, |du)^p\} \\ &\leq C_3(p)\{\mathbb{E}|X_0|^p + C_4\mathbb{E}\int_0^{t\wedge\tau_n}(1 + |X_u|^p)du\}\end{aligned}$$

where C_4 depends on p, d, K_T.

Let $\psi(t) = \mathbb{E} \sup_{u \le t} |X_{u\wedge\tau_n}|^p$. Then we have for some constant $C_5(p)$,

$$\psi(t) \le C_5(p)\{\mathbb{E}|X_0|^p + t + \int_0^t \psi(u)du\}.$$

Since $\psi(t) \le n^p$, it follows from Lemma 4.1 that

$$\psi(t) \le C_5(p)(E|X_0|^p + T)\exp(C_3(p)T)$$

Thus

$$\mathbb{E}[\sup_{t\le T} |X_{t\wedge\tau_n}|^p] \le K_1(E|X_0|^p + K_2)$$

for suitable constants K_1 and K_2. Taking the limit as $n \longrightarrow \infty$ and using Fatou's lemma, the result follows. □

4.3 The Feynman–Kac formula

Earlier we saw the connection between the solution to the Cauchy problem for the heat equation and Brownian motion. Similarly, we can represent the solution to the Cauchy problem for $\frac{\partial}{\partial t} + A_t$ and the process (X_t) (where A_t, (X_t) are as in the previous section). We will consider a more general equation, with a potential c.

The representation formula is known as the Feynman–Kac formula.

We will continue to use the notation from the previous section. Thus, σ, b satisfy (4.8) and (4.9), $a(t,x) = \sigma(t,x)\sigma^*(t,x)$, A_t is the differential operator acting on $C^{1,2}([0,T]\times\mathbb{R}^d)$, $\{X_t^{s,x} : s \le t \le T\}$ is the solution to the SDE (4.5) for $t \in [s,T]$ with the initial condition $X_s^{s,x} = x$. $P_{s,x}$ denotes the law of $X^{s,x}$ on $C([s,T],\mathbb{R}^d)$ and ξ_t is the coordinate process on it. Let $c : [0,T]\times\mathbb{R}^d \longrightarrow [0,\infty)$ be a continuous function. Consider the PDE

$$A_t v = cv$$

on $(0,T)\times\mathbb{R}^d$, i.e., for $(t,x) \in (0,T)\times\mathbb{R}^d$

$$\frac{\partial v}{\partial t}(t,x) + \frac{1}{2}\sum_{i,j=1}^d a^{ij}(t,x)\frac{\partial^2 v}{\partial x_i \partial x_j}(t,x) + \sum_{i=1}^d b^i(t,x)\frac{\partial v}{\partial x_i}(t,x) = c(t,x)v(t,x). \tag{4.43}$$

Theorem 4.6 (Feynman–Kac formula) *Suppose*

$$v \in C([0,T]\times\mathbb{R}^d) \cap C^{1,2}((0,T)\times\mathbb{R}^d)$$

is a solution to the PDE (4.43) satisfying

$$|v(t,x)| \le C(1+|x|^{\lambda}) \tag{4.44}$$

for some $C < \infty, \lambda < \infty$. *Then for* $(s,x) \in [0,T) \times \mathbb{R}^d$,

$$v(s,x) = \mathbb{E}_{P_{s.x}}[v(T,\xi_T)\exp\left(\int_s^T -c(v,\xi_u)du\right)]. \tag{4.45}$$

In particular, uniqueness holds for the Cauchy problem for the equation (4.43) with boundary condition $v(T,x) = \varphi(x)$.

Proof Fix $(s,x) \in [0,T) \times \mathbb{R}^d$. Let

$$B_t = \exp(-\int_s^t c(u, X_u^{s,x})du).$$

Let

$$g(t,x,z) = v(t,x)\exp(z).$$

Using Ito's formula, it follows that

$$M_t = g(t, X_t^{s,x}, B_t) = v(t, X_t^{s,x})\exp(-\int_s^t c(u, X_u^{s,x})du)$$

is a local martingale since it is a stochastic integral with respect to a Brownian motion (the remaining terms add up to zero as v is assumed to be a solution of the PDE (4.43)).

Since $c \ge 0$ and v is assumed to satisfy (4.44), it follows that

$$|M_t| \le C(1+|X_t^{s,x}|^{\lambda}).$$

Hence, using Theorem 4.5, we get

$$\begin{aligned} \mathbb{E}[\sup_{s\le t\le T} |M_t|] &\le C(1+\mathbb{E}\sup_{0\le t\le T}|X_t^{s,x}|^{\lambda}) \\ &\le C(1+K_1+K_2|x|^{\lambda}). \end{aligned}$$

As a consequence, the local martingale M_t is a martingale. Thus, $\mathbb{E}M_s = \mathbb{E}M_T$, i.e.,

$$v(s,x) = \mathbb{E}\left(v(T, X_T^{s,x})\exp(-\int_s^T c(u, X_u^{s,x})du\right)].$$

This is the same as (4.45) since $P_{s,x}$ is the law of $(X^{s,x})$.

Since we have obtained the representation (4.45) for an arbitrary solution v, it follows that if the boundary values (at $t = T$) of two solutions v_t, v_2 agree, i.e., $v_1(T,x) = v_2(T,x)\ \forall x$, then

$$v_1(s,x) = v_2(s,x) \quad \forall\,(s,x) \in [0,T] \times \mathbb{R}^d.$$

□

Remark Note that the key ingredient in the proof of the Feynman–Kac formula is to show that the local martingale (M_t) is a martingale. For other sets of conditions on a, b, c under which the representation is valid, see (Kallianpur and Karandikar, 1988).

4.4 The Ornstein–Uhlenbeck process (O.U.P)

An important special case of the stochastic differential equation discussed in Section 4.2 is the one whose solution is a Gaussian–Markov process called the Ornstein–Uhlenbeck process. We present a brief account of it here, leaving the details to the reader.

The n-dimensional O.U. process $X_t = \left(X_t^1, \ldots, X_t^n\right)$ is the unique solution of the SDE

$$dX_t^j = -\lambda_j X_t^j dt + \sigma_j dW_t^j, \quad j = 1, \ldots, n \quad , \quad \sigma_j, \lambda_j > 0 \text{ constants,} \tag{4.46}$$

with $X_0^j = \xi_j$, independent of the Brownian motion $\left(W_t^1, \ldots, W_t^n\right)$. For $s < t$, writing $\mathcal{G}_s$ for the filtration generated by $\left\{X_u^j, 0 \le u \le s, j = 1, \ldots, n\right\}$, it is easy to verify that the conditional distribution of $\left(X_t^1, \ldots, X_t^n\right)$ given $\mathcal{G}_s$ is Gaussian with (conditional) mean

$$m = \left(e^{-\lambda_1(t-s)}X_s^1, \ldots, e^{-\lambda_n(t-s)}X_s^n\right) \tag{4.47}$$

and diagonal variance matrix with entries

$$\gamma_j^2 = \frac{\sigma_j^2}{2\lambda_j}\left[1 - e^{-2\lambda_j(t-s)}\right] \tag{4.48}$$

for $j = 1, \ldots, n$.

The Markov property is easily checked, and the transition probability density of X_t given $X_s = x$ is given by

$$p(t-s; x, y) = \prod_{j=1}^{n} \frac{1}{\sqrt{2\pi}\gamma_j} \exp\left\{\frac{\left(y_j - m_j x_j\right)^2}{2\gamma_j^2}\right\}. \tag{4.49}$$

If the initial value ξ_j is taken to be a Gaussian random variable with zero mean and variance $\frac{\sigma_j^2}{2\lambda_j}$, then X_t is stationary with

$$E\left(X_t^j X_s^k\right) = \begin{cases} \frac{\sigma_j^2}{2\lambda_j} e^{-\lambda_j|t-s|} & , \quad j = k \\ 0 & , \quad j \neq k. \end{cases} \tag{4.50}$$

The generator (or differential operator) is

$$Af = \frac{1}{2}\sum_{j=1}^{n} \sigma_j^2 \frac{\partial^2 f}{\partial x_j^2} - \sum_{j=1}^{n} \lambda_j x_j \frac{\partial f}{\partial x_j}. \tag{4.51}$$

5
Girsanov's Theorem

An important issue in mathematical finance is that of putting conditions on a semimartingale X (defined on $(\Omega, \mathcal{F}, P)$) which ensure the existence of a probability measure Q equivalent to P such that X is a local martingale on $(\Omega, \mathcal{F}, Q)$. We will discuss this in detail in later chapters. Here, we will consider probability measures Q equivalent to P, and show that in general, X is a semimartingale on $(\Omega, \mathcal{F}, Q)$ as well. Also, one can obtain the decomposition of the semimartingale X on $(\Omega, \mathcal{F}, Q)$ into a Q-local martingale N and a process with bounded variation paths B, and relate N, B to M, A, where $X = X_0 + M + A$ is the decomposition of X on $(\Omega, \mathcal{F}, P)$ into a P-local martingale M and a process with bounded variation paths A. The classical Girsanov's theorem is a consequence of this.

5.1 Auxiliary results

The cross-quadratic variation (also called the *square bracket function* $[X, Y]$) was introduced in Chapter 1. It has the property that if X, Y are local martingales vanishing at 0, then $Z = XY - [X, Y]$ is also a local martingale. If X, Y are L^2-local martingales, the predictable cross quadratic variation (also called their *sharp bracket* $\langle X, Y\rangle$) has been introduced as the unique *predictable* process such that $W = XY - \langle X, Y\rangle$ is a local martingale. In this chapter we will consider the two probability measures P and Q which are mutually absolutely continuous, $Q \equiv P$. To avoid confusion in what follows, it is important to know how both these brackets behave when the measure is changed from P to Q. It is easy to see from Theorem 1.39 that $[X, Y]$ is the same under P and Q since P and Q are equivalent. However, such is not the case in general for $\langle X, Y\rangle$. When X and Y are continuous, of course, we have $[X, Y] = \langle X, Y\rangle$.

Throughout the remainder of this section, we fix probability measures P, Q with $Q \equiv P$, i.e., Q and P are mutually absolutely continuous. Let $L = \frac{dQ}{dP}$ and $L_t = \mathbb{E}_P(L|\mathcal{F}_t)$. Let us note that if P_t and Q_t are respectively the restrictions of P, Q to $\mathcal{F}_t$, then $L_t = \frac{dQ_t}{dP_t}$.

We need the following simple lemma.

Lemma 5.1 *Let Y be an r.c.l.l. adapted process and let σ be a bounded stopping time. If either $Y_\sigma L_\sigma$ is P-integrable or if Y_σ is Q-integrable, then*

$$\mathbb{E}_P[Y_\sigma L_\sigma] = \mathbb{E}_Q[Y_\sigma] \tag{5.1}$$

Proof First assume that Y is bounded. Let σ be bounded by T. Consider $f = Y_\sigma 1_{(\sigma,T]}$. Then, f is predictable (Lemma 1.5) and $Z_t = \int_0^t f dL$ is a P-martingale and $Z_0 = 0$. So $\mathbb{E}_P[Z_T] = 0$, i.e.,

$$\mathbb{E}_P[Y_\sigma(L_T - L_\sigma)] = 0$$

Since Y_σ is $\mathcal{F}_T$-measurable, $\mathbb{E}_Q[Y_\sigma] = \mathbb{E}_P[Y_\sigma L_T]$. From these observations it follows that (5.1) is valid for bounded processes Y. Using the monotone convergence theorem, we can conclude that (5.1) is true for any positive process Y. The general case follows by writing any Y as $Y = Y^+ - Y^-$. □

The following result admits a more direct proof using the notion of a stopped σ-field $\mathcal{F}_\tau$ for a stopping time τ. We have avoided introducing this in order to keep technicalities to a minimum, hence this slightly indirect proof.

Lemma 5.2 *An adapted r.c.l.l. process Y_t is a Q-local martingale if and only if $Y_t L_t$ is a P-local martingale.*

Proof Assume YL is a P-local martingale. Let σ_n be a sequence of stopping times increasing to infinity $P{-}a.s.$ (and hence Q-a.s.), such that $Z_t^n = Y_{t\wedge\sigma_n} L_{t\wedge\sigma_n}$ is a P-martingale. Thus for any bounded stopping time τ,

$$\mathbb{E}_P[Y_{\tau\wedge\sigma_n} L_{\tau\wedge\sigma_n}] = \mathbb{E}_P[Y_0 L_0].$$

Using Lemma 5.1, this implies that

$$\mathbb{E}_Q[Y_{\tau\wedge\sigma_n}] = \mathbb{E}_Q[Y_0].$$

This holds for all bounded stopping times τ and hence $Y_{t\wedge\sigma_n}$ is a martingale, or Y is a Q-local martingale. The reverse assertion follows similarly. □

Note that $L_t > 0$ a.s. P (as well as Q). Thus it follows from the result given above that if Z_t is a P-martingale, then $Z_t L_t^{-1}$ is a Q-martingale.

Theorem 5.3 *Let X be a P-semimartingale. If $Q \equiv P$, then X is a Q-semimartingale.*

Proof

$$\begin{aligned} L_t X_t &= \int_0^t L_{s-} dX_s + \int_0^t X_{s-} dL_s + [X, L]_t \\ &= N_t + B_t, \end{aligned}$$

where $N_t = \int_0^t L_{s-} dX_s + \int_0^t X_{s-} dL_s$ and $B = [X, L]$. Theorem 1.40 implies that N is a P-local martingale and $B \in \mathcal{V}$. This implies that $M_t = L_t^{-1} N_t$ is a Q-martingale. Also, $R_t = L_t^{-1}$ is a Q martingale and hence by Ito's formula, $R_t B_t$ is a Q-semimartingale. We then have that

$$X_t = M_t + R_t B_t \tag{5.2}$$

is a sum of two Q-semimartingales and hence is a Q-semimartingale. □

We will now consider a continuous P semimartingale and obtain a relation between its canonical decompositions under P and under Q.

Theorem 5.4 *Let X be a continuous P-semimartingale. Let $Q \equiv P$. Let $X_t = X_0 + M_t + A_t$ be the canonical decomposition of X under P and let $X_t = X_0 + \tilde{M}_t + \tilde{A}_t$ be the canonical decomposition of X under Q. Then*

$$\tilde{M}_t = M_t - \int_0^t \frac{1}{L_{s-}} d[M, L]_s \tag{5.3}$$

and

$$\tilde{A}_t = A_t + \int_0^t \frac{1}{L_{s-}} d[M, L]_s \tag{5.4}$$

where L_t denotes the P-martingale $L_t = \mathbb{E}[\frac{dQ}{dP} | \mathcal{F}_t]$.

Proof Since X is continuous, M,A,$\tilde{M}$ and $\tilde{A}$ are all continuous processes. To complete the proof, it suffices to show that $\tilde{M}_t$ defined by (5.3) is a Q-local martingale. In view of Lemma 5.2, we need to prove that $\tilde{M}_t L_t$ is a P-local martingale. Note that

$$\begin{aligned} \tilde{M}_t L_t &= \int_0^t \tilde{M}_s dL_s + \int_0^t L_{s-} d\tilde{M}_s + [\tilde{M}, L]_t \\ &= \int_0^t \tilde{M}_s dL_s + \int_0^t L_{s-} dM_s - \int_0^t L_{s-} L_{s-}^{-1} d[M, L]_s + [\tilde{M}, L]_t \\ &= \int_0^t \tilde{M}_s dL_s + \int_0^t L_{s-} dM_s, \end{aligned} \tag{5.5}$$

where we have used the fact that $[M, L]_s$ is continuous since M is, and also that $[\tilde{M}, L]_s = [M, L]_s$ since M and $\tilde{M}$ differ by a continuous process with bounded variation paths. As L and M are P-local martingales, (5.5) implies that $\tilde{M}_t L_t$ is a P-local martingale. As noted earlier, this completes the proof. □

Remark Let U be defined by

$$U_t = \int_0^t \frac{1}{L_{s-}} dL_s. \tag{5.6}$$

Then it follows from the result proved above that if M is a P-local martingale, then $\tilde{M}$ defined by

$$\tilde{M}_t = M_t - [M, U]_t$$

is a Q-local martingale.

5.2 Girsanov's Theorem

As a consequence of the general results given in the previous section, we have the following important result which is known as Girsanov's theorem. Throughout this section, we will consider $t \in [0, T]$

Theorem 5.5 (Girsanov) *Let $\{W_t^1, W_t^2, \dots, W_t^d : 0 \le t \le T\}$ be independent Brownian motions on $(\Omega, \mathcal{F}, P)$. Let $f^1, f^2, \dots, f^d)$ be predictable processes such that*

$$\int_0^T \sum_{j=1}^d |f_s^j|^2 ds < \infty. \tag{5.7}$$

Let

$$L_t := \exp\left\{ \sum_{j=1}^d \int_0^t f^j dW^j - \frac{1}{2} \sum_{j=1}^d \int_0^t |f_s^j|^2 ds \right\}.$$

Suppose that

$$\mathbb{E}_P[L_T] = 1. \tag{5.8}$$

Let Q be the probability measure defined by $dQ = L_T dP$. Then the processes

$$\hat{W}_t^j := W_t^j - \int_0^t f_s^j ds, \quad j = 1, \dots, d, \ 0 \le t \le T$$

are independent Brownian motions under Q.

Proof Let us note that U_t defined by (5.6) is given here by

$$U_t = \sum_{j=1}^{d} \int_0^t f_j dW^j,$$

and so

$$[W^j, U]_t = \int_0^t f_s^j ds.$$

Hence by the previous result, $\hat{W}_t^j = W_t^j - [W^j, U]_t$ is a Q-local martingale. Since W^j and $\hat{W}^j$ differ only by a continuous process with bounded variation paths, it follows that

$$[\hat{W}^j, \hat{W}^k]_t = [W^j, W^k]_t.$$

Thus $[\hat{W}^j, \hat{W}^j]_t$=t and $[\hat{W}^j, \hat{W}^k]_t = 0$ for $1 \leq j, k \leq d$, $j \neq k$. The result now follows from Theorem 2.9. □

For an application of the previous result, it is important to know if, for a given choice of f^j satisfying (5.7), the condition (5.8) holds. Sufficient conditions for this were obtained in Theorem 2.8. Also, see the remark following this theorem.

A natural question arises: Are there any other processes $A \in \mathcal{V}$ such that $\tilde{W}_t = W_t + A_t$ is a local martingale under an equivalent measure Q? The next result answers this in the negative.

Theorem 5.6 *Let (X_t) be an $\mathbb{R}^d$-valued continuous semimartingale with canonical decomposition*

$$X_t = X_0 + M_t + A_t, \tag{5.9}$$

where M is a continuous local martingale (under P) and $A \in \mathcal{V}$. Suppose Q is a probability measure equivalent to P such that (X_t) is a local martingale under Q. Then $A_t = (A_t^1, \ldots, A_t^d)$ admits a representation

$$A_t^i = \sum_{j=1}^{d} \int_0^t h_s^j d\langle X^i, X^j\rangle_s \tag{5.10}$$

where $h = (h^1, \ldots, h^d)$ is a predictable process satisfying

$$\int_0^T \sum_{i=1}^{d} |h_s^i|^2 d\langle X^i, X^i\rangle_s \Big] < \infty \ a.s. \tag{5.11}$$

Proof Let $L = \frac{dQ}{dP}$ and $L_t = \mathbb{E}[L \mid \mathcal{F}_t]$. Let

$$U_t = \int_0^t (L_{s-})^{-1} dL_s.$$

Then

$$[M, U]_t = \int_0^t (L_{s-})^{-1} d[L, M]_s.$$

Since X is a continuous, the decomposition $X_t = X_0 + M_t + A_t$ is unique, and hence it follows from Theorem 5.4 that X is a Q-martingale if and only if

$$A_t = -[M, U]_t. \tag{5.12}$$

Here, M is a continuous local martingale. So we can get a decomposition of U as follows (see Theorem 3.5)

$$U_t = U_0 + \sum_{i=1}^{d} \int_0^t f^i(s) dM_s^i + Z_t, \tag{5.13}$$

where Z is a local martingale such that $[M^i, Z] = 0$ and f is an $\mathbb{R}^d$-valued predictable process such that

$$\int_0^T \sum_{i=1}^{d} |f_s^i|^2 d\langle M^i, M^i\rangle_s \Big] < \infty \ a.s. \tag{5.14}$$

It now follows that

$$[M^i, U]_t = \sum_{j=1}^{d} \int_0^t f_s^j d\langle M^i, M^j\rangle_s.$$

Note that since X, M, A are continuous, $\langle M^i, M^j\rangle_s = \langle X^i, X^i\rangle_s$, and hence

$$[M^i, U]_t = \sum_{j=1}^{d} \int_0^t f_s^j d[X^i, X^j]_s.$$

The result follows from this by taking $h^i = -f^i$. □

Remark The measure Q as in the previous result, if it exists, is called an equivalent martingale measure (EMM). We have seen that if an EMM Q exists, then A must satisfy (5.10)–(5.11). When A satisfies these conditions, we have a natural choice of Q. See the next result for conditions under which this Q turns out to be an EMM.

Here is an extension of the Girsanov Theorem. Its proof is exactly along the lines of the proof of Girsanov's theorem and hence is omitted.

Theorem 5.7 *Let (X_t) be an $\mathbb{R}^d$-valued continuous semimartingale with canonical decomposition*

$$X_t = X_0 + M_t + A_t, \tag{5.15}$$

where M is a continuous local martingale (under P) and $A \in \mathcal{V}$. Suppose that A satisfies

$$A_t^i = \sum_{j=1}^{d} \int_0^t h_s^j d\langle X^i, X^j \rangle_s, \tag{5.16}$$

where $h = (h^1, \dots, h^d)$ is a predictable process satisfying

$$\int_0^T \sum_{i=1}^{d} |h_s^i|^2 d\langle X^i, X^i \rangle_s \Big] < \infty \ a.s. \tag{5.17}$$

Let

$$\rho_t = \exp\left\{ \sum_{j=1}^{d} - \int_0^t h_s^j dM_s^j - \frac{1}{2} \sum_{j=1}^{d} \int_0^t |h_s^j|^2 d[M^j, M^j]_s \right\}. \tag{5.18}$$

Suppose $\mathbb{E}[\rho_T] = 1$ and let Q^* be defined by $\frac{dQ^*}{dP} = \rho_T$. Then we have that $X = (X^1, X^2, \dots, X^d)$ is an $\mathbb{R}^d$-valued Q-local martingale.

6
Option Pricing in Discrete Time

In this chapter, we consider the problem of pricing an option in discrete time trading. We will introduce and discuss various important notions from stochastic finance, such as *investment strategy, arbitrage opportunity, complete markets*, and the role of *equivalent martingale measures* in discrete time. We assume a discrete model for the underlying stock.

6.1 Arbitrage opportunities

We begin with an informal description of some of the technical terms used in finance. We fix a unit of time Δ (which may be a day, an hour or ten minutes) and stipulate that all transactions take place at times $\{k\Delta : k \geq 0\}$.

A *bond* is a riskless security, earning a fixed rate of interest r in each unit of time. Thus an investment B_0 at time 0 in the bond is worth $B_0(1+r)^k$ at time k. It is convenient to write $R = (1+r)$. It is assumed that bonds can be bought or sold which means an investor can invest or borrow at the rate of interest r.

Shares of the *stock* of a specified company are traded in the stock market. The price S_k at which one share of the stock can be bought or sold is modelled as a random process. Bonds and stock are together known as *securities* (sometimes also referred to as *primary securities*).

Suppose that the price process $\{S_k\}$ of a stock is such that

$$P(S_1 \geq RS_0) = 1, \quad P(S_1 > RS_0) > 0.$$

Then an investor can borrow an amount S_0 and buy one share of the stock at time 0. At time 1 he can sell the stock at price S_1, settle his debts by paying RS_0, and his profit $P = S_1 - RS_0$ is non-negative with probability one and is strictly positive with positive probability.

It can be argued that if such a stock is available in the market, all investors would like to invest large amounts of money (by borrowing) into the stock—since there is nothing to lose and something to be gained. This will disturb the equilibrium and push the price of the stock (at time 0) up.

The situation described above is an example of an *arbitrage opportunity*. In general, an *arbitrage opportunity* in a market consisting of several securities is a strategy of buying and selling these securities *without any investment* by the investor, such that it leads to profit (strictly positive) with positive probability *without any risk of a loss.*

Market analysts agree that if an arbitrage opportunity exists, all the investors would like to follow that strategy which would thus disturb the equilibrium, pushing up the price of the security being purchased. *So we impose a blanket assumption that arbitrage opportunities do not exist (written in short as* no arbitrage *or NA).*

We begin by examining the restrictions that the principle of no arbitrage puts on the model for stock prices. We are considering here trading in discrete time. Without loss of generality, let us assume that the unit of time is a day and that stock prices change every afternoon at 2:00 pm. Investors are allowed to trade in the morning, at 11:00 am, at the prevailing prices, namely those of the previous evening.

The price of the stock on the kth afternoon is denoted by S_k. The price S_0 of the stock on day zero is assumed to be deterministic, $S_0 = s_0$. Also the face value of the bond is 1 on the morning of day zero. The interest rate is r per day, due at 2:00 pm, so that the value of the bond on the kth day at 2:00 pm is R^k where $R = (1 + r)$.

For $k \geq 0$, let ξ_k^1 denote the number of shares a specified investor decides to hold on the morning of the kth day and let ξ_k^0 denote the number of bonds he decides to hold. Thus, if for $k \geq 1$ $\xi_k^1 \geq \xi_{k-1}^1$, he buys $\xi_k^1 - \xi_{k-1}^1$ shares and if $\xi_k^1 < \xi_{k-1}^1$, then he sells $\xi_{k-1}^1 - \xi_k^1$ shares. We have a similar interpretation for bonds.

Clearly, ξ_k^1, ξ_k^0 should depend only on $\{S_0, S_1, S_2, \ldots, S_{k-1}\}$ for $k \geq 1$. When choosing ξ_k^1, ξ_k^0, the only information the investor has is $\{S_i : i \leq k-1\}$. ξ_0^1, ξ_0^0 are required to be constants. We express this as

$$\xi_k^1 = \pi_k^1(S_0, S_1, \ldots, S_{k-1}) \tag{6.1}$$

$$\xi_k^0 = \pi_k^0(S_0, S_1, \ldots, S_{k-1}), \tag{6.2}$$

where π_k^0, π_k^1 are real-valued functions on $\mathbb{R}^k$, $k \geq 1$. Of course to implement such a trading strategy, the investor may have to put in extra money on certain days while he may have surplus on other days. We are going to consider a special class of trading strategies, called *self-financing strategies*. These are trading strategies where there is no money put in and there is no surplus on any day except for the initial investment x. Thus, on any given day, the investor only moves his money from shares to bonds or vice-versa. The shares and bonds held by an investor together are known as his *portfolio*. On the kth morning ($k \geq 1$), the investor's portfolio is worth $\xi_{k-1}^1 S_{k-1} + \xi_{k-1}^0 R^{k-1}$ and he needs $\xi_k^1 S_{k-1} + \xi_k^0 R^{k-1}$ to implement his trading strategy. Since the strategy is assumed to be self-financing, it follows that these two quantities must be equal, i.e.,

$$\xi_{k-1}^1 S_{k-1} + \xi_{k-1}^0 R^{k-1} = \xi_k^1 S_{k-1} + \xi_k^0 R^{k-1}, \quad k \geq 1, \tag{6.3}$$

so that for any $j \geq 1$ (writing $\beta = R^{-1}$),

$$\xi_j^0 = \xi_{j-1}^0 + (\xi_{j-1}^1 - \xi_j^1) S_{j-1} \beta^{j-1}. \tag{6.4}$$

Also, $x = \xi_0^0 + \xi_0^1 S_0$ is the initial investment. We conclude using (6.4) that

$$\xi_j^0 = x - \xi_0^1 S_0 + \sum_{i=1}^{j} (\xi_{i-1}^1 - \xi_i^1) S_{i-1} \beta^{i-1}. \tag{6.5}$$

It is clear that in a self-financing strategy, the investor only chooses ξ_k^1 for $k \geq 0$ and together with x, this determines ξ_k^0 via (6.5). Let V_k denote the worth of the portfolio on the evening of the kth day. Then $V_k = \xi_k^1 S_k + \xi_k^0 R^k$. Let us rewrite this as

$$\beta^k V_k = \xi_k^1 S_k \beta^k + \xi_k^0,$$

and using (6.5) we get

$$\begin{aligned} \beta^k V_k &= \xi_k^1 \beta^k S_k + x - \xi_0^1 S_0 + \textstyle\sum_{j=1}^k (\xi_{j-1}^1 - \xi_j^1) S_{j-1} \beta^{j-1} \\ &= x + \textstyle\sum_{j=1}^k \xi_j^1 (S_j \beta^j - S_{j-1} \beta^{j-1}). \end{aligned} \tag{6.6}$$

Here $\tilde{V}_k = \beta^k V_k$ is the discounted value process and if we define $\tilde{G}_k = \sum_{j=1}^k \xi_j^1 (S_j \beta^j - S_{j-1} \beta^{j-1})$, then $\tilde{G}_k$ represents the discounted gains process. The equation (6.6) can be recast as

$$\tilde{V}_k = x + \tilde{G}_k,$$

that is, the (discounted) value of the portfolio from a self-financing strategy is equal to the initial investment plus the (discounted) gain from the strategy.

We will further assume that each S_k takes only finitely many values and that we are considering a time horizon of N days. Let

$$\mathcal{S} = \left\{ (s_0, s_1, \ldots, s_N) \in \mathbb{R}^{N+1} : P(S_0 = s_0, \ldots, S_N = s_N) > 0 \right\}.$$

Then $\mathcal{S}$ is a finite set and

$$P((S_0, \ldots, S_N) \in \mathcal{S}) = 1.$$

We will now assume without loss of generality that the underlying probability space is $\mathcal{S}$, P is a probability measure on $\mathcal{S}$ and $S_0, S_1, \ldots, S_N$ are given by

$$S_i(s_0, \ldots, s_N) = s_i.$$

A self-financing strategy is represented by $\theta = \{x, \pi_1^1, \ldots, \pi_k^1\}$, where π_k^1 is a function on $\mathbb{R}^k$. θ determines ξ_j^1, ξ_j^0 via (6.1) and (6.5). For a self-financing strategy θ, the worth of the portfolio $V_k(\theta)$ on the kth evening corresponding to the self-financing strategy is given by

$$V_k(\theta)(s_0, \ldots, s_N) = \left[x + \sum_{j=1}^{k} \pi_j^1(s_0, \ldots, s_{j-1})(s_j\beta^j - s_{j-1}\beta^{j-1})\right] R^k. \tag{6.7}$$

In this context, an *arbitrage opportunity* is a self-financing strategy $\theta = (0, \pi^1)$ such that

$$V_N(\theta)(s_0, \ldots, s_N) \geq 0 \quad \text{for all } (s_0, \ldots, s_N) \in \mathcal{S} \tag{6.8}$$

and

$$V_N(\theta)(s_0^1, \ldots, s_N^1) > 0 \quad \text{for some } (s_0^1, \ldots, s_N^1) \in \mathcal{S}. \tag{6.9}$$

The principle of no arbitrage here means that if a stategy θ satisfies (6.8), then it cannot satisfy (6.9).

Suppose for some i, $(\tilde{s}_0, \ldots, \tilde{s}_i)$

$$P(S_{i+1} \geq RS_i | S_0 = \tilde{s}_0, \ldots, S_i = \tilde{s}_i) = 1. \tag{6.10}$$

If

$$P(S_{i+1} > RS_i | S_0 = \tilde{s}_0, \ldots, S_i = \tilde{s}_i) > 0, \tag{6.11}$$

take $\theta = \{0, \pi_1^1, \ldots, \pi_N^1\}$ defined as follows: π_j^1 is identically equal to zero for $j \neq (i+1)$ and

$$\pi_{i+1}^1(s_0, \ldots, s_N) = \begin{cases} 1 & \text{if } (s_0, \ldots, s_i) = (\tilde{s}_0, \ldots, \tilde{s}_i) \\ 0 & \text{otherwise.} \end{cases}$$

Then one has

$$V_N(\theta)(s_0, \ldots, s_N) = R^{N-i}\left(\frac{s_{i+1}}{R} - \tilde{s}_i\right) 1_{\{s_0 = \tilde{s}_0, \ldots, s_i = \tilde{s}_i\}}.$$

In view of (6.10) and (6.11) this is an arbitrage opportunity. Thus (6.11) cannot be true. Similarly, we can show that if for any i, $\tilde{s}_0, \ldots, \tilde{s}_i$

$$P(S_{i+1} \leq RS_i | S_0 = \tilde{s}_0, \ldots, S_i = \tilde{s}_i) = 1, \tag{6.12}$$

then

$$P(S_i < RS_i | S_0 = \tilde{\beta}_0, \ldots , S_i = \tilde{s}_i) = 0. \tag{6.13}$$

In other words, we have that for any i, $(\tilde{s}_0, \ldots , \tilde{s}_i)$ such that $P(S_0 = \tilde{s}_0, \ldots , S_i = \tilde{s}_i) > 0$,

$$P(S_{i+1} = RS_i | S_0 = \tilde{s}_0, \ldots , S_i = \tilde{s}_i) < 1 \tag{6.14}$$

implies

$$P(S_{i+1} > RS_i | S_0 = \tilde{s}_0, \ldots , S_i = \tilde{s}_i) > 0 \tag{6.15}$$

and

$$P(S_{i+1} < RS_i | S_0 = \tilde{s}_0, \ldots , S_i = \tilde{s}_i) > 0. \tag{6.16}$$

We will now prove the main result of this section. Let $\mathcal{F}_i = \sigma\{S_j : 0 \leq j \leq i\}$. Note that each $\mathcal{F}_i$ is a finite field.

Theorem 6.1 *The following are equivalent.*

(i) No arbitrage.

(ii) There exists a probability measure Q on $\mathcal{S}$ such that $\{S_i \beta^i, \mathcal{F}_i\}$ is a Q-martingale and

$$Q(s_0, \ldots , s_N) > 0 \quad \forall (s_0, \ldots , s_N) \in \mathcal{S}. \tag{6.17}$$

Proof Suppose (ii) holds. Then it is easy to see using (6.7) that for every strategy θ,

$$V_k(\theta)(S_0, \ldots , S_N)$$

is a Q-martingale, and hence

$$\mathbb{E}^Q[V_k(\theta)(S_0, \ldots , S_N)]\beta^k = V_0(\theta). \tag{6.18}$$

Let θ be a strategy such that $V_0(\theta) = 0$ and

$$V_N(\theta)(s_0, \ldots , s_N) \geq 0 \quad \text{for all } (s_0, \ldots , s_N) \in \mathcal{S}. \tag{6.19}$$

Then (6.18), $V_0(\theta) = 0$, (6.19) and the assumption on Q that

$$Q((s_0, \ldots , s_N)) > 0 \;\; \forall (s_0, \ldots , s_N) \in \mathcal{S}$$

implies that

$$V_N(\theta)(s_0, \ldots , s_N) = 0 \;\; \forall (s_0, \ldots , s_N) \in \mathcal{S}.$$

Thus arbitrage opportunities, i.e., strategies satisfying (6.8) and (6.9), do not exist.

Now suppose (i) holds. We will now construct Q. For $0 \leq M \leq N$, let

$$\mathcal{S}^{(M)} = \{(s_0, \ldots, s_M) : P(S_0 = s_0, \ldots, S_M = s_M) > 0\}$$

and

$$\mathcal{S}^* = \cup_{M=0}^{N-1} \mathcal{S}^{(M)}.$$

We will denote elements of $\mathcal{S}^*$ by α (so that $\alpha = (s_0, \ldots, s_i)$ for some $i, 0 \leq i < N$). For $\alpha = (s_0, \ldots, s_i) \in \mathcal{S}^*$, let $\ell_\alpha = s_i$ and

$$C_\alpha = \left\{s : (s_0, \ldots, s_i, s) \in \mathcal{S}^{(i+1)}\right\}.$$

We will choose $p_\alpha(s), \alpha \in \mathcal{S}^*, s \in C_\alpha$ such that $0 < p_\alpha(s) \leq 1$,

$$\sum_{\alpha \in C_\alpha} p_\alpha(s) = 1 \tag{6.20}$$

and

$$\sum_{\alpha \in C_\alpha} s p_\alpha(s) = \ell_\alpha R. \tag{6.21}$$

Having chosen $(p_\alpha(s))$ satisfying 6.20)–(6.21), defining

$$Q((s_0, \ldots, s_N)) = p_{(s_0, \ldots, s_{N-1})}(s_N) \cdot p_{(s_0, \ldots, s_{N-2})}(s_{N-1}) \ldots p_{s_0}(s_1), \tag{6.22}$$

we can check that for $\alpha = (s_0, \ldots, s_i)$

$$\mathbb{E}[S_{i+1} | S_0 = s_0, \ldots, S_i = s_i] = \sum_{s \in C_\alpha} s p_\alpha(s) = s_i R$$

and hence conclude that $\{S_j \beta^j, \mathcal{F}_j\}$ is a Q-martingale.

It remains to choose p_α satisfying (6.20) and (6.21). If C_α is a singleton, then taking $C_\alpha = \{\ell_\alpha R\}$, $p_\alpha(\ell_\alpha R) = 1$ we can check that (6.20) and (6.21) hold. If C_α has more than one element, then (6.1) and (6.16) hold and we can choose $a_\alpha, b_\alpha \in C_\alpha, a_\alpha > \ell_\alpha R$ and $b_\alpha < \ell_\alpha R$. Let $D_\alpha = C_\alpha - \{a_\alpha, b_\alpha\}$. For $0 \leq \epsilon < 1$, take $p_\alpha^\epsilon(s) = \epsilon$ for $s \in D_\alpha$. Let $p_\alpha^\epsilon(a_\alpha), p_\alpha^\epsilon(b_\alpha)$ be solutions to the equations

$$p_\alpha^\epsilon(a_\alpha) + p_\alpha^\epsilon(b_\alpha) = 1 - \epsilon(\#D_\alpha) \tag{6.23}$$

$$a_\alpha p_\alpha^\epsilon(a_\alpha) + b_\alpha p_\alpha^\epsilon(b_\alpha) = \ell_\alpha R - \epsilon \cdot \sum_{s \in D_\alpha} s. \tag{6.24}$$

Since $a_\alpha > b_\alpha$, these equations admit a unique solution given by

$$p_\alpha^\epsilon(a_\alpha) = \frac{\ell_\alpha R - b_\alpha - \epsilon \sum_{s \in D_\alpha}(s - b_\alpha)}{a_\alpha - b_\alpha} \tag{6.25}$$

$$p_\alpha^\epsilon(b_\alpha) = \frac{a_\alpha - \ell_\alpha R - \epsilon \sum_{s \in D_\alpha}(a_\alpha - s)}{a_\alpha - b_\alpha}. \tag{6.26}$$

It is clear from these expressions that for suitably small $\epsilon > 0$, $p_\alpha^\epsilon(a_\alpha) > 0$ and $p_\alpha^\epsilon(b_\alpha) > 0$. Thus for sufficiently small $\epsilon > 0$, $\{p_\alpha^\epsilon(s) : \alpha \in C_\alpha\}$ satisfies (6.20) and (6.21). As noted earlier, this completes the proof. □

Remark Let us note here that when the number of elements in any C_α is more than two, we have two (in fact infinitely many) distinct choices of $\epsilon > 0$ for which $p_\alpha^\epsilon(a_\alpha) > 0$, $p_\alpha^\epsilon(b_\alpha) > 0$ and we get two distinct measures Q^1 and Q^2, satisfying requirements in part (ii). Conversely, it is easy to see that if the cardinality of each C_α is at most two, then for $\alpha = (s_0, \ldots, s_1)$

$$p_\alpha(s) = Q(S_{i+1} = s | S_0 = s_0, \ldots, S_i = s_i)$$

is uniquely determined by (6.20) and (6.21) and hence the probability measure Q is uniquely determined by (6.22)

6.2 Option pricing: an example

We begin with the stock of a specified company and a (stochastic) model $\{S_k : k \geq 0\}$ for the price of the stock. Here it is assumed that trading is possible only at an integer multiple of a fixed time interval Δ (Δ may be an hour or a day) and S_k is the price of the stock at $k\Delta$.

A stock broker is selling a coupon that entitles the holder to buy one share of the specified company stock at time N (called the *terminal time*) at a price K (called the *striking price*), if the holder so desires. Thus the holder has an option of buying the stock and if he so demands, the broker is committed to sell. Such a one-sided contract is known as an option coupon (or more precisely, a European call option).

The issue we are going to address is that of determining the market price of the option coupon.

We will assume that there are no transaction costs and that one can borrow or lend at the same rate of interest r (per unit time, namely Δ). Let $R = (1 + r)$, $\beta = R^{-1}$.

The holder of the option coupon can make a profit of $(S_N - K)^+$ at time N : if $S_N > K$, he can buy at a price K and sell at the price S_N; if $S_N \leq K$, the option coupon is worthless. The present (discounted) value of the potential profit is $(S_N - K)^+ R^{-N}$ where $R = (1 + r)$. One may be tempted to conclude that the price of the coupon should be an expected discounted gain namely,

$$E[(S_N - K)^+ R^{-N}].$$

This need not be true, as the following example shows. The example is artificial but it illustrates several interesting points.

Suppose that Δ = one year, $S_0 = \$10$ and $S_1 = \$16$ with probability 0.6 and $S_1 = \$6$ with probability 0.4. Suppose $r = 10\%$ so that $R = 1.1$. Consider an option coupon with $N = 1$ and $K = 10.5$. Here, the discounted expected gain is

$$\mathbb{E}(S_1 - K)^+ \cdot R^{-1} = 5.5 \times 0.6 \times \frac{1}{1.1} = 3.$$

Suppose the market price of the coupon is \$3, i.e., there are sellers as well as buyers of the coupon at this price. An investor is considering investing \$300 by buying 100 option coupons at a price of \$3 per coupon. A friend of his suggests that instead of buying the coupons, he should do the following: borrow an additional \$230 (at 10% interest rate) and invest \$530 to buy 53 shares at the price of \$10 each. His friend explains that at the end of one year, if the stock has gone up, namely $S_1 = 16$, then the 100 option coupons will be worth $5.5 \times 100 = 550$, whereas if he follows the alternative, the 53 shares would be worth $53 \times 16 = 848$, out of which he has to settle a debt of $\$230 \times 1.1 = \$253$, so the worth of his investments would be \$595. On the other hand, if $S_1 = 6$, the option coupons are worthless, whereas the worth of the portfolio if the alternate strategy is followed would be $53 \times 6 - 230 \times 1.1 = 65$. So regardless of whether the stock goes up or down, it is better to borrow \$230 and buy 53 shares than to buy 100 option coupons at \$3 per coupon.

His friend goes on to advise him that if indeed the market price is \$3, he should sell option coupons at the rate of \$3 per coupon, and for every 100 option coupons sold, buy 53 shares by borrowing \$230. Then, as seen earlier, he will definitely make a profit of \$45 if the stock goes up ($S_1 = 16$) or of \$65 if the stock goes down ($S_1 = 6$) per 100 option coupons sold. This is a strategy which leads to profit without taking any risk, in other words an arbitrage opportunity.

So, in our example, if the option coupons are priced at \$3, soon there would be no buyer! Market forces would thus bring down the price. The strategy given above is an arbitrage opportunity if the price of the option coupon is $\geq \$2.55$; at the price \$2.55 it leads to a profit of \$20 (per 100 coupons sold) if $S_1 = 6$ and to a zero profit if $S_1 = 16$ (but still there is no loss). So we can conclude that the market price should be $\leq \$2.55$.

Consider instead the following strategy with an initial investment of \$250: borrow \$300 and buy 55 shares of the stock at time 0. At time 1, the investor's holdings are worth $55 \times 16 - 300 \times 1.1 = 550$ if the stock goes up to \$16; 55 $\times 6 - 300 \times 1.1 = 0$ if the stock goes down. This is exactly the worth of 100 option coupons. He can conclude that the price of the option coupon must be \$2.50, since if the price p is strictly greater than \$2.50, he can sell 100 option coupons, borrow \$300 and buy 55 shares. This would lead to a pofit of $\$100(p - 2.5)$ regardless of whether the stock goes up or down and is thus an arbitrage opportunity. If the price p were strictly less than \$2.50, he could buy 100 option coupons, sell 55 shares and invest \$300 on the bonds. This is also an arbitrage opportunity since once again our profit would be $\$100(2.5 - p)$.

Thus the principle of no arbitrage implies that the price must be \$2.50. The strategy described in the previous paragraph which exactly replicates the outcomes from the option coupons is called a *hedging strategy*.

We can use arguments similar to the ones in this example to get upper and lower bounds on the price of the option. We will also consider a special case (as in this example) where the upper and lower bounds coincide, determining the option price uniquely.

Note that we are considering an ideal market where there are no transaction costs. This is never true in a real market, but one can understand the underlying notions and get a price in an ideal market first and then correct it for transaction costs later.

6.3 European call option

Let us consider the discrete model for stock prices as in section 6.1 and let us assume that arbitrage opportunities do not exist. Consider a European call option with striking price K and terminal time N. The holder of the option coupon makes a profit of $(S_N - K)^+$ at the terminal time N. Suppose p is the price of such an option coupon in the marketplace, so that there are buyers as well as sellers at this price.

Suppose x is such that there exists an investment strategy $\theta = (x, \pi^1)$ such that

$$V_N(\theta)(s_0, \dots, s_N) \geq (s_N - K)^+ \quad \forall (s_0, \dots, s_N) \in \mathcal{S}. \tag{6.27}$$

If $x < p$, consider the strategy of selling an option coupon at price p, and following strategy $\Phi = (p, \pi^1)$ for investing in shares and bonds. At time N, the assets are $V_N(\Phi)$ and the liabilities are $(S_N - K)^+$. Thus the net worth is

$$\begin{aligned} V_N(\Phi)(s_0, \dots, s_N) - (s_N - K)^+ & \\ &= V_N(\theta)(s_0, \dots, s_N) - (s_N - K)^+ + p - x \\ &\geq p - x \\ &> 0. \end{aligned} \tag{6.28}$$

This yields an arbitrage opportunity and hence, as argued earlier, x must be greater than or equal to p.

Let A^+ be the set of all x such that there exists $\theta = (x, \pi^1)$ satisfying (6.27) and let

$$x^+ = \inf A^+.$$

We then have

$$x^+ \geq p. \tag{6.29}$$

Similarly, taking A^- to be the set of all x such that there exists $\theta = (x, \pi^1)$ satisfying

$$V_N(\theta)(s_0, \dots, s_N) \leq (s_N - K)^+ \; \forall (s_0, \dots, s_N) \in \mathcal{S} \tag{6.30}$$

and $x^- = \sup A^-$, we can conclude that

$$x^- \leq p. \tag{6.31}$$

We have thus obtained upper and lower bounds for the price of the option using only the stipulation that arbitrage opportunities do not exist. Following Merton, we call these the bounds for the *rational price* of the option. When the upper and lower bounds coincide, they determine the rational price uniquely. We first observe that these constraints are consistent, namely $x^+ \geq x^-$. Indeed, if Q is a probability measure such that $S_i\beta^i$ is a Q-martingale, then for a self-financing strategy $\theta = (x, \pi^1)$

$$\mathbb{E}_Q\big[V_N(\theta)(S_0, \ldots, S_N)\beta^N\big] = x. \tag{6.32}$$

and hence it follows that if $x_1 \in A^+$, $x_2 \in A^-$, then

$$x_2 \leq \mathbb{E}^Q\big[(S_N - K)^+\beta^N\big] \leq x_1.$$

Thus,

$$x^- \leq \mathbb{E}^Q\big[(S_N - K)^+\beta^N\big] \leq x^+.$$

We will now analyze a special case where $x^- = x^+$ and, as a consequence, the rational price is completely determined by the principle of no arbitrage.

Theorem 6.2 *Suppose that arbitrage opportunities do not exist and that*

$$\#C_\alpha = 1 \text{ or } 2 \quad \forall \alpha \in \mathcal{S}^*. \tag{6.33}$$

Let Q be the probability measure in (ii), Theorem 6.1 and let

$$\hat{x} = \mathbb{E}^Q[(S_N - K)^+R^{-N}]. \tag{6.34}$$

Then there exists a (self-financing) strategy $\theta = (x, \pi^1)$ such that

$$V_N(\theta)(s_0, \ldots, s_N) = (s_N - K)^+ \quad \forall (s_0, \ldots, s_N) \in \mathcal{S}. \tag{6.35}$$

As a consequence, $x^+ = x^- = \hat{x}$.

Proof For $\alpha \in \mathcal{S} = \mathcal{S}^{(N)}$, let

$$g(\alpha) = (S_N(\alpha) - K)^+. \tag{6.36}$$

We define π_N^1, π_N^0 and g on $\mathcal{S}^{(N-1)}$ as follows. Fix $\alpha = (s_0, \ldots, s_{N-1}) \in \mathcal{S}^{(N-1)}$. If $\#C_\alpha = 1$, then $C_\alpha = \{s_{N-1}R\}$ and define

$$\begin{aligned} \pi_N^1(\alpha) &= 0, \\ \pi_N^0(\alpha) &= R^{-N}g((s_0, \ldots, s_{N-1}, Rs_{N-1})). \end{aligned}$$

If $\#C_\alpha = 2$, then $C_\alpha = \{a_\alpha, b_\alpha\}$ with $a_\alpha > s_{N-1}R > b_\alpha$ (recall our observation that (6.14) implies (6.1) and (6.16)). Consider the equations

$$ya_\alpha + zR^N = g((s_0, \dots, s_{N-1}, a_\alpha))$$

$$yb_\alpha + zR^N = g((s_0, \dots, s_{N-1}, b_\alpha))$$

in variables y, z. These equations admit a unique solution as $a_\alpha > b_\alpha$. Let $\pi_N^1(\alpha) = y, \pi_N(\alpha) = z$. Take

$$g(\alpha) = \pi_N^1(\alpha)\mathcal{S}_{N-1} + \pi_N^0(\alpha)R^{N-1}.$$

it is clear that if at time $N-1$ an investor had an amount $g((s_0, \dots, S_{N-1}))$, he/she can follow a strategy of investing $\pi_N^1((s_0, \dots, S_{N-1}))$ on the stock and $\pi_N^0((S_0, \dots, S_{N-1}))$ on the bonds and end up at time N with exactly the same reward as the one from the option.

We now define $\pi_{i+1}^1(\alpha), \pi_{i+1}^0(\alpha)$ and $g(\alpha)$ for $\alpha \in \mathcal{S}^{(i)}, 0 \le i \le N-2$ by backward induction.

Having defined π_{j+1}^1, π_{j+1}^0 and g for $j > i, i \le N-2$, we define these for $j = i$ as follows.

Fix $\alpha = (s_0, \dots, s_i) \in \mathcal{S}^{(i)}$. If $C_\alpha = \{Rs_i\}$, then $\pi_{i+1}^1(\alpha) = 0, \pi_{i+1}^0(\alpha) = R^{-(i+1)}g((s_0, \dots, s_i, Rs_i))$. On the other hand, if $C_\alpha = \{a_\alpha, b_\alpha\}$ (with $a_\alpha > b_\alpha$), then $\pi_{i+1}^1(\alpha)$ and $\pi_{i+1}^0(\alpha)$ are the unique solutions to the equations

$$\pi_{i+1}^1(\alpha)a_\alpha + \pi_{i+1}^0(\alpha)R^{i+1} = g((s_0, \dots, s_i, a_\alpha)) \tag{6.37}$$

$$\pi_{i+1}^1(\alpha)b_\alpha + \pi_{i+1}^0(\alpha)R^{i+1} = g((s_0, \dots, s_i, b_\alpha)). \tag{6.38}$$

Then define

$$g(\alpha) = \pi_{i+1}^1(\alpha)s_i + \pi_{i+1}^0(\alpha)R^i. \tag{6.39}$$

Take $\theta = \{\pi_i^1 : 1 \le i \le N\}$ and $\hat{x} = g((s_0))$.

By construction, one has, for $(s_0, \dots, s_i, s_{i+1}) \in \mathcal{S}^{(i+1)}$,

$$\pi_{i+1}^1((s_0, \dots, s_i))s_{i+1} + \pi_{i+1}^0((s_0, \dots, s_i))R^{i+1} = g((s_0, \dots, s_i, s_{i+1}))$$

and

$$\pi_{i+1}^1((s_0, \dots, s_i))s_i + \pi_{i+1}^0((s_0, \dots, s_i))R^i = g((s_0, \dots, s_i)).$$

Thus (recall $\beta = R^{-1}$)

$$\begin{aligned}\pi_{i+1}^1((s_0, \dots, s_i))(s_{i+1}\beta^{i+1} - s_i\beta^i) \\ &= g((s_0, \dots, s_i, s_{i+1}))\beta^{i+1} - g((s_0, \dots, s_1))\beta^i.\end{aligned}$$

Now it is easy to check that (6.34) and (6.35) hold with $\hat{x} = g((s_0))$.

It follows that $\hat{x} \in A^+$ so that $x^+ \le \hat{x}$, and $\hat{x} \in A^-$ so that $x^- \ge \hat{x}$. Since $x^- \le x^+$, it follows that $x^- = x^+ = \hat{x}$. □

6.4 Complete markets

Consider a coupon that entitles the holder of the coupon to receive a payoff of $f(S_0, \dots, S_N)$ at time N, where f is a specified non-negative function. This coupon is called a contingent claim. We will denote this claim by $CC(f; N)$. Thus a European call option is the contingent claim with $f(s_0, \dots, s_N) = (s_N - K)^+$.

We can consider other examples of derivative securities such as: the payoff to the holder of the coupon at time N is

(a) $(K - S_N)^+$

(b) S_N^2

(c) $S_1 + S_2 + \dots + S_N$

(d) $\max_{i \le N} |S_i - K|^+$.

All of the above are examples of contingent claims. The contingent claim in (a) above is a European put option with terminal time N and striking price K and is traded in European markets: it entitles the holder to sell one share of the specified company, at the striking price K at terminal time N if he so wishes, making a profit of $(K - S_N)^+$. The other contingent claims corresponding to the other examples are not traded on the markets.

In analogy with the European option case, let us define $A^+(f; N)$ to be the set of $x \in \mathbb{R}$ such that there exists a self-financing strategy θ such that

$$V_N(x, \theta)(s_0, \dots, s_N) \ge f(s_0, \dots, s_N) \tag{6.40}$$

for all $(s_0, \dots, s_N) \in \mathcal{S}$ and let

$$x^+(f; N) = \inf A^+(f; N).$$

$A^-(f; N)$ is defined similarly with the $\ge$ in (6.40) replaced by $\le$; and then define

$$x^-(f; N) = \sup A^-(f; N).$$

Here again, the principle of *no arbitrage* implies that $x^+(f; N)$ is an upper bound and $x^-(f; N)$ is a lower bound for the rational price of the contingent claim $CC(f; N)$.

A contingent claim $CC(f; N)$ is said to be *attainable* if there exists a strategy θ and $\hat{x}$ such that

$$V_N(\hat{x}, \pi)(s_0, \dots, s_N) = f(s_0, \dots, s_N) \quad \forall (s_0, \dots, s_N) \in \mathcal{S}. \tag{6.41}$$

This means having the contingent claim $CC(f : N)$ at time 0 is the same as having an amount $\hat{x}$ at time 0—since starting with an amount $\hat{x}$ at time 0, following strategy θ, the portfolio can be matched with the contingent claim for every outcome of the stock process. Note that if (6.41) holds, then $\hat{x}$ belongs to $A^+(f; N)$

as well as to $A^-(f; N)$. Thus the price of the claim $CC(f; N)$ must be $\hat{x}$. Further, using (6.18) it follows that the price $\hat{x}$ of an attainable claim $CC(f; N)$ is given by

$$\hat{x} = R^{-N}\mathbb{E}^Q[f(S_0, \dots, S_N)], \tag{6.42}$$

where Q is any probability measure on $\mathcal{S}^{(N)}$ such that $\beta^i S_i$ is a Q-martingale.

Definition The market consisting of the bond and the stock $\{S_i\}$ is said to be *complete* if every contingent claim is attainable.

As observed above, in a complete market, prices of all contingent claims are completely determined. The following theorem characterizes completeness of a market consisting of a bond and a stock.

Theorem 6.3 *Consider a market consisting of a bond and a stock $\{S_k\}$. Assume that arbitrage opportunities do not exist. Let*

$$\mathcal{E}(P) = \{Q \text{ on } \mathcal{S}^{(N)} : \beta^i S_i \text{ is a } Q\text{-martingale}, Q\{\alpha\} > 0\ \forall \alpha \in \mathcal{S}^{(N)}\}.$$

Then the following are equivalent

(a) The market is complete.

(b) $\mathcal{E}(P)$ is a singleton.

(c) $\forall \alpha \in \mathcal{S}^*, \#C_\alpha \in \{1, 2\}$.

Proof We have seen earlier that $\mathcal{E}(P)$ is nonempty in view of our assumption of *no arbitrage*.

First, we will prove (a) $\Rightarrow$ (b). Suppose the market is complete. Let $Q^1, Q \in \mathcal{E}(P)$. Define $g : \mathcal{S} \to \mathbb{R}$ as follows:

$$g(s_0, \dots, s_N) = \frac{Q^1((s_0, \dots, s_N))}{Q((s_0, \dots, s_N))} \quad (s_0, \dots, s_N) \in \mathcal{S}. \tag{6.43}$$

By completeness of the market, there exists a strategy $\theta = (x, \pi^1)$ such that the contingent claim g is attained at time N, i.e.,

$$g(s_0, \dots, s_N) = [x + \sum_{j=0}^{N-1} \pi^1_{j+1}(s_0, \dots, s_j)(\beta^{j+1}s_{j+1} - \beta^j s_j)]R^N. \tag{6.44}$$

Let $\mathcal{F}_i = \sigma(S_0, \dots, S_i)$. For $i \geq 0$, define

$$h^i(s_0, \dots, s_i) = \frac{Q^1(\{S_0 = s_0, \dots, S_i = s_i\})}{Q(\{S_0 = s_0, \dots, S_i = s_i\})}.$$

It is easy to verify that $Z_i = h^i(S_0, \ldots, S_i)$ is a Q-martingale w.r.t. $(\mathcal{F}_i)$. Since $g = h^N$ one has, in view of (6.44),

$$Z_N = [x + \sum_{j=0}^{N-1} \pi^1_{j+1}(S_0, \ldots, S_j)(\beta^{j+1} S_{j+1} - \beta^j S_j)]R^N. \tag{6.45}$$

Using the fact that Z_j, $\beta^j S_j$ are Q-martingales (w.r.t. $(\mathcal{F}_i)$), it follows that

$$Z_{i+1} - Z_i = \pi^1_{i+1}(s_0, \ldots, s_i)(\beta^{i+1} S_{i+1} - \beta^i S_i) \cdot R^N. \tag{6.46}$$

We will now show that $\beta^j S_j Z_j$ is a Q-martingale. Now

$$\begin{aligned}
&\mathbb{E}^Q[\beta^{i+1} S_{i+1} Z_{i+1} | S_0 = s_0, \ldots, S_i = s_i] \cdot Q(S_0 = s_0, \ldots, S_i = s_i) \\
&= \sum_{s_{i+1}, \ldots, s_N} \beta^{i+1} s_{i+1} h^{i+1}(s_0, \ldots, s_i) Q((s_0, \ldots, s_N)) \\
&= \sum_{s_{i+1}} \beta^{i+1} s_{i+1} h^{i+1}(s_0, \ldots, s_i) Q(S_0 = s_0, \ldots, S_i = s_i) \\
&= h^i(s_0, \ldots, s_i) \cdot \sum_{s_{i+1}} \beta^{i+1} s_{i+1} Q^1(S_0 = s_0, \ldots, S_i = s_i) \\
&= h^i(s_0, \ldots, s_i) \beta^i s_i \cdot Q(S_0 = s_0, \ldots, S_i = s_i).
\end{aligned}$$

Here, we have used the definition of h^i, h^{i+1} and the fact that $\beta^j S_j$ is a Q^1-martingale. We have thus proved

$$\mathbb{E}^Q \left[\beta^{i+1} S_{i+1} Z_{i+1} | S_0, \ldots, S_i\right] = \beta^i S_i Z_i. \tag{6.47}$$

In the steps that follow, we use (6.46), (6.47) and the fact that $\beta^j S_j$, Z_j are Q-martingales w.r.t. $(\mathcal{F}_i)$. Let us write $\xi_{i+1} = \pi^1_{i+1}(S_0, \ldots, S_i) \cdot R^N$ for convenience.

$$\begin{aligned}
\{\mathbb{E}^Q((Z_{i+1} - Z_i)^2) &= \mathbb{E}^Q((Z_{i+1} - Z_i)\xi_{i+1}(\beta^{i+1} S_{i+1} - \beta^i S_i))\} \\
&= \mathbb{E}^Q[\xi_{i+1} \mathbb{E}^Q((Z_{i+1} - Z_i)(\beta^{i+1} S_{i+1} - \beta^i S_i) | \mathcal{F}_i)] \\
&= \mathbb{E}^Q[\xi_{i+1}(\beta^i S_i - \beta^i S_i - \beta^i S_i + \beta^i S_i) Z_i] \\
&= 0.
\end{aligned}$$

We thus have $Z_{i+1} = Z_i$ Q-a.s. for all i, and as a consequence, $Z_N = Z_0$ Q-a.s. By definition of $\mathcal{E}(P)$, every singleton $\alpha \in \mathcal{S}$ has positive Q-probability. It thus follows that Z_N is a constant, which means g is a constant function. In turn, this yields $Q^1 = Q$. Thus we have proved, (a) $\Rightarrow$ (b).

Next, we will prove (b) $\Rightarrow$ (c). Suppose that for some $\alpha \in \mathcal{S}^*$, $\#C_\alpha > 2$. Going back to the proof of Theorem 6.2, we see that we can get *two* distinct solutions to the equations (6.20) and (6.21) (given by $p^\epsilon_\alpha(s)$ for $\epsilon = \epsilon_1$ and $\epsilon = \epsilon_2$ for suitably small ϵ_1, ϵ_2). This in turn gives two distinct probability measures Q^1 and

Q^2 (defined by (6.22)) in $\mathcal{E}(P)$. We conclude that if (b) holds, then $\#C_\alpha \leq 2$ for all α.

It remains to prove (c) $\Rightarrow$ (a). This is essentially the same as the proof of Theorem 6.2. Given a contingent claim $f(S_0, \dots, S_N)$, define

$$g(s_0, \dots, s_N) = f(s_0, \dots, s_N)$$

(instead of (6.36)). Proceeding as in the proof of Theorem 6.2, we obtain a strategy π for which

$$V_N(x, \pi)(s_0, \dots, s_N) = f(s_0, \dots, s_N).$$

Thus if (c) holds, every contingent claim is attainable.

6.5 The American option

An American option can be exercised by the holder *at any time before the terminal time.*

Thus, in case of an American call option with terminal time N and striking price K, the holder of the coupon can, if he so wishes, exercise his option at time $n \leq N$ and buy one share at price K, making a profit of $(S_n - K)^+$ at time n. Of course, the decision to exercise the option at time n or not has to be based on actual information available at that time, namely $S_0, \dots, S_n$.

If τ denotes the (random) time at which the option is exercised, the event $\{\tau = n\}$ should depend only on $S_1, \dots, S_n$ and then τ must be a stopping time, i.e., $\{\tau = n\}$ should belong to the σ-field $\mathcal{F}_n = \sigma(S_0, S_1, \dots, S_n)$. Since we are considering the case of discrete random variables $S_0, S_1, \dots, S_N$, each taking only finitely many values, $\mathcal{F}_n$ is actually a field generated by atoms

$$\{S_) = s_0, S_1 = s_1, \dots, S_n = s_n\}.$$

Thus the American call option can be described as follows: the holder can exercise his option at any stopping time $\tau \leq N$ and make a profit of $(S_\tau - K)^+$.

In the case of an American put option, the holder can sell one share at price K at any stopping time $\tau \leq N$ making a profit of $(K - S_\tau)^+$. In order to consider American call and put options at the same time, we will consider the following American type security:

The holder of the security can exercise it at any stopping time $\tau \leq N$ to make a profit of Z_τ, where $Z_n = h_n(S_0, S_1, \dots, S_n)$. The security is denoted by $(\{h_n\}; N)$.

We will regard h_n as a function on $\mathcal{S}$ defined by

$$h_n(s_0, \dots, s_N) := h_n(s_0, \dots, s_n).$$

For a stopping time $\tau \leq N$ (on $\mathcal{S}$),

$$h_\tau(s_0, \dots, s_N) := h_{\tau(s_0, \dots, s_N)}(s_0, \dots, s_N).$$

We assume that arbitrage opportunities do not exist, so that $\exists Q \in \mathcal{E}(P)$. Let $B^+(\{h_n\}; N)$ consist of $y \in [0, \infty)$, such that there exists a strategy (y, π^1) with

$$V_k(y, \pi^1)(s_0, \ldots, s_N) \geq h_k(s_0, \ldots, s_N) \quad \forall (s_0, \ldots, s_N) \in \mathcal{S}, \tag{6.48}$$

and let $B^-(\{h_n\}; N)$ consist of $y \in [0, \infty)$, such that there exists a strategy (y, π^1) and a stopping time τ such that

$$V_\tau(y, \pi^1)(s_0, \ldots, s_N) \leq h_\tau(s_0, \ldots, s_N) \quad \forall (s_0, \ldots, s_N) \in \mathcal{S}. \tag{6.49}$$

Let

$$y^+(\{h_n\}; N) = \inf B^+(\{h_n\}; N)$$
$$y^-(\{h_n\}; N) = \sup B^-(\{h_n\}; N).$$

If the price p of the American type security $(\{h_n\}; N)$ is less than $y^-(\{h_n\}; N)$, an investor can buy the security at price p and invest the amount $-p$ on the stock market consisting of the bond and the stock $\{s_k\}$ following the strategy $-\pi$. At time τ, he would exercise the American security, and liquidate his investments on the stock and bond. Thus, starting from zero investment, his net assets at time τ would be

$$\begin{aligned} h_\tau + V_\tau(-p, -\pi) &= h_\tau - V_\tau(p, \pi) \\ &= h_\tau - \{V_\tau(y, \pi) + (p - y)R^\tau\} \\ &\geq (y - p)R^\tau. \end{aligned}$$

This would be an arbitrage opportunity. Hence, $y^-(\{h_n\}; N)$ is a lower bound for the rational price.

On the other hand, if $p > y$ for $y \in B^+(\{h_n\}; N)$, let π be such that (6.48) holds. An investor can sell the American security at price p and invest p on the stock $\{S_k\}$ per bond following the strategy π. If the buyer exercises his option at time τ, the investor can liquidate his investments at time τ as well, and his net assets then are

$$\begin{aligned} V_\tau(p, \pi) - h_\tau &= V_\tau(y, \pi) + (p - y) \cdot R^\tau - h_\tau \\ &\geq (p - y)R^\tau, \end{aligned}$$

making this an arbitrage opportunity. Thus $y^+(\{h_n\}; N)$ is an upper bound for the rational price of the American security. The following lemma implies that these constraints on the price are consistent.

Lemma 6.4 *Let $Q \in \mathcal{E}(P)$. Then for any stopping time $\tau \leq T$,*

$$y^-(\{h_n\}; N) \leq E^Q[Z_\tau \beta^\tau] \leq y^+(\{h_n\}; N).$$

Proof The assertion follows from the observation that for any y, π

$$V_n(y, \pi)(S_0, \ldots, S_N) \cdot \beta^n$$

is a Q-martingale (with mean y) and hence

$$y = E^Q(V_\tau(y, \pi)(S_0, \dots, S_N)\beta^\tau). \qquad \square$$

Remark Let us note that when an investor buys the American option, he can choose the stopping time τ at which he can exercise his option, whereas the seller of the American option has to be prepared for any choice of τ made by the buyer. This asymmetry is reflected in the definition of $B^+(\{h_n\}; N)$ and $B^-(\{h_n\}; N)$.

Let us briefly consider the corresponding European security, namely the contingent claim $h_N(S_0, \dots, S_N)$. As we have seen earlier, the upper bound $x^+ = x^+(h_N, N)$ and the lower bound $x^- = x^-(h_N, N)$ for the rational price of the contingent claim $h_N(S_0, \dots, S_N)$ are given by $x^+ = \inf A^+(h_N, N)$ and $x^- = \sup A^-(h_N, N)$, where $A^+(h_N, N)$ consists of $y \in [0, \infty)$ such that there exists π such that

$$V_N(y, \pi)(s_0, \dots, s_N) \geq h_N(s_0, \dots, s_N) \quad \forall (s_0, \dots, s_N) \in \mathcal{S} \tag{6.50}$$

and $A^-(h_N, N)$ consists of $y \in [0, \infty)$ such that there exists π such that

$$V_N(y, \pi)(s_0, \dots, s_N) \leq h_N(s_0, \dots, s_N) \quad \forall (s_0, \dots, s_N) \in \mathcal{S}. \tag{6.51}$$

If y, π are such that (6.48) holds, then clearly (6.50) holds so that

$$B^+(\{h_n\}; N) \subseteq A^+(h_N, N)$$

and, as a consequence

$$y^+(\{h_n\}; N) \geq x^+(h_N, N). \tag{6.52}$$

If y, π are such that (6.51) holds, then (6.49) holds for $\tau = N$ and $A^-(h_N, N) \subseteq B^-(\{h_n\}; N)$ implying that

$$x^-(h_N, N) \leq y^-(\{h_n\}; N) \tag{6.53}$$

We will prove later that in a complete market, $y^+(\{h_n\}; N) = y^-(\{h_n\}; N)$. First we will consider an American call option $(h_n(s_0, \dots, s_n) = (s_n - K)^+)$ and show that if $x^+(h_N, N) = x^-(h_N, N)$ for the corresponding European call option, then

$$y^+(\{h_n\}; N) = y^-(\{h_n\}; N) = x^+(h_N, N) = x^-(h_N, N).$$

Theorem 6.5 *Suppose that*

$$x^+((s_N - K)^+, N) = x^-((s_N - K)^+, N) = \tilde{x}. \tag{6.54}$$

Then

$$y^+(\{(s_n - K)^+\}; N) = y^-(\{(s_n - K)^+\}; N). \tag{6.55}$$

Proof Let us write x^+, y^+, x^-, y^- for the upper and lower bounds of the European call option and the American call option. Likewise we will drop $(\{h_n\}; N)$ and (h_N, N) from the notation A^+, A^-, B^+, B^-.

We will prove that $A^+ = B^+$ and as a consequence that

$$x^+((s_N - K)^+, N) = y^+(\{(s_n - K)^+\}, N). \tag{6.56}$$

This along with (6.53) and (6.54), would give the required equality (6.55), completing the proof.

We have seen earlier that $B^+ \subseteq A^+$. Let $y \in A^+$ and π be such that (6.50) holds, (with $h_N(s_0, \ldots, s_N) = (s_N - K)^+$). Then

$$V_N(y, \pi)(S_0, \ldots, S_N) \geq (S_N - K)^+.$$

Let $Q \in \mathcal{E}(P)$. Recall that $\mathcal{E}(P)$ is nonempty because of our assumption of no arbitrage. Since $V_n(y, \pi)(S_0, \ldots, S_n)\beta^n$ is a Q-martingale, we get

$$V_n(y, \pi)(S_0, \ldots, S_n)\beta^n \geq E^Q[(S_N - K)^+ | \mathcal{F}_n]\beta^n.$$

Using Jensen's inequality for conditional expectations and the fact that $S_n\beta^n$ is a Q martingale, we get

$$\begin{aligned} V_n(y, \pi)(S_0, \ldots, S_n)\beta^n &\geq (E^Q(S_N | \mathcal{F}_n) - K)^+ \\ &= (S_n \beta^{N-n} - K)^+ \beta^N \\ &= (S_n - K R^{N-n})^+ \beta^n \\ &\geq (S_n - K)^+ \beta^n \end{aligned}$$

as $R > 1$. We thus have proved that

$$V_n(y, \pi)(S_0, \ldots, S_N) \geq (S_n - K)^+ \quad \text{a.s. } Q. \tag{6.57}$$

By the definition of $\mathcal{E}(P)$, $Q(S_0 = s_0, \ldots, S_N = s_N) > 0$ for all $(s_0, \ldots, s_N) \in \mathcal{S}$, and hence (6.48) follows from (6.57) and so $y \in B^+$.

Thus $A^+ = B^+$. As observed earlier, this completes the proof. □

Let us return to an American type security $(\{h_n, \}, N)$. The following is the main result of this section. The following result shows that $B^+ = B^-$ and hence the price of the American security $(\{h_n\}, N)$ is uniquely determined by the no arbitrage principle. Recall our notation:

$$Z_n = h_n(S_0, \ldots, S_n) = h_n(S_0, \ldots, S_N).$$

Theorem 6.6 *Suppose that arbitrage opportunities do not exist and that the market is complete. Let*

$$\tilde{y} = \sup_{\tau \leq T} E^Q[Z_\tau \beta^\tau], \tag{6.58}$$

where the supremum is taken over all stopping times $\tau \leq T$. Then

$$y^+(\{h_n\}, N) = y^-(\{h_n\}, N) = \tilde{y}. \tag{6.59}$$

Proof In view of our assuptions, $\mathcal{E}(P) = \{Q\}$ (see Theorems 6.1, 6.3). Let us define $\{Y_n : n \leq N\}$ by backward induction as follows:

$$Y_N = Z_N \beta^N \tag{6.60}$$

and having defined $\{Y_i : n+1 \leq i \leq N\}$, define

$$Y_n = \max\{Z_n \beta^n, E(Y_{n+1}|\mathcal{F}_n)\}. \tag{6.61}$$

By construction $\{Y_n, \mathcal{F}_n\}$ is a supermartingale, i.e.,

$$Y_n \geq E(Y_{n+1}|\mathcal{F}_n).$$

For $n \geq 1$, define

$$D_n = \sum_{i=0}^{n+1} \{Y_i - E^Q(Y_{i+1}|\mathcal{F}_i)\}.$$

Then

$$Y_n = Y_0 + M_n - D_n \tag{6.62}$$

with $D_n \geq D_{n-1} \geq 0$, where

$$M_n = \sum_{i=1}^{n} (Y_i - E^Q(Y_i|\mathcal{F}_i)). \tag{6.63}$$

Then M_n is a Q-martingale. Consider the contingent claim

$$(Y_0 + M_N)R^N.$$

Using completeness of the market, get y, θ such that

$$V_N(y, \pi)(S_0, \ldots, S_N) = (Y_0 + M_N)R^N. \tag{6.64}$$

Since S_0 is a constant and $\mathcal{F}_0 = \sigma(S_0)$, Y_0 is a constant. Multiplying both sides in (6.64) by β^N (recall $\beta = R^{-1}$) we get $y = Y_0$. Further, using the fact that $V_n(Y_0, \pi)(S_0, \ldots, S_n)\beta^n$ and M_n are Q-martingales, we conclude that

$$V_n(Y_0, \pi)(S_0, \ldots, S_n)\beta^n = Y_0 + M_n. \tag{6.65}$$

The identities (6.62) and (6.65) along with the observation that $D_n \geq 0$ imply that for all n

$$\begin{aligned} V_n(Y_0, \pi)(S_0, \ldots, S_n)\beta^n &\leq Y_n \\ &\geq Z_n \beta^n. \end{aligned} \tag{6.66}$$

We thus conclude that $Y_0 = y$ and θ chosen above satisfy (6.48) so that $Y_0 \in B^+$. Also for any stopping time τ

$$\mathbb{E} V_\tau(Y_0, \pi)(S_0, \ldots, S_n)\beta^\tau = Y_0,$$

and so (6.66) gives

$$\mathbb{E}(Z_\tau \beta^\tau) \leq Y_0. \tag{6.67}$$

Define

$$\sigma = \inf\{i < N : Y_i > E(Y_{i+1}|\mathcal{F}_i\} \min N.$$

Then for $i < N$,

$$\{\sigma = i\} = \{Y_j = E(Y_{j+1}|\mathcal{F}_j), \quad \forall j < i; \quad Y_i > E(Y_{i+1}|\mathcal{F}_i)\}$$

and

$$\{\sigma = N\} = \{Y_j = E(Y_{j+1}|\mathcal{F}_j), \quad \forall j < N\}.$$

It follows that σ is a stopping time. By the definition of σ, $D_i 1_{\{\sigma=i\}} = 0$ and so

$$Y_i 1_{\{\sigma=i\}} = (Y_0 + M_i) 1_{\{\sigma=i\}}. \tag{6.68}$$

Also, if for $i < N$, $Y_i > E(Y_{i+1}|\mathcal{F}_i)$, then $Y_i = Z_i \beta^i$ and $Y_N = Z_N \beta^N$. It follows that

$$Y_i 1_{\{\sigma=i\}} = Z_i \beta^i 1_{\{\sigma=i\}}. \tag{6.69}$$

Together, (6.68) and (6.69) imply

$$Z_\sigma \beta^\sigma = Y_0 = (Y_0 + M_\sigma), \tag{6.70}$$

and as a consequence

$$E^Q(Z_\sigma \beta^\sigma) = Y_0. \tag{6.71}$$

From (6.67) and (6.71) we conclude that $\tilde{y} = Y_0$ and (6.65) and (6.71) yield

$$V_\sigma(Y_0, \pi(S_0, \ldots, S_N)) = Z_\sigma \beta^\sigma, \tag{6.72}$$

and thus $Y_0 \in B^-$. We have proved $Y_0 = \tilde{y} \in B^+ \cap B^-$. Hence $\tilde{y} = y^+ = y^-$. □

7
Introduction to Continuous Time Trading

In this chapter, we begin with an informal description of the technical terms used in finance in the context of continuous time trading. We saw these terms in the previous chapter, which was devoid of technicalities. When it comes to continuous time, we cannot escape these technicalities which is why they were first introduced.

We will define most of the technical terms in this chapter and elaborate on them later.

7.1 Introduction

Markets usually consist of two types of securities: the first security is the stock of various companies. Shares of these stocks can be bought or sold in the market. Their price is subject to a large number of factors and it can go up or go down. Stocks are considered *risky* assests since we cannot be sure if the price will go up or go down. The price of the stock at time t, (S_t), is modelled as a random variable. The other security is a bond. This is a riskless security, where one always gets back the investment, plus interest which can be fixed, or which can vary with time. Thus, the price of the bond (B_t) at time t satisfies, $B_s \leq B_t$ if $s < t$.

We are going to consider options on the stocks. They are of the following types:

A *European call option* on the stock of a specified company gives its owner the right (but not the obligation) to buy one share of stock (from the seller of the option contract, also called the writer of the option) at a specific time $T > 0$ (called the terminal time) and at a fixed price K called the exercise price or striking price. The writer of the option contract is committed to sell at time T at a price K if the owner so desires.

It is clear that the owner of the option contract would like to exercise his option if the price of the stock at time T is more than K. Otherwise, the option contract is worthless. It is possible that options may be available in the market for the same stock with different combinations of T and K.

An *European put option* on the stock of a specified company gives its owner the right (but not the obligation) to sell one share of stock (to the seller of the option contract) at terminal time $T > 0$ and at striking price K.

An *American call option* on the stock of a specified company gives its owner the right (but not the obligation) to buy one share of stock *at any time up to the terminal time* T at the striking price K. The *American put option* is defined similarly by replacing *buy* above by *sell.*

More generally, a *contingent claim* is a positive random variable X; if the time period under consideration is $[0, T]$, then it is required that X is $\sigma\{S_u, B_u : 0 \leq u \leq T\}$ measurable. Often, it is also required that $\mathbb{E}(X) < \infty$ or $\mathbb{E}(X^2) < \infty$. It is to be regarded as follows: the buyer of this claim gets a reward of X at time T; the actual amount is random and depends upon the contingency, namely the price evolution in the market of the underlying stock. When $X = (S_T - K)^+$, the contingent claim X is the European call option on a stock (S_t) with terminal time T and striking price K.

In order to study the market consisting of stocks and bonds and options based on the stocks, we have to mathematically model the trading strategies available to investors in the stocks and bonds markets. First, let us consider a market consisting of one stock whose price at t is denoted by S_t and one bond B_t. Consider a strategy of an investor consisting of buying and selling at fixed times, $t_0 < t_1 < \ldots < t_m$. Such strategies will be called *simple strategies.* Let π_{t_i} be the amount of the stock held (or owned) during the interval $(t_i, t_{i+1}]$, and ψ_{t_i} the amount (in some units) of the bond held during the same interval. We cannot allow an investor to forsee the future and hence π_{t_i} and ψ_{t_i} can only depend upon the observed stock and bond prices over the interval $[0, t_i]$. Mathematically, this is described as

$$\pi_{t_i} \text{ is } \sigma\{S_u, B_u; 0 \leq u \leq t_i\} \text{ measurable.}$$

$\{\pi_t, \psi_t\}$ is called the *portfolio* of the investor at time t and (π, ψ) is called the investor's trading strategy. The value V_t of the investor's portfolio at time t is

$$V_t = \pi_t S_t + \psi_t B_t.$$

Suppose there are k stocks whose price processes are denoted by $S_t^i, i = 1, \ldots, k$. It is convienient to work with discounted prices: $\tilde{S}_t^i = \frac{S_t^i}{B_t}$ and the discounted value $\tilde{V}_t(x, \pi) = \frac{V_t(x,\pi)}{B_t} = x + \sum_{i=1}^k \pi_t^i \tilde{S}_t^i$. We would be considering strategies that involve an initial investment x but no further investment or consumption. Thus any money earned by selling stocks is invested in bonds and any money needed to buy more stocks is obtained by selling the bonds. Such strategies are called *self-financing*. Note here that for a self-financing strategy, the initial investment x and π_t determine ψ_t. For a self-financing strategy, when computing in terms of discounted prices, the only change can be through the holding in stocks since the discounted price of the bond is a constant. Hence, when there is only one stock, the discounted gain up to time t, $\tilde{G}_t$, is given by

$$\tilde{G}_t(\pi) = \sum_{t_j \le t} \pi_{t_j}(\tilde{S}_{t_j} - \tilde{S}_{t_{j-1}}).$$

Thus, for a self-financing strategy, the discounted value equals the intial investment plus the discounted gain, giving

$$\tilde{V}_t = x + \sum_{i=0}^{m} \pi_{t_i}(\tilde{S}_{t_{i+1}\wedge t} - \tilde{S}_{t_i \wedge t}). \tag{7.1}$$

7.2 A general model

More generally, let us consider a market consisting of k stocks, whose price at time $t \in [0, T]$ is given by $(S_t^1, \ldots, S_t^k)$, and a bond whose price we denote by S_t^0. It is assumed that $S_u^0 \le S_t^0$ for $u < t$. Further, we also assume that $S_t^0, S_t^1, \ldots, S_t^k$ are r.c.l.l. processes.

The discounted stock prices are $\tilde{S}_t^i = S_t^i(S_t^0)^{-1}$. The stock price processes are assumed to be defined on some probability space $\{\Omega, \mathcal{F}, P\}$. Let $\mathcal{G}_t$ denote the smallest σ field with respect to which S_u^i $0 \le i \le k, 0 \le u \le t$ are measurable. $\mathcal{G}_t$ denotes the information available up to time t. A self-financing *simple* strategy is characterized by an initial investment x, $0 \le t_0 < t_1 < \ldots < t_m \le T$ (these are times at which the portfolio changes) and $(a_{t_j}^i)$, $1 \le i \le k, 0 \le j \le m$, where $a_{t_j}^i$ is a bounded $\mathcal{G}_{t_j}$ measurable random variable and it denotes the number of shares of the ith stock held by the investor during $(t_j, t_{j+1}]$ (where $t_{m+1} = T$ if we are considering a finite horizon T or ∞ if we are considering an infinite horizon). Let us write

$$\pi_t^i = a_{t_j}^i \quad \text{for } t_j < t \le t_{j+1};\ 0 \le i \le k.$$

In other words, $\pi_t^i = \sum_j \pi_{t_j}^i \mathbf{1}_{(t_j, t_{j+1}]}(t)$. π_t^i is hence $\mathcal{F}_t$-measurable and left-continuous. We will denote the strategy as

$$\theta = (x, \pi),$$

where $\pi = (\pi^1, \ldots, \pi^k)$. The discounted value process of the self-financing simple strategy $\theta = (x, \pi)$ is (in analogy with (7.1)) given by

$$\tilde{V}_t(\theta) = x + \sum_{j=1}^{k} \sum_{i=0}^{m-1} a_{t_i}^j(\tilde{S}_{t_{i+1}\wedge t}^j - \tilde{S}_{t_i \wedge t}^j). \tag{7.2}$$

Note that the value $V_t(\theta)$ of the portfolio at time t is S_t^0 times the discounted value $\tilde{V}_t(\theta)$. We will write θ or (x, π) to denote a strategy. We will restrict ourselves to self-financing strategies from now on and drop the adjective self-financing.

A contingent claim X is said to be *attainable* via a simple strategy at time T if there exists a strategy $\theta = (x, \pi)$, such that

$$\begin{aligned} X &= V_T(\theta) \\ &= S_T^0 \tilde{V}_T(\theta). \end{aligned}$$

A simple strategy $\theta = (x, \pi)$ is said to be an *arbitrage opportunity* if

$$P(\tilde{V}_T(\theta) \geq 0) = 1 \text{ and } P(\tilde{V}_T(\theta) > 0) > 0. \tag{7.3}$$

It is usual to assume that

$$P[\tilde{V}_0(\theta) = x] = 1.$$

Definition The market consisting of $S_t^0, S_t^1, \ldots, S_t^k$ is said to satisfy the *No Arbitrage* (NA) property in the class of simple strategies if there does not exist any simple strategy θ such that (7.3) holds.

It is natural to require that any model of stock prices satisfy the no arbitrage property, since, if an arbitrage opportunity exists, everyone would try to replicate that strategy which would destabilize the market and push up prices of the stocks bought through this strategy.

A probability measure Q is said to be an *Equivalent Martingale Measure* (written as EMM) for $\tilde{S} = (\tilde{S}^1, \ldots, \tilde{S}^k)$ if Q is equivalent to P and each of $\tilde{S}_t^1, \ldots, \tilde{S}_t^k$ are Q-local martingales. In continuous time finance, the existence of an EMM is not equivalent to NA. However, it is essentially equivalent and the precise relationship between these two concepts constitutes the Fundamental Theorem of Asset Pricing whose statement and proof will be given in the next chapter. We only note here that we have to rule out approximate arbitrage opportunities (suitably defined) to get an equivalent martingale measure Q. If an EMM Q exists, then it follows from Theorem 5.3 that $S_t^i, \tilde{S}_t^i$ are P-semimartingales for $1 \leq i \leq k$.

This justifies our considering only semimartingales as plausible models of stock prices.

7.3 Trading strategies and arbitrage opportunities

In the rest of the chapter we will assume that the *stock price processes S_t^i are continuous semimartingales.*

Let us note that, as in the discrete case, we assume that we are considering an ideal situation where there are no transaction costs and short selling of bonds (which amounts to obtaining a loan at the same rate of interest as is available on investments) as well as short selling of stock is permitted, though an overall limit may be placed on an individual investor's debt. This assumption is often referred to as a *frictionless market.*

Consider a market consisting of a bond, whose price at time t is S_t^0, and k stocks, whose prices are S_t^i, $1 \leq i \leq k$. We will consider a finite time horizon T, thus $t \in [0, T]$. Let

$$\tilde{S}_t^i = S_t^i (S_t^0)^{-1}, \quad S_0^i = 1,$$

denote the discounted price of the ith stock. Let $\mathcal{F}_t^S$ be the filtration generated by $S = (S^0, \ldots, S^k)$. We have so far considered the class of simple (self-financing) strategies in the previous section. A trading strategy is defined in general as follows.

Definition $\theta = (\pi^0, \pi^1, \ldots, \pi^k)$ is said to be a trading strategy if (writing $\mathcal{F}_t$ for $\mathcal{F}_t^S$)

(a) each π_t^i is $(\mathcal{F}_t)$-predictable,

(b) The stochastic integral $\int_0^T \pi_t^i d\tilde{S}_t^i$ exists for $i = 0, \ldots, k$.

As seen earlier, π_t^i is to be interpreted as the number or the amount of the ith stock held by the investor at time t ($i = 0$ corresponds to the bond). Thus, $\theta = (\pi_t^0, \pi_t^1, \ldots, \pi_t^k)$ represents the holding of the investor at time t and is also known as the investor's *portfolio.*

The trading strategies that depend on a continuous time parameter cannot really be implemented in the market—an investor can really trade only at finitely many time points. These strategies are limits of strategies that an investor can pursue.

Definition For a given portfolio $\theta = (\pi^0, \pi^1, \ldots, \pi^k)$, its *value* or *wealth process* is defined as

$$V_t(\theta) := \sum_{i=0}^{k} \pi_t^i S_t^i, \quad (t > 0).$$

Definition The accumulated gains or losses up to (and including) the instant t are called the *gains* process and is given by

$$G_t(\theta) := \sum_{i=0}^{k} \int_0^t \pi_u^i dS_u^i.$$

The *discounted* value process $\tilde{V}_t(\theta)$ and the *discounted* gains process $\tilde{G}_t(\theta)$ are respectively given by

$$\tilde{V}_t(\theta) := \sum_{i=0}^{k} \pi_t^i \tilde{S}_t^i = \pi_t^0 + \sum_{i=1}^{k} \pi_t^i \tilde{S}_t^i,$$

$$\tilde{G}_t(\theta) := \sum_{i=1}^{k} \int_0^t \pi_u^i d\tilde{S}_u^i.$$

Note that $\tilde{G}_t(\theta) \neq \frac{1}{S_t^0} G_t(\theta)$ in general.

Definition $\theta = (\pi^0, \pi^1, \ldots, \pi^k)$ is said to be a *self-financing* strategy if there is no investment or consumption at any time $t > 0$. That is, $\theta = (\pi^0, \pi^1, \ldots, \pi^k)$ is a self-financing strategy if

$$\tilde{V}_t(\theta) = \tilde{V}_0(\theta) + \tilde{G}_t(\theta) \quad \text{a.s. } 0 \le t \le T.$$

If, for a self-financing strategy θ, $\tilde{V}_0(\theta)$, $\pi^1, \pi^2, \ldots, \pi^k$ are given, then one has

$$\pi_t^0 + \sum_{i=1}^k \pi_t^i \tilde{S}_t^i = \tilde{V}_0(\theta) + \sum_{i=1}^k \int_0^t \pi_u^i d\tilde{S}_u^i. \tag{7.4}$$

Thus π_t^0 is determined by (7.4). In view of this observation, we will denote a self-financing strategy $\theta = (\pi^0, \pi^1, \ldots, \pi^k)$ by

$$\theta = (x, \pi),$$

where

$$\pi = (\pi^1, \pi^2, \ldots, \pi^k)$$

and $x = \tilde{V}_0(\theta)$ is the initial investment. Also, we have

$$\tilde{V}_t(\theta) = x + \sum_{i=1}^k \int_0^t \pi_u^i d\tilde{S}_u^i \quad \forall t \ge 0. \tag{7.5}$$

Let us define

$$\hat{S}_t^i = \int_0^t (\tilde{S}_t^i)^{-1} d\tilde{S}_t^i. \tag{7.6}$$

Then $\hat{S}_t^i$ is also a semimartingale and

$$\tilde{S}_t^i = 1 + \int_0^t \tilde{S}_u^i d\hat{S}_u^i,$$

and, as a consequence,

$$\int_0^T \pi_t^i d\tilde{S}_t^i = \int_0^T \pi_t^i \tilde{S}_t^i d\hat{S}_t^i.$$

Let

$$\hat{S}_t^i = M_t^i + A_t^i \tag{7.7}$$

be the canonical decomposition of the continuous semimartingale $\hat{S}^i$. Then by recasting the integrability condition in the definition of a trading strategy, it can be seen that $\theta = (x, \pi^1, \dots, \pi^k)$ is a (self-financing) trading strategy if $\pi^1, \dots, \pi^k$ are predictable processes such that

$$\int_0^T (\pi_t^i)^2 (\tilde{S}_t^i)^2 d\langle M, M\rangle_t < \infty \quad a.s. \tag{7.8}$$

$$\int_0^T |\pi_t^i||\tilde{S}_t^i| d|A|_t < \infty \quad a.s. \tag{7.9}$$

Let us now introduce the important concept of an admissible strategy. It is reasonable to expect that markets will allow an investor to borrow (i.e., short sell a bond) or short sell a stock (i.e., $\pi_t^i < 0$) as long as his net worth is positive. In other words, if m is the investor's net worth, then he must ensure that

$$\tilde{V}_t(\theta) \geq -m.$$

This leads us to the following definition.

Definition A self-financing strategy $\theta = (x, \pi^1, \dots, \pi^k)$ is admissible (or tame) if for some $m < \infty$,

$$P\big\{\tilde{V}_t(\theta) \geq -m \quad \forall t\big\} = 1. \tag{7.10}$$

The no arbitrage property is defined as follows.

Definition An admissible strategy $\theta = (x, \pi^1, \dots, \pi^k)$ is said to be an *arbitrage opportunity* if $x = 0$,

$$\tilde{V}_T(\theta) \geq 0 \quad P - a.s. \tag{7.11}$$

and

$$P[\tilde{V}_T(\theta) > 0] > 0. \tag{7.12}$$

Definition $\tilde{S} = (\tilde{S}^1, \dots, \tilde{S}^k)$ has the *no arbitrage property* (NA) if there does not exist an admissible strategy $\theta = (0, \pi)$ such that (7.11) and (7.12) hold.

Thus, NA holds if and only if for any predictable process $(\pi_t^1, \dots, \pi_t^k)$, such that π^i is S^i-integrable, and for some $m < \infty$,

$$P\Bigg(\sum_{i=1}^k \int_0^t \pi_u^i d\tilde{S}_u^i \geq -m \quad \forall t\Bigg) = 1, \tag{7.13}$$

$$P\Bigg(\sum_{i=1}^k \int_0^T \pi_u^i d\tilde{S}_u^i \geq 0\Bigg) = 1 \tag{7.14}$$

implies that

$$P\left(\sum_{i=1}^{k}\int_0^T \pi_u^i d\tilde{S}_u^i = 0\right) = 1. \tag{7.15}$$

Let

$$\mathcal{M}(P) = \left\{Q : Q \equiv P \text{ and } \tilde{S}_t^i \text{ a } Q\text{-local martingale}, \quad 1 \le i \le k\right\}$$

be the class of equivalent (local) martingale measures (abbreviated as EMM). We have seen that in discrete time, NA holds iff $\mathcal{M}(P)$ is non-empty at this level of generality. Here we note that $\mathcal{M}(P) \neq \emptyset$ implies NA. The reverse implication holds under some additional conditions, and this is dealt with in the next chapter.

Theorem 7.1 *Let $Q \in \mathcal{M}(P)$. For an admissible strategy $\theta = (x, \pi^1, \ldots, \pi^k)$,*

$$U_t := \sum_{i=1}^{k}\int_0^t \pi_u^i d\tilde{S}_u^i \tag{7.16}$$

is a Q-local martingale and a Q-super martingale.

Thus $\mathcal{M}(P) \neq \emptyset$ implies NA.

Proof Let us note that under Q, $\tilde{S}^i$ is a local martingale and hence U, which is a stochastic integral with respect to $\tilde{S}$, is also a local martingale.

If $\{\tau_n\}$ is an increasing sequence of stopping times such that $P(\tau_n = T) \to 1$ and $U_t^n = U_{t\wedge\tau_n}$ is a Q-martingale, then for $s \le t$, $E^Q(U_t^n|\mathcal{F}_s^S) = U_s^n$. In view of the admissibility of π, $U_t^n \ge -m$ for some m and hence, by Fatou's lemma for conditional expectation, we get

$$\begin{aligned} E^Q(U_t|\mathcal{F}_s^S) &= E^Q(\liminf U_t^n|\mathcal{F}_s^S) \\ &\le \liminf E^Q(U_t^n|\mathcal{F}_s^S) \\ &= \liminf U_s^n = U_s. \end{aligned}$$

This proves that U_t is a Q-supermartingale. In particular, $E^Q(U_t) \le E^Q(U_0) = 0$. Thus if $P(U_T \ge 0) = 1$, then $Q(U_T \ge 0) = 1$. Hence $E^Q(U_t) \le 0$ implies $Q(U_T = 0) = 1$ and therefore, $P(U_T = 0) = 1$. Thus NA holds. □

Let us examine the requirement of admissibility, namely that

$$U_t = \sum_{i=1}^{k}\int_0^t \pi^i d\tilde{S}^i \ge -m \tag{7.17}$$

on a trading strategy π. This enables us to conclude that U_t is a Q-supermartingale and hence $E_Q(U_T) \le 0$. If all predictable strategies π for which $\sum_i \int \pi^i d\tilde{S}^i$ is defined are allowed, then arbitrage opportunities may exist even in a market consisting of a single stock which is a martingale (and so trivially the EMM property holds). Here is an example. Consider a single stock whose price is

$$dS_t = S_t(dW_t + r dt), \quad S_0 = 1,$$

where (W_t) is a Brownian motion on $(\Omega, \mathcal{F}, Q)$ along with a bond with interest rate r. Here the discounted stock price is

$$\tilde{S}_t = \exp(W_t - \frac{1}{2}t)$$

and hence the discounted stock price is a martingale so that (as mentioned above), the EMM property holds. We give below an example of a strategy which is not admissible and which is an arbitrage opportunity. Take $f_t = \frac{1}{\sqrt{T-t}}$ and note that for $s < T$

$$\int_0^s f_t^2 dt = \int_0^s \frac{1}{T-t} dt = \log\left(\frac{T}{T-s}\right).$$

So the integral

$$Y_s = \int_0^s f_t dW_t$$

is defined for $s < T$. It is a Gaussian process with mean 0 and covariance

$$E^Q(Y_{s_1} Y_{s_2}) = \log\left(\frac{T}{T - s_1 \wedge s_2}\right).$$

If we define

$$\beta_u = Y_s \quad \text{if} \quad u = \log\left(\frac{T}{T-s}\right),$$

then β_u is a Brownian motion under Q. Thus

$$\tau_1 = \inf\{u : \beta_u = 2\}$$

satisfies $Q(\tau_1 < \infty) = 1$. As a consequence,

$$\tau = \inf\{s : Y_s = 2\}$$

satisfies $Q(\tau < T) = 1$. Define

$$\pi_s^1 = 1_{(s \leq \tau]} f_s (\tilde{S}_s)^{-1}.$$

Since τ is a stopping time, it follows that π^1 is predictable. Now

$$\int_0^T \pi_s^1 d\tilde{S}_s = \sigma \int_0^\tau f_s dW_s = 2; \tag{7.18}$$

so it is an arbitrage opportunity. Further, $(0, \pi^1)$ cannot be an admissible strategy, for if it were, $\int_0^t \pi_s^1 d\tilde{S}_s$ would be a supermartingale and thus

$$E^Q\left[\int_0^T \pi_s^1 d\tilde{S}_s\right] \leq 0,$$

contradicts (7.18).

Thus to rule out arbitrage opportunities, we have to put some restriction on the risk that an investor is allowed to take. Putting a uniform lower bound is one way of doing it, in which case, we get admissible strategies. There are other ways to do so. For this, suppose the stock price model satisfies the EMM property, i.e., an EMM exists. Let Q denote a fixed EMM. Let us note that in this case the integrability condition (7.8) implies (7.9) since, when an EMM exists, the canonical decomposition of $\tilde{S}^i$ satisfies (5.10) in Theorem 5.6. Thus, $\theta = (x, \pi^1, \pi^2, \ldots, \pi^k)$ is a trading strategy if π^i are predictable processes satisfying

$$\int_0^T (\pi_t^i)^2 (\tilde{S}_t^i)^2 d[\hat{S}^i, \hat{S}^i]_t < \infty \quad a.s. \tag{7.19}$$

Here we have used the fact that $[\hat{S}^i, \hat{S}^i]_s = \langle M^i, M^i \rangle_s$, where M^i is the local martingale part of the canonical decomposition of $\hat{S}^i$.

Definition A trading strategy $\theta = (x, \pi^1, \pi^2, \ldots, \pi^k)$ is said to be p-admissible (for $p > 1$) if

$$E^Q[\{\int_0^T \sum_{i=1}^k |\pi_s^i|^2 d\langle \tilde{S}^i, \tilde{S}^i \rangle\}^{p/2}] < \infty. \tag{7.20}$$

We have the following result on p-admissible strategies which will also be needed in the sequel.

Lemma 7.2 *Let $\theta = (x, \pi^1, \ldots, \pi^k)$ be a p-admissible strategy. Then its discounted value function $\tilde{V}_t(\theta)$ is a Q-martingale such that $\mathbb{E}|\tilde{V}_T(\theta)|^p < \infty$.*

Proof If θ satisfies (7.20), then $\tilde{V}_t(\theta)$ is a local martingale with

$$E^Q(\langle \tilde{V}_t(\theta), \tilde{V}_t(\theta) \rangle^{p/2}) < \infty.$$

Hence by Burkholder's inequality (4.42),

$$\begin{aligned} E^Q(\sup_{t \leq T} |\tilde{V}_t(\theta)|^p) &\leq C_p E(\langle \tilde{V}_t(\phi), \tilde{V}_t(\phi) \rangle^{p/2}) \\ &< \infty. \end{aligned}$$

It follows from this that $\tilde{V}_t(\phi)$ is indeed a Q-martingale. □

It follows that when $Q \in \mathcal{M}(P)$, arbitrage opportunities do not exist in the class of p-admissible strategies. So depending upon the context, we would require a trading strategy to be admissible or p-admissible. Or, we could just require that the discounted value process $\tilde{V}_t(\theta)$ be a Q-martingale.

7.4 Examples

We will consider a couple of examples of stock price models.

Example 1. Geometric Brownian Motion (GBM) Consider a market consisting of one stock whose price is given by

$$dS_t = S_t(\sigma W_t + \mu dt), \quad S_0 = 1,$$

where $\sigma > 0$ and $\mu \in \mathbb{R}$ are constants and W is standard Brownian motion. Then

$$S_t = \exp\{\sigma W_t + (\mu - \frac{1}{2}\sigma^2)t\}, \quad t > 0.$$

Suppose that the bond price B_t is given by $B_t = \exp\{rt\}$. Then the discounted stock price $\tilde{S}_t = S_t(B_t)^{-1}$ is given by

$$\tilde{S}_t = \exp\{\sigma W_t + (\mu - \frac{1}{2}\sigma^2 - r)t\}, \quad t > 0.$$

and thus from (7.6), we get $\hat{S}_t = \sigma W_t + (\mu - r)t$. By Girsanov's theorem (Theorem 5.5), it follows that Q, defined by

$$\frac{dQ}{dP} = \exp\left\{-\frac{\mu - r}{\sigma} W_T - \frac{T}{2}\frac{(\mu - r)^2}{\sigma^2}\right\},$$

is a probability measure and that $\tilde{S}_t$ is a Q-martingale. Indeed, $\hat{S}_t$ is a Brownian motion under Q. Thus, this model satisfies the NA property.

Example 2. Diffusion model for k stocks. Suppose $S^1, S^2, \ldots, S_t^k$ are the prices of k stocks given by

$$dS_t^i = S_t^i\{\sum_{j=1}^d \sigma_{ij}(S_t, t)dW_t^j + b^i(S_t, t)dt\}, \quad i = 1, \ldots, k,$$

where

$$a_{ij}(x, t) := \sum_m \sigma_{im}(x, t)\sigma_{jm}(x, t)$$

is assumed to be positive definite for each (x, t) and $(W^1, W^2, \ldots, W^d)$ are independent Brownian motions. Suppose the bond price $S_t^0 = \exp\{rt\}$. Then in this case $\tilde{S}_t^i$ and $\hat{S}_t^i$ are given by

$$\hat{S}_t^i = \int_0^t \{\sum_{j=1}^d \sigma_{ij}(S_u, u)dW_u^j + (b^i(S_u, u) - r)du\}$$

$$\tilde{S}_t^i = 1 + \int_0^t \tilde{S}_u^i d\hat{S}_u^i.$$

Here, if $\hat{S}_t^i = M_t^i + A_t^i$ is the canonical decomposition of $\hat{S}_t^i$, then

$$\langle M^i, M^i \rangle_t = \int_0^t a_{ii}(S_u, u)du$$

and

$$A_t^i = \int_0^t (b^i(S_u, u) - r)du.$$

Thus defining $h^i(x, u) = (a_{ii}(x, u))^{-1}(b^i(x, u) - r)$, we have

$$A_t^i = \int_0^t h^i(S_u, u)d\langle M^i, M^i \rangle_u.$$

If we assume that h is bounded, then $\mathbb{E}[\rho_T] = 1$ where

$$\begin{aligned} \rho_T \quad = \quad & \exp\Big\{-\sum_{i=1}^k \int_0^T h^i(S_u, u)dM_u^i \\ & - \frac{1}{2}\sum_{i=1}^k \sum_{j=1}^k h^i(S_u, u)h^j(S_u, u)\, d\langle M^i, M^j \rangle_u\Big\} \end{aligned}$$

and Q, defined by

$$\frac{dQ}{dP} = \rho_T,$$

is a probability measure under which $\hat{S}_t^i$ and $\tilde{S}_t^i$ are local martingales. Thus Q is an EMM. As seen earlier, this imples that NA holds in this case.

7.5 Contingent claims and complete markets

Let us consider the stock price model as in Section 7.3. As we have remarked earlier, it is natural to impose the condition that NA hold. We will in fact assume that EMM exists. We remark here that it is economically meaningful to assume that EMM exists. Thus we fix a probability measure Q such that under Q, the discounted stock prices, $\tilde{S}_t^i$ are martingales.

Definition A *contingent claim* is an $\mathcal{F}_T^S$-measurable random variable Z_T satisfying

$$Z_T \geq 0 \quad \text{a.s.,} \quad E^Q(Z_T) < \infty.$$

Examples of contingent claims: Let S_t denote the price of a given stock at time t, and for $K < \infty$ and $T < \infty$, let $Z_T = (S_T - K)^+$, $Y_T = (K - S_T)^+$. Then Z_T and Y_T are contingent claims; the first one is the European call option and the second one is the European put option on the stock S_t—both with terminal time T and striking price K. We will discuss them in detail in a later chapter.

Definition A *contingent claim* is said to be *attainable* if there exists a strategy $\theta = (x, \pi^1, \dots, \pi^k)$ such that $\tilde{V}_t(\pi)$ is a Q-martingale and

$$V_T(\theta) = Z_T \quad \text{a.s.} \tag{7.21}$$

Definition The strategy $\theta = (x, \pi^1, \dots, \pi^k)$ satisfying (7.21) is said to be a *hedging strategy* for the contingent claim Z_T.

Definition of Complete Markets A market consisting of k-stocks $(S^1, S^2, \dots, S^k)$ and a bond (S_t^0) is said to be *complete* if every contingent claim is attainable.

Equivalently, the market is complete if, for every positive Q-integrable $\mathcal{F}_T^S$ measurable random variable, Z, there exists a strategy $\theta = (x, \pi^1, \dots, \pi^k)$ such that $\tilde{V}_t(\theta)$ is a Q-martingale and

$$\tilde{V}_t(\theta) = x + \sum_{i=1}^{k} \int_0^T \pi^i d\tilde{S}^i = Z_T. \tag{7.22}$$

From our definitions, it appears that the notion of completeness depends upon the choice of the EMM Q. We will see in Chapter 9 that this is not the case.

8
Arbitrage and Equivalent Martingale Measures

As we have seen in the discrete case, the NA property plays a central role in option pricing and in discrete time; it is equivalent to the existence of an equivalent martingale measure. We will explore this relation and show that a suitable extension of the NA property is equivalent to NA.

8.1 Introduction

Borrowed from the French, the word *arbitrage* has acquired the following special meaning in (stochastic) finance. Let us suppose that there exists a (self-financing) trading strategy such that with a zero initial investment (or even with borrowed money for an initial investment), the strategy enables an investor to obtain a profit at the final time T *without any possibility of loss*. We then say that there is *arbitrage* or, more precisely, that there is a possibility for arbitrage. If no such strategy exists, we say that there is *no arbitrage* (and the model is said to satisfy the NA property).

Clearly, if an arbitrage opportunity exists, it will be the preferred strategy for an investor. So many investors would try to enter the market and follow this strategy, which would disturb the equilibrium of the market so that the prices of the stocks being traded would change. It is important, therefore, to exclude such an unrealistic situation by imposing a general enough condition on the underlying model of stock prices in an option pricing theory.

The question then is to find a necessary and sufficient condition on our stochastic model that ensures no arbitrage. In Chapter 6, we saw that in the special case considered, namely that there is one stock that takes finitely many values and we are considering finitely many time points at which trading is allowed, that the NA property is equivalent to the EMM property (existence of an equivalent (local) martingale measure). As we indicated in Chapter 7, the equivalence breaks down in general when trading is allowed at infinitely many points. We saw an example (in continuous time trading) where EMM exists but NA does not hold unless we restrict the class of trading strategies by controlling the associated risk. Later in this chapter, we will see an example in discrete time trading over an infinite horizon which shows that NA does not imply EMM in general.

In the next section, we will consider the implications of NA in the model discussed in the previous chapter, where we model the stock prices as continuous semimartingales. Surprisingly, the search for the answer has led to rather deep problems of stochastic analysis. We will see that the requirement of NA implies that the semimartingales must be of a specific structure.

In the rest of the chapter, we will go one step back and consider a very general stochastic model for the prices $(S_t^1, \ldots, S_t^k)$ of the stocks, assuming only that they are r.c.l.l. processes without assuming that they are semimartingales. We can still consider simple trading strategies and the associated value function. We will see that the property of no arbitrage in the class of simple strategies is *essentially equivalent* to the existence of an equivalent martingale measure. We have to qualify the previous statement by *essentially* because the EMM property is equivalent to NA in the *closure* of the class of simple strategies. The closure is to be taken in an appropriate topology. We will discuss this point in detail later in this chapter.

The equivalence between NA (in the extended class) and the EMM property is essentially a functional analytic result. It is an infinite dimensional analogue of the theorem of an alternative for a system of linear equations. It crucially uses a separation theorem, versions of which were independently proved by Kreps (Kreps, 1981) and Yan (Yan, 1980). (Yan proved this in a different but related context—that of characterizing semimartingales).

In section 8.4, we will prove the Kreps–Yan separation theorem and then give its consequences, when the closure described above is taken in L^p for $1 \leq p < \infty$. In this case, we get equivalence of the NA (in the closure) and existence of an EMM with density in L^q, the dual of L^p.

The case $p = \infty$ is the most interesting since then we would get equivalence with the EMM property itself (since every density belongs to L^1). However, in this case one is forced to use the weak* topology on L^∞ while taking closure. This raises questions about the interpretation since this topology is not metrizable. The weak* closure can also be described via Orlicz spaces. We present some of our recent work, which uses Orlicz spaces and gives an economically meaningful condition that characterizes the EMM property. The definitions and required results on Orlicz spaces are given in a separate section.

8.2 Necessary and sufficient conditions for NA

In the stock model considered in the previous chapter, we saw that a market consisting of a bond (S_t^0) and stocks ($S_t^1, \dots, S_t^k$) satisfies no arbitrage (NA) if and only if the discounted stock price process $\tilde{S}_t^i = S_t^i[S_t^0]^{-1}$ satisfies (7.13)–(7.15). This leads us to the following.

Definition We say that a ($\mathbb{R}^k$-valued) continuous semimartingale $(X_t^1, \dots, X_t^k)$ satisfies NA if for $(\mathcal{F}_t^X)$-predictable processes $(f^1, \dots, f^k)$, such that $\int_0^T f^i dX^j$ is defined,

$$\sum_{j=1}^k \int_0^t f^j dX^j \geq -1 \text{ for all } t \text{ a.s.}$$

and

$$\sum_{j=1}^k \int_0^T f^j dX^j \geq 0 \text{ a.s.}$$

imply that

$$\sum_{j=1}^k \int_0^T f^j dX^j = 0 \text{ a.s.}$$

Thus, as noted above, the market consisting of $\{S_t^0, S_t^1, \dots, S_t^k\}$ satisfies NA if and only if the semimartingale $\{\tilde{S}_t^1, \dots, \tilde{S}_t^k\}$ satisfies NA. Let

$$\hat{S}_t^i = \int_0^t (\tilde{S}_t^i)^{-1} d\tilde{S}_t^i.$$

Note that for $(f^1, \dots, f^k)$ predictable processes, $\sum_{j=1}^k \int_0^t f^j d\tilde{S}^j$ is defined if and only if $\sum_{j=1}^k \int_0^t f^j \tilde{S}^j d\hat{S}^j$ is defined, and then the two integrals are equal. It follows that $(\tilde{S}_t^1, \dots, \tilde{S}_t^k)$ satisfies NA if and only if $(\hat{S}_t^1, \dots, \hat{S}_t^k)$ satisfies NA.

In the rest of the section, we will get necessary and sufficient conditions for $(\hat{S}_t^1, \dots, \hat{S}_t^k)$ to satisfy NA, which in turn is the same as saying that the market consisting of $(S_t^0, \dots, S_t^k)$ satisfies NA.

For $1 \leq i \leq k$, let

$$\hat{S}_t^i = M_t^i + A_t$$

be the canonical decomposition of the continuous semimartingale $\hat{S}^i$ with M^i being a local martingale and $A^i \in \mathcal{V}$.

When $(\hat{S}^1, \dots, \hat{S}_k)$ admits an equivalent martingale measure (EMM), then Theorem 5.6 implies that A^i admits a representation (5.10). We will see that the same is true if $(\hat{S}_t^1, \dots, \hat{S}_t^k)$ satisfies NA.

We need a technical lemma that says that for the continuous processes $A, B \in \mathcal{V}$, absolute continuity implies that the Radon–Nikodym derivative can be chosen to be predictable. In what follows, we will be working with the filtration $(\mathcal{F}_t^{\hat{S}})$, which is also the filtration $(\mathcal{F}_t^S)$. The term *predictable* in the rest of the section refers to this filtration. The underlying probability space is $(\Omega, \mathcal{F}, P)$.

Lemma 8.1 *Let A_t, B_t be continuous processes $A, B \in \mathcal{V}^+$. Suppose that for a P-null set N*

$$dA(\omega) << dB(\omega) \quad \forall \omega \notin N. \tag{8.1}$$

Then there exists a predictable process f such that $\int_0^T |f_s| d|B|_s < \infty$ a.s., and for P-null set N_1

$$A_t(\omega) = \int_0^t f_u(\omega) dB_u(\omega) \ \forall t \leq T, \ \forall \omega \notin N_1. \tag{8.2}$$

Proof Let $D_t = |A|_t + |B|_t$. D_t is a continuous process and $D \in \mathcal{V}$. Let P' be a probability measure defined by

$$dP' = ce^{-D_T} dP,$$

where c is a constant, chosen so that $P'(\Omega) = 1$. Let $\mathcal{H}$ be the class of predictable ϕ such that

$$\|\phi\|_D^2 = E_{P'}\left[\int_0^T |\phi_u|^2 dD_u\right] < \infty.$$

$\mathcal{H}$ is a Hilbert space with norm $\|\cdot\|_D$ and inner product

$$(\phi^1, \phi^2)_D = \mathbb{E}_{P'} \int_0^T \phi_u^1 \phi_u^2 dD_u.$$

Note that for $\phi \in \mathcal{H}$,

$$\begin{aligned}
\mathbb{E}_{P'}\left[\int_0^T |\phi_u| d|A|_u\right] &\leq \mathbb{E}_{P'}\left[\int_0^T |\phi_u| dD_u\right] \\
&\leq \left(\mathbb{E}_{P'}\left[\int_0^T |\phi_u|^2 dD_u\right]\right)^{\frac{1}{2}} (\mathbb{E}_{P'}[D_T])^{\frac{1}{2}} \\
&\leq c_1 \|\phi\|_D.
\end{aligned}$$

Thus, Λ, defined by

$$\Lambda(\phi) = \mathbb{E}_{P'}\left[\int_0^T \phi_u dA_u\right], \ \phi \in \mathcal{H},$$

is a continuous linear functional on $\mathcal{H}$. Hence there exists $\psi \in \mathcal{H}$ such that

$$\Lambda(\phi) = (\phi, \psi)_D.$$

It follows that

$$\mathbb{E}_{P'}\left[\int_0^T \phi_u dA_u\right] = \mathbb{E}_{P'}\left[\int_0^T \phi_u \psi_u dD_u\right]. \tag{8.3}$$

Let $N_t = A_t - \int_0^t \psi_u dD_u$. Then $N \in \mathcal{V}$ and N is continuous. Further, taking $\phi_u = 1_C 1_{(s,t]}(u)$, $C \in \mathcal{F}_s^S$, in (8.3), we conclude that

$$\mathbb{E}_{P'}[1_C(N_t - N_S)] = 0.$$

So N is a martingale on $(\Omega, \mathcal{H}, P')$. By Theorem 1.9, it follows that $N_t \equiv 0$. Thus,

$$A_t = \int_0^t \psi_u dD_u. \tag{8.4}$$

Similarly, we conclude that $B_t = \int_0^t \gamma_u dD_u$ for some predictable γ. From the assumption (8.1) on A and B, it follows that

$$A_t = \int_0^t \psi_u 1_{(\gamma_u > 0)} dD_u,$$

and hence

$$A_t = \int_0^t \psi_u \cdot 1_{(\gamma_u > 0)} \cdot (\gamma_u)^{-1} dB_u.$$

(By convention, $1_{(\gamma_u>0)}(\gamma_u)^{-1}$ is defined to be 0 if $\gamma_u = 0$). The result follows by taking

$$f_u(\omega) = \psi_u(\omega) \cdot 1_{(\gamma_u(\omega)>0)} \cdot (\gamma_u(\omega))^{-1}. \qquad \square$$

Remark When $A, B \in \mathcal{V}^+$, f can be taken to be $[0, \infty)$-valued—just replace f by $\max(f, 0)$.

Lemma 8.2 *Let $M^1, M^2, \ldots, M^k$ be continuous local martingales and let $\Lambda \in \mathcal{V}^+$ be such that*

$$d\langle M^i, M^i\rangle(\omega) << d\Lambda(\omega) \;\; \forall \omega \notin N_0, \tag{8.5}$$

where N_0 is a null set. Then there exists σ^{ij} predictable processes such that $\sigma_u(\omega) = (\sigma_u^{ij}(\omega))$ is non-negative definite for all u, ω and

$$\langle M^i, M^j\rangle_t(\omega) = \int_0^t \sigma_u^{ij}(\omega) d\Lambda_u(\omega) \;\; \forall t, \; \forall \omega \notin N_1$$

for some P-null set N_1.

Proof It follows that $d\langle M^i, M^j\rangle << d\Lambda$ a.s. for all i, j. Invoking Lemma 8.1, one can get predictable processes ϕ^{ij} such that using the Kunita–Watanabe inequality (1.67)

$$\langle M^i, M^j\rangle_t = \int_0^t \phi_u^{ij} d\Lambda_u \text{ a.s.}$$

For $a = (a^1, \dots, a^k) \in \mathbb{R}^k$, where a_i is rational, let

$$Y_t^a = \sum_{i=1}^k a^i M_t^i.$$

Then

$$\langle Y^a, Y^a\rangle_t = \int_0^t (\sum_{i=1}^k \sum_{j=1}^k a^i a^j \phi_u^{ij}) d\Lambda_u \ a.s.$$

Since $\langle Y^a, Y^a\rangle_t$ is an increasing process, it follows that

$$\int_0^T 1_{(\sum_{i=1}^k \sum_{j=1}^k a^i a^j \phi_u^{ij} \leq 0)} d\Lambda_u = 0 \text{ a.s.}$$

As a consequence, if

$$\Gamma_u(\omega) = \{\omega : \sum_{i=1}^k \sum_{j=1}^k a^i a^j \phi_u^{ij}(\omega) \geq 0 \text{ for all } a_1, \dots, a_k \text{ rational}\},$$

then

$$\int_0^T 1_{\Gamma_u^c(\omega)} d\Lambda_u = 0 \text{ a.s.}$$

Define

$$\sigma_u^{ij}(\omega) = \psi_u^{ij}(\omega) 1_{\Gamma_u(\omega)}.$$

This choice of σ satisfies the required conditions. □

We will now give an important structural result on semimartingales. The proof closely follows the ideas in Delbaen and Schachermayer (Delbaen and Schachermayer, 1994).

Theorem 8.3 *Suppose $(\hat{S}_t^1, \dots, \hat{S}_t^k)$ is a continuous semimartingale satisfying NA and*

$$\hat{S}_t^i = M_t^i + A_t^i$$

is the canonical decomposition of $\hat{S}^i$. Then A^i admits a representation

$$A_t^i = \sum_{j=1}^k \int_0^t h_u^j d\langle M^j, M^i\rangle_u \ a.s. \tag{8.6}$$

Proof Let $\Lambda_t = \sum_{i=1}^k (|A^i|_t + \langle M^i, M^i \rangle_t)$. Let g_i, σ^{ij} be predictable processes such that

$$A_t^i = \int_0^t g_u^i d\Lambda_u \quad \text{a.s.}$$

$$\langle M^i, M^j \rangle_t = \int_0^t \sigma_u^{ij} d\Lambda_u \quad \text{a.s.},$$

where σ is assumed to be non-negative definite and matrix-valued (see Lemma 8.2).

Let

$$\xi_u(\omega) = \lim_{M \longrightarrow \infty} \sigma_u \cdot (\sigma_u + \frac{1}{M} I)^{-2}.$$

Since σ_u is a non-negative definite, $(\sigma_u + \frac{1}{M} I)^{-1}$ exists. It can be shown that the limit above exists and that

$$\sigma_u(\omega)\xi_u(\omega)\sigma_u(\omega) = \sigma_u(\omega), \tag{8.7}$$

i.e., ξ_u is a generalized inverse of σ_u. Then the construction given above ensures that ξ_u is predictable.

Define

$$\eta_u^{ij}(\omega) = 1_{\{i=j\}}(u) - \sum_{m=1}^k \sigma_u^{im} \xi_u^{mj}.$$

Using (8.7), note that

$$\begin{aligned} \sum_{j=1}^k \eta_u^{ij} \sigma_u^{jn}(\omega) &= \sigma_u^{in}(\omega) - \sum_{j=1}^k \sum_{m=1}^k \sigma_u^{im} \xi_u^{mj} \sigma_u^{jn}(\omega) \\ &= \sigma_u^{in}(\omega) - \sigma_u^{in}(\omega) \\ &= 0. \end{aligned} \tag{8.8}$$

Let $h_u^i = \sum_{r=1}^k \xi_u^{ir} g_u^r$ and $\alpha_u^i = \sum_{j=1}^k \eta_u^{ij} g_u^j$. It follows from the definitions of η, h, α that

$$\alpha_u^i = g_u^i - \sum_{j=1}^k \sigma_u^{ij} h_u^j. \tag{8.9}$$

Now fix i, $1 \leq i \leq k$. Let

$$f_u^j(\omega) = \eta_u^{ij}(\omega). \ (\text{sgn}\ (\alpha_u^i(\omega))).$$

Then

$$\begin{aligned} \sum_{j=1}^k \int_0^t f_u^j dA_u^j &= \int_0^t \left(\sum_{j=1}^k \eta_u^{ij} g_u^j \right) \text{sgn}\ (\alpha_u^i) d\Lambda_u \\ &= \int_0^t |\alpha_u^i| d\Lambda_u \\ &\geq 0. \end{aligned}$$

Let $N_t = \sum_{j=1}^{k} \int_0^t f^j dM^j$. Then

$$\langle N, N \rangle_t = \sum_{j,n} \int_0^t \eta_u^{ij} \sigma_u^{jn} \eta_u^{in} du.$$

Using (8.8), it follows that $\langle N, N \rangle_t = 0$ and so $N_t \equiv 0$. Thus,

$$\begin{aligned} \int_0^t \sum_{j=1}^{k} f_u^j d\hat{S}_u^j &= \sum_{j=1}^{k} \int_0^t f_u^i dA_u^j \\ &= \int_0^t |\alpha_u^i| d\Lambda_u. \end{aligned}$$

The assumption that $(\hat{S}_t^1, \ldots, \hat{S}_t^k)$ satisfies NA implies that $\int_0^T |\alpha_u^i| d\Lambda_u = 0$. It follows that

$$\int_0^t g_u^i d\Lambda_u = \int_0^t \sum_{j=1}^{k} \sigma_u^{ij} h_u^j d\Lambda_u.$$

This completes the proof. □

Remark In the preceding result, we do not need to assume NA, only the following weaker condition suffices:

$$\int_0^t \sum_{j=1}^{k} f^j d\hat{S}^j \geq 0 \text{ a.s. } \quad \forall t \leq T \Rightarrow \int_0^T \sum_{j=1}^{k} f^j d\hat{S}^j = 0 \text{ a.s.} \tag{8.10}$$

Let us now assume that the discounted stock prices $(\tilde{S}_t^i)$ satisfy the following: there exist predictable processes $h^1, \ldots, h^k$ such that

$$A_t^j(\omega) = \sum_{i=1}^{k} \int_0^t h_u^i(\omega) d\langle M^i, M^j \rangle_u(\omega), \quad t \leq T \tag{8.11}$$

and

$$\sum_{i=1}^{k} \int_0^T (h_u^i(\omega))^2 d\langle M^i, M^i \rangle_u(\omega) < \infty, \tag{8.12}$$

where $\tilde{S}_t^i = M^i + A^i$ is the canonical decomposition of $\tilde{S}$. We have seen above that if the market satisfies NA, then (8.11) holds. So we are now assuming (8.12) in addition. We will make one more assumption, namely that every $(\mathcal{F}_t^S)$-martingale (N_t) admits an integral representation

$$N_t = N_0 + \int_0^t \sum_{i=1}^{k} \varphi_u^i dM_u^i \tag{8.13}$$

for some predictable process φ satisfying

$$\sum_{i=1}^{k} \int_0^T (\varphi_u^i(\omega))^2 d\langle M^i, M^i \rangle_u(\omega) < \infty. \tag{8.14}$$

Before we proceed, let us note that these conditions are satisfied by the stock price model $S_t^0 = e^{rt}$ and

$$dS_t^i = S_t^i \left\{ \sum_{j=1}^{k} \sigma^{ij}(S_t) dW_t^j + b^i(S_t) dt \right\}$$

for $i = 1, \ldots, k$, where $(W_t^1, \ldots, W_t^k)$ is an $\mathbb{R}^k$-valued standard Brownian motion and σ^{ij}, b^i are bounded Lipschitz continuous functions and $\sigma(x) = (\sigma^{ij}(x))$ is (strictly) positive definite for each x. Here,

$$d\tilde{S}_t^i = \tilde{S}_t^i \left\{ \sum_{j=1}^{k} \sigma^{ij}(S_t) dW_t^j + (b^i(S_t) - r) dt \right\},$$

and we set

$$d\hat{S}_t^i = \sum_{j=1}^{k} \sigma^{ij}(S_t) dW_t^j + (b^i(S_t) - r) dt.$$

Thus,

$$M_t^i = \int_0^t \sum_0^k \sum_{j=1}^{k} \sigma^{ij}(S_u) dW_u^j$$

and

$$A_t^i = \int_0^t h_u^i du,$$

where

$$h_t^i = \sum_{j=1}^{k} [\sigma^{-1}(S_t)]^{ij} (b^j(S_t) - r).$$

Invertibility of σ implies that $\sigma(S_u^i : u \leq t, 1 \leq i \leq k) = \sigma(W_u^i : u \leq t, i \leq k)$, i.e., $\mathcal{F}_t^S = \mathcal{F}_t^W$. Thus, every martingale N (w.r.t. $(\mathcal{F}_t^S)$) admits a representation

$$N_t = N_0 + \int_0^t \sum_{j=1}^{k} g_u^j dW_u^j$$

with $\displaystyle\sum_{j=1}^{k} \int_0^T (g_u^j)^2 du < \infty$. Defining $\varphi_u^i = \sum_{j=1}^{k} g_u^j (\sigma^{-1}(S_u))^{ji}$, it can be seen that N, φ satisfy (8.13) and (8.14).

Theorem 8.4 *Suppose that the stock price model satisfies (8.11)–(8.14). Let*

$$Z_t = \exp\{-\sum_{j=1}^{k}\int_0^t h_u^j dM_u^j - \frac{1}{2}\sum_{i=1}^{k}\sum_{j=1}^{k}\int_0^t h_u^i h_u^j d\langle M^i, M^j\rangle_u\}.$$

Then the market consisting of $\{S_t^0, S_t^1, \dots, S_t^k\}$ *satisfies NA if and only if*

$$\mathbb{E}_P(Z_T) = 1.$$

Proof Suppose $\mathbb{E}_P(Z_T) = 1$. Then Q, defined by $dQ = Z_T dP$, is a probability measure. Theorem 5.7 implies that $\hat{S}^i$ are Q-local martingales. Hence, Q is an equivalent martingale measure (EMM); so Theorem 7.1 implies that NA holds. For the other part, we will prove that if $\mathbb{E}_P(Z_T) < 1$, then it leads to an arbitrage opportunity.

Suppose $c = \mathbb{E}_P(Z_T) < 1$. Choose $\delta > 0$ such that $c + \delta < 1$. Consider the martingale

$$Z_t^* = \mathbb{E}_P[(Z_T + \delta)|\mathcal{F}_t^S].$$

In view of the assumption of integral representation, it follows that there exists predictable processes $\psi^1, \dots, \psi^k$ such that

$$Z_t^* = (c+\delta) + \int_0^t \sum_{i=1}^{k} \psi_u^i dM_u^i.$$

It follows that Z_t^* is continuous. Further, $Z_t^* \geq \delta > 0$. Thus, defining $\phi_t^i = (Z_t^*)^{-1}\psi_t^i$, it follows that

$$Z_t^* = (c+\delta) + \int_0^t \sum_{i=1}^{k} Z_u^* \phi_u^i dM_u^i. \tag{8.15}$$

Let $Y_t = (Z_t)^{-1}$ and $X_t = Z_t^* Y_t$. Using Ito's formula, we have the relations

$$\begin{aligned}
dZ_t &= Z_t \left\{\sum_{j=1}^{k}(-h_u^j)dM_u^j\right\} \\
dZ_t^* &= Z_t^* \left\{\sum_{j=1}^{k}\varphi_u^j dM_u^j\right\} \\
dY_t &= (-1)(Z_t)^{-1}dZ_t + (-1)\cdot(-2)\cdot(\frac{1}{2})\cdot(Z_t)^{-3}\cdot d\langle Z, Z\rangle_t \\
&= Y_t\left\{\sum_{j=1}^{k} h_u^j dM_u^j\right\} + Y_t\left\{\sum_{i=1}^{k}\sum_{j=1}^{k} h_u^i h_u^j d\langle M^i, M^j\rangle_u\right\} \\
&= Y_t\left\{\sum_{j=1}^{k} h_u^j d\hat{S}_u^j\right\}.
\end{aligned}$$

Hence

$$
\begin{aligned}
dX_t &= Z_t^* dY_t + Y_t dZ_t^* + d\langle Z^*, Y\rangle_t \\
&= X_t \left\{ \sum_{j=1}^{k} h_u^j d\hat{S}_u^j \right\} + X_t \left\{ \sum_{j=1}^{k} \varphi_u^j dM_u^j \right\} \\
&\quad + X_t \left\{ \sum_{j=1}^{k} \sum_{i=1}^{k} h_u^i \varphi_u^j d\langle M^j, M^i\rangle_u \right\} \\
&= X_t \left\{ \sum_{j=1}^{k} (h_u^j + \varphi_u^j) d\hat{S}_u^j \right\},
\end{aligned}
$$

using (8.11) in the last step. Thus, taking

$$\pi_u^j = (h_u^j + \varphi_u^j) X_u \cdot (\tilde{S}_u)^{-1}$$

and $\theta = (0, \pi^1, \ldots, \pi^k)$ as a trading strategy with a zero initial investment, the value function satisfies

$$
\begin{aligned}
\tilde{V}_t(\theta) &= \int_0^t \sum_{j=}^{k} \pi_u^j d\tilde{S}_u^j \\
&= \int_0^t \sum_{j=1}^{k} (h_u^j + \varphi_u^j) X_u d\hat{S}_u^j \\
&= X_t - X_0.
\end{aligned}
$$

Since $X_t \geq 0$ a.s. and $X_0 = (c + \delta)$, it follows that $\tilde{V}_t(\theta) \geq -1$ and hence θ is an admissible strategy. Also, $\tilde{V}_T(\theta) = X_T - X_0 > 1 - c - \delta$. This is an arbitrage opportunity since $c + \delta$ was assumed to be less than 1. Hence NA is violated. Thus NA implies $\mathbb{E}(Z_T) = 1$. This completes the proof. □

The above theorem as well as its proof is based on the ideas of Levental and Skorohod (Levental and Skorohod, 1995).

8.3 A general model of stock prices

We now consider the general model of Section 7.2. Let us recall the market consisting of k stocks, whose price at time t is given by $(S_t^1, \ldots, S_t^k)$, $t \in [0, T]$, and a bond whose price we denote by S_t^0. The only assumption on the processes is that $S_t^0, S_t^1, \ldots, S_t^k$ are r.c.l.l. processes (and that $S_u^0 \leq S_t^0$ for $u < t$). The discounted stock prices are given by $\tilde{S}_t^i = S_t^i (S_t^0)^{-1}$. The stock price processes are assumed to be defined on some probability space $\{\Omega, \mathcal{F}, P\}$. Let $\mathcal{G}_t$ denote the smallest σ-field with respect to which S_u^i, $0 \leq i \leq k$, $0 \leq u \leq t$ are measurable. $\mathcal{G}_t$ denotes the information available to an investor up to time t.

The (self-financing) simple trading strategies are given by $\theta = (x, \pi)$ where $x \geq 0$ is the initial investment and π is a simple predictable $\mathbb{R}^k$-valued process given by

$$\pi_t^i = a_{t_j}^i \quad \text{for } t_j < t \leq t_{j+1};\ 1 \leq i \leq k,$$

and where $a_{t_j}^i$ is a bounded $\mathcal{G}_{t_j}$-measurable random variable which denotes the number of shares of the ith stock held by the investor during $(t_j, t_{j+1}]$. The discounted value process of the self-financing simple strategy $\theta = (x, \pi)$ is given by (see (7.2))

$$\tilde{V}_t(\theta) = x + \sum_{j=1}^{k} \sum_{i=0}^{m-1} a_{t_i}^j (\tilde{S}_{t_{i+1}\wedge t}^j - \tilde{S}_{t_i \wedge t}^j). \tag{8.16}$$

The model satisfies the NA property in the class of simple strategies, if for a simple strategy $\theta = (0, \pi)$ and some $t \geq 0$, $P(\tilde{V}_t(\theta) \geq 0) = 1$ implies that $P(\tilde{V}_t(\theta) = 0) = 1$.

Recall that a probability measure Q equivalent to P is an equivalent martingale measure (EMM) if $(\tilde{S}_t^1, \ldots, \tilde{S}_t^k)$ are Q-local martingales. The model is said to satisfy the EMM property if an EMM Q exists.

The connection between NA and EMM is apparent from the two observations that follow. First, let

$$\mathcal{K}_s = \{\tilde{V}_t(\theta) : 0 \leq t < \infty, \theta \text{ a simple strategy }\} \tag{8.17}$$

be the class of all discounted contingent claims that are attainable over a finite time horizon via simple strategies and

$$L_+^\infty(P) = \{Z \in L^\infty(P) : P(Z \geq 0) = 1\}. \tag{8.18}$$

It is easy to see that $S_t = (S_t^0, S_t^1, \ldots, S_t^k)$ satisfies the NA (no arbitrage) property in the class of simple strategies if and only if

$$\mathcal{K}_s \cap L_+^\infty(P) = \{0\}. \tag{8.19}$$

Regarding the EMM property, observe that if $g \in L^1(P)$ is such that

$$P(g > 0) = 1, \tag{8.20}$$

and for all $f \in \mathcal{K}_s$, $fg \in L^1(P)$ with

$$\int fg dP = 0, \tag{8.21}$$

then Q, defined by $dQ = gdP$, is an EMM. In fact, in this case, $(\tilde{S}_t^1, \ldots, \tilde{S}_t^k)$ are Q-martingales because $1_A(\tilde{S}_t^i - \tilde{S}_s^i) \in \mathcal{K}_s$ for $A \in \mathcal{G}_s$ and $0 \leq s \leq t$.

It is immediate from these observations (and also as seen in Chapter 7) that (8.21) implies (8.19).

Example 1 Let $\Omega = \{-1, 1\}^{\mathbb{N}}$ and let γ_i be the coordinate mappings on Ω. Let P be the probability measure on Ω such that γ_i's are independent, and for $n \geq 1$, $P(\gamma_i = 1) = \frac{1}{2} + \frac{1}{2\sqrt{i}}$, $P(\gamma_i = -1) = \frac{1}{2} - \frac{1}{2\sqrt{i}}$. Let the stock price process $\{S_n : n \geq 1\}$ be given by $S_0 = 1$ and

$$S_n = [\prod_{i=1}^{n}(1 + \frac{1}{2}\gamma_i)](1 + r)^n,$$

where r is the rate of interest (on the bond). The discounted stock price $\tilde{S}_n$ is given by

$$\tilde{S}_n = \prod_{i=1}^{n}(1 + \frac{1}{2}\gamma_i).$$

It is easy to see that on $(\Omega, \mathcal{F})$ (where $\mathcal{F} = \sigma(\gamma_i : i \geq 1)$) there is a unique probability measure Q under which $\tilde{S}_n$ is a martingale; it is the one under which γ_i's are independent with $Q(\gamma_i = 1) = Q(\gamma_i = -1) = \frac{1}{2}$. There is no other probability measure on Ω under which $\tilde{S}_n$ is a local martingale.

Writing $\rho_i(\gamma_i) = \frac{dP_i}{dQ_i}(\gamma_i)$ and P_i, Q_i being the restrictions to $\{-1, 1\}$ of the infinite product measures P and Q, we have $\rho_i(\gamma_i) = 1 + \frac{\gamma_i}{\sqrt{i}}$ and $\int \sqrt{\rho_i(\gamma_i)}\, dP_i(\gamma_i) = \frac{1}{2}\left(1 + \frac{\gamma_i}{\sqrt{i}}\right)^{3/2}$. Hence

$$\prod_{i=1}^{n} \int \sqrt{\rho_i(\gamma_i)}\, dP_i(\gamma_i) = \frac{1}{2^n} \prod_{i=1}^{n}\left(1 + \frac{\gamma_i}{\sqrt{i}}\right)^{3/2}.$$

The right hand side is of order $2^{-n} \cdot e^{3\sqrt{n}}$ which tends to 0 as $n \to \infty$.

From a well-known theorem of Kakutani (Kakutani, 1948), it follows that P and Q are mutually singular or orthogonal, $P \perp Q$. Hence there does not exist an EMM or an equivalent local martingale measure for P. However, NA holds as shown by the following argument:

Let $\mathcal{F}_n = \sigma(\gamma_i : 1 \leq i \leq n)$ and P^n, Q^n be restrictions of P, Q on $\mathcal{F}_n$. It is easy to see that P^n and Q^n are equivalent. Here, $\mathcal{K}_s$—the class of contingent claims expressed in discounted prices that are attainable via (simple) trading strategies over finite time—is given by

$$\mathcal{K} = \left\{ \sum_{j=0}^{n-1} a_j(\tilde{S}_{j+1} - \tilde{S}_j) \; : \; a_j \text{ bounded } \mathcal{F}_j\text{-measurable}, n \geq 1 \right\}.$$

Note that every $Z \in \mathcal{K}$ is $\mathcal{F}_m$-measurable for some m. Hence, it follows that $\mathcal{K} \cap L_+^\infty(P) = \{0\}$. Indeed, if $W \in \mathcal{K} \cap L_+^\infty(P)$, then W is $\mathcal{F}_m$-measurable for some m and hence $W \in \mathcal{K} \cap L_+^\infty(Q)$. But under Q, $\tilde{S}_n$ is a martingale and hence $\mathbb{E}_Q(W) = 0$, so $Q(W = 0) = 1$. Hence $P(W = 0) = 1$. So NA holds.

As noted above, an equivalent (local) martingale measure does not exist. Let us note that if $g_n = \frac{dP^n}{dQ^n}$, then

$$g_n = \prod_{i=1}^{n} \left(1 + \frac{\gamma_i}{\sqrt{i}}\right).$$

It can be verified that $\tilde{S}_{n+1} - \tilde{S}_n = \tilde{S}_n \frac{\gamma_{n+1}}{2}$ and

$$g_n - 1 = \sum_{m=0}^{n-1} \frac{2g_m}{\sqrt{m}\tilde{S}_m} (\tilde{S}_{m+1} - \tilde{S}_m).$$

Hence $Z_n = g_n - 1 \in \mathcal{K}$. Since $Q \perp P$, $g_n \to \infty$ a.s. P. Thus,

$$P(Z_n \geq 1) \longrightarrow 1.$$

Thus, although there is no arbitrage opportunity in the class $\mathcal{K}$ of attainable claims, there is a sequence $\{Z_n\} \subset \mathcal{K}$ such that $P(Z_n \geq 1) \longrightarrow 1$.

The example discussed above suggests that in order to have an equivalent (local) martingale measure, one should rule out the existence of sequences $\{Z_n\} \subset \mathcal{K}$ such that $P(Z_n \geq Z) \longrightarrow 1$, $Z \in L_+^\infty$ and $P(Z = 0) < 1$. Let us tentatively call such sequences $\{Z_n\}$ *approximate arbitrage opportunities*. However, the existence of an equivalent martingale measure does not rule out approximate arbitrage opportunities as the following example shows.

Example 2 In the setup of Example 1, consider the stock prices $\tilde{S}_n$ on the probability space $(\Omega, \mathcal{F}, Q)$. Since $\tilde{S}_n$ is a Q-martingale, the EMM property trivially holds. Let $f_0 = 1$ and

$$f_n = \prod_{i=1}^{n} (1 + \gamma_i).$$

Then $f_m - f_{m-1} = 2f_{m-1}\frac{(\tilde{S}_m - \tilde{S}_{m-1})}{\tilde{S}_{m-1}}$ and hence $Z_n = 1 - f_n$ can be written as

$$\begin{aligned} Z_n &= -\sum_{i=1}^{n} (f_i - f_{i-1}) \\ &= -\sum_{i=1}^{n} \frac{2f_{i-1}}{\tilde{S}_{i-1}} (\tilde{S}_i - \tilde{S}_{i-1}); \end{aligned}$$

hence $Z_n \in \mathcal{K}$. Note that $Q(f_n = 2^n) = 2^{-n}$ and $Q(f_n = 0) = 1 - 2^{-n}$. Thus $Z_n \longrightarrow 1$ a.s. $[P]$. Thus $\{Z_n\}$ is an approximate arbitrage opportunity $Q(Z_n \geq 1 - \varepsilon) \longrightarrow 1$ for every $\varepsilon > 0$.

In the first example, let us note that $Z_n \geq -1$, i.e., the risk associated with the approximate arbitrage opportunity $\{Z_n\}$ (namely (Z_n^-) is bounded by 1. In the second example, Z_n^- is not bounded. Indeed $Q(Z_n^- = 2^n - 1) = 2^{-n}$ for all n.

These comments suggest that to characterize the EMM property, one should rule out those approximate arbitrage opportunities for which the associated risks are controlled (in some appropriate sense).

Thus, in general, the equivalence between NA and EMM breaks down. One needs to strengthen (8.19) in order that it be equivalent to (8.21). Kreps (Kreps, 1981) (and independently, Yan (Yan, 1980) in a different context) came up with the idea that one should consider

$$\mathcal{U} = \{f - h : \ f \in \mathcal{K}_s, \ h \in L^\infty_+\}.$$

Note that one still has

$$\mathcal{U} \cap L^\infty_+(P) = \{0\}.$$

Let $\mathcal{V}$ be the closure of $\mathcal{U}$ in an appropriate topology. Consider

$$\mathcal{V} \cap L^\infty_+(P) = \{0\} \tag{8.22}$$

which is stronger than (8.19). We will now prove a result called the Kreps–Yan separation theorem, the proof of which uses the Hahn–Banach separation theorem. While Kreps considered closure in the weak*-topology on L^∞, Yan considered closure in the norm topology in $L^1(P)$. The next section deals with these ideas.

8.4 The separation theorem

The following theorem can be attributed to Kreps (Kreps, 1981) and Yan (Yan, 1980). While Kreps proved the result for $p = \infty$, Yan proved it for $p = 1$ in another context. Stricker (Stricker, 1990) and Ansel–Stricker (Ansel and Stricker, 1990) noted the relevance of Yan's result in the context of mathematical finance and obtained the result for closure in L^p, $1 < p < \infty$. The proof given below follows Schachermayer (Schachermayer, 1994). We need to invoke the Hahn–Banach theorem which is stated below for the convenience of the reader.

Theorem 8.5 (Hahn-Banach separation theorem) *Let $\mathcal{X}$ be a Banach space and let $\mathcal{X}^*$ be its dual. Suppose that C is a non-empty convex subset of $\mathcal{X}$ closed in the weak* topology on $\mathcal{X}$ and K is a non-empty convex subset of $\mathcal{X}$ compact in the weak* topology on $\mathcal{X}$. Suppose that*

$$C \cap K = \emptyset.$$

Then there exists $\lambda \in \mathcal{X}^*$ such that

$$\sup\{\lambda(x) : x \in C\} < \inf\{\lambda(x) : x \in K\}.$$

See, e.g., Rudin (Rudin, 1974) Theorem 3.4 for a proof.

Theorem 8.6 *Let $1 \leq p, q \leq \infty$ be such that $\frac{1}{p} + \frac{1}{q} = 1$. Let $\mathcal{U} \subseteq L^p(P)$ be a convex cone such that $0 \in \mathcal{U}$. Let $\mathcal{V}_p$ be the norm closure of $\mathcal{U}$ in L^p if $1 < p < \infty$ and the weak*-closure of $\mathcal{U}$ in L^∞ if $p = \infty$. Suppose that $\mathcal{U}$ satisfies*

$$f - h \in \mathcal{U} \quad \forall f \in \mathcal{U}, h \in L^\infty_+ \tag{8.23}$$

and

$$\mathcal{V}_p \cap L^\infty_+ = 0. \tag{8.24}$$

Then there exists $g \in L^q$ such that $P(g > 0) = 1$ and

$$\int fgdP \leq 0 \quad \forall f \in \mathcal{U}. \tag{8.25}$$

Proof Note that for $1 < p < \infty$, $\mathcal{V}_p$ is a convex set, closed in the $\sigma(L^p, L^q)$ topology, since it is norm-closed (by definition) and L^p is reflexive. (See (Rudin, 1974) Theorem 3.12.)

Fix $A \in \mathcal{F}$ such that $P(A) > 0$. We will first prove that there exists $g^A \in L^q(P)$ with $P(g^A \geq 0) = 1$ such that $\int 1_A g^A dP > 0$ and

$$\int fg^A dP \leq 0 \quad \forall f \in \mathcal{U}. \tag{8.26}$$

Here 1_A does not belong to $\mathcal{V}_p$ and $\mathcal{V}_p$ is a convex set, closed in the $\sigma(L^p, L^q)$ topology. It follows from the Hahn–Banach separation theorem that there exists $g^A \in L^q$ such that

$$\sup\left\{\int fg^A dp \ : \ f \in \mathcal{V}_p\right\} = a < \int 1_A g^A.$$

Since $0 \in \mathcal{V}_p$, it follows that $a \geq 0$. On the other hand, $\mathcal{V}_p$ is also a cone, the closure of the cone $\mathcal{U}$. Hence $a = 0$, for if $\int fgdP > 0$ for some $f \in \mathcal{V}_p$, then $uf \in \mathcal{V}_p$ for all $u > 0$ and this will force $a = \infty$. On the other hand, $a < \int 1_A g^A$ and hence $a < \infty$. Therefore $a = 0$. Since $0 \in \mathcal{U}$, $-1_B \in \mathcal{U}$ for all $B \in \mathcal{F}$ and so

$$-\int 1_B g^A dP \leq 0$$

for all $B \in \mathcal{F}$. Hence $P(g^A \geq 0) = 1$. We have already seen that $0 < \int 1_A g^A dP$.

Now let

$$\mathcal{N} = \{g \in L^q(P) : \int fgdP \leq 0 \ \forall f \in \mathcal{U}\} \tag{8.27}$$

and

$$b = \sup\{P(g > 0) : \ g \in \mathcal{N}\}. \tag{8.28}$$

We now claim that the supremum in (8.28) is attained. To see this, let $g_n \in \mathcal{N}$ be such that $P(g_n > 0) \to b$. Take

$$g = \sum_{n=1}^{\infty} (2^n(1 + ||g_n||_p))^{-1} g_n$$

(where the series converges in $L^q(P)$) and note that $g \in \mathcal{N}$. Also, $P(g > 0) = b$, proving the claim. The result will follow once we prove that $b = 1$.

Suppose $b < 1$ and let $g \in \mathcal{N}$ be such that $P(g > 0) = b$. Let

$$A = \{g = 0\}$$

and let $g^A \in \mathcal{N}$ such that $\int 1_A g^A dP > 0$. This can be done since $P(A) > 0$. Hence

$$P(\{g^A > 0\} \cap A) > 0. \tag{8.29}$$

Now let

$$\bar{g} = g + g^A.$$

Then $\bar{g} \in \mathcal{N}$ and, in view of (8.29), $P(\bar{g} > 0) > b$. This contradicts (8.28). Hence we conclude that $b = 1$. □

Remark Let us note that we only used

$$A \in \mathcal{F}, P(A) > 0 \Rightarrow 1_A \notin \mathcal{V}_p \tag{8.30}$$

instead of (8.24) in the proof.

Remark In the proof of the separation theorem given above, we used that for $1 \le p < \infty$, the dual of $L^p(P)$ (with the norm topology) is $L^q(P)$ and hence we took norm closure while defining $\mathcal{V}_p$ for $1 \le p < \infty$. However, the dual of L^∞ is not L^1 and hence we had to resort to the weak* closure, while defining $\mathcal{V}_\infty$ for the dual of $L^\infty(P)$ in the weak* topology is $L^1(P)$.

We will first deduce a result due to Stricker and Ansel–Stricker on the existence of EMM with density in L^q.

Theorem 8.7 *Let $1 \le p < \infty$ and q be such that $\frac{1}{p} + \frac{1}{q} = 1$. Suppose that*

$$\tilde{S}_t^i \in L^p(P) \quad 1 \le i \le k, \ 0 \le t \le T.$$

Let $\mathcal{K}_s$ be the class of contingent claims attainable via simple strategies and let

$$\mathcal{U} = \{f - h : \ f \in \mathcal{K}_s, \ h \in L_+^\infty\}.$$

Let $\mathcal{V}_p$ be the closure of $\mathcal{U}$ in L^p (norm topology).

The following are equivalent:

1. *There exists a probability measure Q equivalent to P such that*

$$\frac{dQ}{dP} \in L^q \tag{8.31}$$

and $(\tilde{S}_t^1, \ldots, \tilde{S}^k)$ are Q-martingales.

2. *For all $A \in \mathcal{F}$ such that $P(A) > 0$, we have $1_A \notin \mathcal{V}_p$.*

3. *$\mathcal{V}_p \cap L_+^\infty = 0$.*

Proof Note that (2) is a special case of (3). From the separation theorem (8.6) and the remark following its proof, it follows that (2) implies the existence of a $g \in L^q$ satisfying (8.25) and such that $P(g > 0) = 1$. Take $dQ = gdP$. We had seen earlier that (8.25) implies that $(\tilde{S}_t^1, \ldots, \tilde{S}^k)$ are Q-martingales. It remains to be shown that (1) implies (3). If (1) holds, then $\tilde{V}_t(\theta)$ is itself a mean zero Q-martingale for all trading strategies, so that $\mathbb{E}_Q[f] = 0$ for all $f \in \mathcal{K}_s$. Hence $\mathbb{E}_Q[f] \leq 0$ for all $f \in \mathcal{U}$, i.e.,

$$\int fgdP = 0 \quad \text{for all } f \in \mathcal{U}. \tag{8.32}$$

Since $\mathcal{U} \subseteq L^p(P)$ and $g \in L^q$, it follows that (8.32) remains valid for f in the closure $\mathcal{V}_p$ (of $\mathcal{U}$) in L^p. □

The aim is to get a characterization of the EMM property. The preceding result gives a characterization of the EMM property with an additional requirement that the density of Q with respect to P is in $L^q(P)$ for $1 < q < \infty$. It would be good to get this result for $q = 1$ since every density belongs to L^1. Here is the result due to Kreps (Kreps, 1981). The formulation is different from the one given in Kreps, who allowed the number of stocks to be infinite. It can be proved exactly along the lines of the previous result, invoking the separation theorem given earlier in the section for $p = \infty$.

Theorem 8.8 *Suppose that*

$$\tilde{S}_t^i \in L^\infty(P) \quad 1 \leq i \leq k,\ 0 \leq t \leq T.$$

Let $\mathcal{K}_s$ be the class of contingent claims attainable via simple strategies and let

$$\mathcal{U} = \{f - h :\ f \in \mathcal{K}_s,\ h \in L_+^\infty\}.$$

Let $\mathcal{V}_\infty$ be the closure of $\mathcal{U}$ in the weak topology on L^∞.*
The following are equivalent:

1. *There exists an EMM Q, i.e., a probability measure Q equivalent to P such that $(\tilde{S}_t^1, \ldots, \tilde{S}^k)$ are Q-martingales.*

2. *For all $A \in \mathcal{F}$ such that $P(A) > 0$, we have $1_A \notin \mathcal{V}_\infty$.*

3. $\mathcal{V}_\infty \cap L_+^\infty = 0$.

The assumption that $\tilde{S}_t^i \in L^\infty(P)$ is too strong. Further, the closure (of $\mathcal{U}$) $\mathcal{V}_\infty$ is the closure in the weak* topology—a topology which is not metrizable and hence the closure cannot be described in terms of sequences, but in terms of nets. This makes it difficult to interpret. In fact, it was in view of this difficulty that Stricker obtained his results under the added condition on the integrability of a suitable power of the density $\frac{dQ}{dP}$. Kusuoka (Kusuoka, 1993) obtained a version of the separation theorem using Orlicz spaces (which generalize the L^p spaces). However, he did not give any interpretation of this condition. We will formulate this result in a way that allows us to express the condition (that characterizes the EMM) in a economically meaningful way.

We define Orlicz spaces and give some basic results in the next section.

8.5 Orlicz spaces

Definition A Young function Φ is a continuous convex increasing function on $[0, \infty)$ with $\Phi(0) = 0$ and $\frac{\Phi(x)}{x} \uparrow \infty$.

With each Young function Φ, one can associate a function space $L(\Phi)$, called the Orlicz space (corresponding to Φ). When $\Phi(x) = x^p, 1 < p < \infty$, the Orlicz space $L(\Phi)$ is the space L^p. If $\frac{\Phi(x)}{x^p} \uparrow \infty$ for every $p > 1$, then

$$L^p \subseteq L(\Phi) \subseteq L^1 \quad \forall p > 1$$

and if $\frac{\Phi(x)}{x^p} \downarrow 0$ then

$$L^\infty \subseteq L(\Phi) \subseteq L^p \quad \forall p > 1.$$

In this sense, the Orlicz spaces fill the gaps between L^p, L^1 and L^∞, L^p $(1 < p < \infty)$.

For a Young function Φ, the function Ψ defined by, for $y \in [0, \infty)$

$$\Psi(y) = \sup\{xy - \Phi(x) : x \in [0, \infty)\} \tag{8.33}$$

is also a Young function. Ψ is called the conjugate function of Φ. From the definition of Φ, it follows that

$$xy \leq \Phi(x) + \Psi(y). \tag{8.34}$$

For any random variable Z on $(\Omega, \mathcal{F}, P)$, let

$$\|Z\|_\Phi = \inf\{c > 0 : \mathbb{E}[\Phi(\frac{|Z|}{c})] \leq 1\}$$

and the Orlicz space $L(\Phi)$ is defined by

$$L(\Phi) = \{Z : \|Z\|_\Phi < \infty\}.$$

Then it is easy to show that $L(\Phi)$ is a Banach space and $L(\Phi) \subseteq L^1$.

Here are some simple observations about Orlicz spaces.

Lemma 8.9 *Suppose $Z_n, Z \in L(\Phi)$ are such that $Z_n \longrightarrow Z$ in $L(\Phi)$. Then $Z_n \longrightarrow Z$ in L^1 as well as $\mathbb{E}\Phi(|Z_n - Z|) \longrightarrow 0$.*

Proof If U is such that $\|U\|_\Phi < 1$, then taking c such that $\|U\|_\Phi < c < 1$, one has $\mathbb{E}\Phi(\frac{|U|}{c}) \leq 1$. Since

$$\begin{aligned} \Phi(|x|) &\leq c\Phi(\frac{|x|}{c}) + (1-c)\Phi(0) \\ &= c\Phi(\frac{|x|}{c}), \end{aligned}$$

it follows that $\mathbb{E}\Phi(|U|) \leq \|U\|_\Phi$. Thus, for $Z_n \longrightarrow Z$ in $L(\Phi)$, we have $\mathbb{E}\Phi(|Z_n - Z|) \longrightarrow 0$.

For the other part, using Jensen's inequality, it follows that

$$\Phi(\mathbb{E}(|Z_n - Z|)) \leq \mathbb{E}\Phi(|Z_n - Z|) \longrightarrow 0.$$

This along with the continuity of Φ at 0 and the fact that $\Phi(0) = 0$ implies that $\mathbb{E}(|Z_n - Z|) \longrightarrow 0$. □

Here is the analogue of Holder's inequality for Orlicz spaces.

Lemma 8.10 *Let Φ, Ψ be conjugate Young's functions. Then for $U \in L(\Phi)$, $V \in L(\Psi)$, one has*

$$\mathbb{E}|UV| \leq 2\|U\|_\Phi \|V\|_\Psi. \tag{8.35}$$

Proof Let $c_1 \geq \|U\|_\Phi$ and $c_2 \geq \|V\|_\Psi$. Then $\mathbb{E}\Phi(\frac{|U|}{c_1}) \leq 1$ and $\mathbb{E}\Psi(\frac{|V|}{c_2}) \leq 1$. Now, the inequality (8.34) gives

$$\frac{|UV|}{c_1 c_2} \leq \Phi\left(\frac{|U|}{c_1}\right) + \Psi\left(\frac{|V|}{c_2}\right).$$

Taking expectations, it follows that $\mathbb{E}(|UV|) \leq 2c_1c_2$. The result follows from this. □

For $V \in L(\Psi)$, the mapping $\lambda_V : L(\Phi) \mapsto \mathbb{R}$, defined by

$$\lambda_V(U) = \mathbb{E}(UV),$$

defines a continuous linear mapping on $L(\Phi)$. Thus $L(\Psi) \subset L(\Phi)^*$. However, this may not be an equality.

Let $E(\Phi)$ be the closure of L^∞ in the $\|\cdot\|_\Phi$ norm. Here is the analogue of the classical result $(L^p)^* = L^q$ for $1 < p < \infty$, $\frac{1}{p} + \frac{1}{q} = 1$:

$$E(\Phi)^* = L(\Psi). \tag{8.36}$$

See Krasnoselskii and Rutickii (Krasnoselskii and Rutickii, 1961) for this result. The proof is along the lines of the proof of the duality result for L^p spaces. We will skip the proof.

8.6 No arbitrage with controlled risk

In this section, we will use Orlicz spaces and get an economically meaningful condition on the stock prices that characterizes the existence of an equivalent (local) martingale measure. In this section, we deviate from our earlier conventions a little bit, and allow for an infinite time horizon. Also, we assume that the discounted stock price processes $\tilde{S}_t^1, \ldots, \tilde{S}_t^k$ are locally bounded, i.e., there exists a sequence $\{\tau_m\}$ of $(\mathcal{F}_t)$-stopping times, τ_m increasing to ∞, such that

$$|\tilde{S}^i_{t\wedge\tau_m}| \leq c_{m,i} \quad t \geq 0 \tag{8.37}$$

for some constants $c_{m,i} < \infty$. By replacing τ_m by $\tau_m \wedge m$ if necessary, we assume that τ_m are bounded stopping times. We will allow investment strategies where the portfolio changes at finitely many times, but these times could themselves be random (stopping times). The strategies thus obtained are called elementary strategies.

An elementary investment strategy on the ith stock is a process π^i of the form

$$\pi_t^i = \sum_{j=0}^{m-1} a_j^i I_{(\sigma_j,\sigma_{j+1}]}(t), \tag{8.38}$$

where a_j^i are $\mathcal{F}_{\sigma_j}$-measurable bounded random variables, $\sigma_0 \leq \sigma_1 \leq \ldots \leq \sigma_m$ are $(\mathcal{F}_t)$ stopping times with $\sigma_m \leq \tau_k$ for some k, τ_k as in (8.37). Thus, π^i is predictable and is constant over each of the intervals $(\sigma_0, \sigma_1]$, $(\sigma_1, \sigma_2]$, $\ldots$, $(\sigma_{m-1}, \sigma_m]$. π_t^i is the number of shares of the ith stock the investor will hold at time t. Let us write $\pi = (\pi^1, \ldots, \pi^k)$. Then as before, an elementary strategy is given by $\theta = (x, \pi)$, where $x \geq 0$ is the initial investment. The discounted value process of the elementary strategy $\theta = (x, \pi)$ is given by (see (7.2))

$$\tilde{V}_t(\theta) = x + \sum_{j=1}^{k} \sum_{i=0}^{m-1} a_{t_i}^j (\tilde{S}^j_{t_{i+1}\wedge t} - \tilde{S}^j_{t_i\wedge t}). \tag{8.39}$$

Let

$$\mathcal{K}_e = \{\tilde{V}_t(\theta) : \ \theta = (0, \pi) \ \text{ is an elementary strategy}, \ t < \infty\}. \tag{8.40}$$

Note that $\mathcal{K}_e \subset L^\infty(\Omega, \mathcal{F}, P)$ and is a linear space. Here is a simple observation needed later in the main result.

Lemma 8.11 *Let Q be given by $\frac{dQ}{dP} = f$ with $P(f > 0) = 1$, $f \in L^1(P)$. Then Q is an EMM for $\tilde{S}$ if and only if*

$$\mathbb{E}_Q[Z] = 0 \quad \forall Z \in \mathcal{K}_e. \tag{8.41}$$

Proof If Q is an EMM, then for a simple strategy π, $\tilde{V}_t(0, \pi)$ is a local martingale on $(\Omega, \mathcal{F}, Q)$ and hence a martingale since it is bounded. In particular,

$$\mathbb{E}_Q(\tilde{V}_t(0, \pi)) = E_Q(\tilde{V}_0(0, \pi)) = 0.$$

Thus,

$$\mathbb{E}_Q(Z) = 0 \quad \forall Z \in \mathcal{K}_e.$$

Conversely, suppose (8.41) is satisfied. Fix $1 \le i \le k$ and a stopping time σ_1. We will show that

$$\mathbb{E}_Q[\tilde{S}^i_{\sigma_1 \wedge \tau_m}] = \mathbb{E}_Q[\tilde{S}^i_0], \tag{8.42}$$

where τ_m are as in (8.37). This will imply that

$$\tilde{S}^i_{t \wedge \tau_m} \text{ is a martingale for all } m$$

and hence that $\tilde{S}^i$ is a local martingale. This will complete the proof.

It remains to prove (8.42). Fix m and let $\sigma = \sigma_1 \wedge \tau_m$, $a^i = 1$ and $a^j = 0$ for $j \neq i$ and

$$\pi^l = a^l 1_{(0,\sigma]}.$$

Let $\pi^i_t = 0$ for $i > 1$ and for all $t \ge 0$. Then for t such that $\tau_m \le t$ (such a t exists since τ_m is bounded),

$$\tilde{V}_t(0, \pi) = \tilde{S}^i_\sigma - \tilde{S}^i_0.$$

Since $\tilde{V}_t(0, \pi) \in \mathcal{K}$, (8.41) implies (8.42). □

In this framework, Delbaen and Schachermayer (Delbaen and Schachermayer, 1994) obtained the following result. $\tilde{S}^1_t, \ldots, \tilde{S}^d_t$ are said to satisfy the condition *No Free Lunch with Bounded Risk* (written as NFLBR) if, for a sequence of elementary strategies θ_n, $\tilde{V}(0, \theta_n)_t \ge -1$ for all t, n and $(\tilde{V}(0, \theta_n)_T)^- \to 0$ in P-probability implies that

$$\tilde{V}(0, \theta_n)_T \to 0 \text{ in } P\text{-probability}.$$

Here is the result from (Delbaen and Schachermayer, 1994).

Theorem 8.12 *Suppose $\tilde{S}^1_t, \ldots, \tilde{S}^k_t$ are locally bounded. Then $\tilde{S}$ satisfies the condition NFLBR if and only if $(\tilde{S}^1_t, \ldots, \tilde{S}^k_t)$ admit an EMM.*

The proof of this result uses fairly deep results from stochastic calculus. Instead, we will give below a variant that admits a purely functional analytic proof.

Remark In what follows, the explicit model of the stock prices and the time horizon do not enter the picture at all. The only thing that matters is the class $\mathcal{K}_e$ of claims that can be attained by the investors over a finite time horizon by trading at finitely many (stopping) times. Thus, if one were to consider infinitely many stocks (as in (Kreps, 1981)), the results can be easily extended. One can define the class $\mathcal{K}_e$ as above, imposing the further condition that one investor can trade infinitely many stocks. The results given below continue to hold for this case also.

Let $\mathcal{R}$ be the class of increasing continuous functions Φ from $[0, \infty)$ onto $[0, \infty)$ such that $\Phi(0) = 0$. $\Phi \in \mathcal{R}$ is to be thought of as a risk function, where the risk associated with the loss W is $\mathbb{E}\Phi(W)$ or the risk associated with the reward R is $\mathbb{E}\Phi(R^-)$.

If $Z \in L_+^\infty$ is such that for every $\Phi \in \mathcal{R}$, there exist $\{W_n\} \subset \mathcal{K}$ with $P(W_n \geq Z - \frac{1}{n}) \longrightarrow 1$ and $\mathbb{E}\Phi(W_n^-) \longrightarrow 0$, then every investor, *whatever his risk function,* can come as close to the position Z as desired using simple strategies over a finite time horizon, keeping this risk as small as desired. Thus Z can be called an approximable arbitrage opportunity. If such a Z (other than Z=0) does not exist, we say that $\tilde{S}$ (or $\mathcal{K}$) satisfies *no approximate arbitrage with controlled risk* property. It is easy to see that given an increasing continuous function Φ on $[0, \infty)$, we can find $\Phi_1 \in \mathcal{E}$ (an increasing continuous convex function Φ_1 with $\Phi_1(0) = 0$ and $\Phi_1(x)/x$ increasing) such that $\Phi(x) \leq \Phi_1(x)$ and thus we can restrict to $\Phi \in \mathcal{E}$.

Definition We say that $\tilde{S}$ (or $\mathcal{K}_e$) satisfies *no approximate arbitrage with controlled risk* (NAACR) if there does not exist $Z \in L_+^\infty - \{0\}$, such that for every $\Phi \in \mathcal{E}$, there exists a sequence $\{W_n : n \geq 1\} \subset \mathcal{K}_e$ with $P(W_n \geq Z - \frac{1}{n}) \longrightarrow 1$ and $\mathbb{E}\Phi(W_n^-) \longrightarrow 0$.

Thus $\tilde{S}$ satisfies NAACR if no non-negative claim Z, other than the zero claim, can be approximated by attainable claims in $\mathcal{K}_e$ with the associated risks going to zero, no matter what the risk function of the individual investor.

For a linear space $\mathcal{H} \subseteq L^\infty(P)$, let $\bar{\mathcal{H}}$ be the class of $Z \in L^\infty$ such that for every $\Phi \in \mathcal{E}$, there exists a sequence $\{W_n : n \geq 1\} \subset \mathcal{H}$ with $P(W_n \geq Z - \frac{1}{n}) \longrightarrow 1$ and $\mathbb{E}\Phi(W_n^-) \longrightarrow 0$.

The NAACR condition on (S^i) can then be rephrased as

$$\bar{\mathcal{K}_e} \cap L_+^\infty = 0. \tag{8.43}$$

In preparation for our main result, here is a version of the separation theorem.

Theorem 8.13 *Let $\mathcal{H} \subseteq L^\infty(P)$ be a linear space. Then the following are equivalent:*

(i) $\bar{\mathcal{H}} \cap L_+^\infty = \{0\}$.

(ii) $\exists f \in L^1(P)$, $P(f > 0) = 1$ *such that*

$$\int f Z dP = 0 \quad \forall Z \in \mathcal{H}.$$

Proof (i) $\Rightarrow$ (ii)
Let $\mathcal{U} = \{V \in L^\infty : V \leq Z \text{ for some } Z \in \mathcal{H}\}$. Let $\mathcal{V}_\infty$ be the closure of $\mathcal{U}$ in the $\sigma(L^\infty, L^1)$ topology on L^∞ (weak*-topology). We will first prove that

$$\mathcal{V}_\infty \subseteq \bar{\mathcal{H}}. \tag{8.44}$$

Fix $Z \in \mathcal{V}_\infty$. Let $\Phi \in \mathcal{E}$. Now $Z \in \mathcal{V}_\infty$ implies that there exists a net $(W_\alpha)_{\alpha \in \Delta} \subset \mathcal{U}$ such that $W_\alpha \longrightarrow Z$ in $\sigma(L^\infty, L^1)$ topology. Since $L^\infty \subset E(\Phi)$ and $E(\Phi)^* =$

$L(\Psi) \subset L^1$ (where Ψ is the conjugate of Φ), it follows that $W_\alpha \longrightarrow Z$ in $\sigma(E(\Phi), L(\Psi))$ topology, i.e., the weak topology on $E(\Phi)$. Thus Z belongs to the weak closure of $\mathcal{C}$ in $E(\Phi)$. Since $\mathcal{C}$ is convex, the weak closure equals the norm closure. (See e.g., Theorem 3.3.12 of (Rudin, 1974).) Thus, there exists a sequence $\{V_n\} \subset \mathcal{U}$ such that $\|V_n - Z\|_\Phi \longrightarrow 0$. Hence $V_n \longrightarrow Z$ in probability and $\mathbb{E}\Phi(|V_n - Z|) \longrightarrow 0$. Since $Z \geq 0$ a.s., $|V_n - Z| \geq V_n^-$ and hence $\mathbb{E}\Phi(V_n^-) \longrightarrow 0$.

Now, let $Z_n \in \mathcal{H}$ be such that $V_n \leq Z_n$. Then $V_n \longrightarrow Z$ in probability and $\mathbb{E}\Phi(V_n^-) \longrightarrow 0$ implies that $P(Z_n \geq Z - \varepsilon) \longrightarrow 1$ for every $\varepsilon > 0$ and $\mathbb{E}\Phi(Z_n^-) \longrightarrow 0$. We can extract a subsequence $U_m = Z_{n_m}$ such that $P(U_m \geq Z - \frac{1}{m}) \longrightarrow 1$ and $\mathbb{E}\Phi(U_m^-) \longrightarrow 0$. Hence, $Z \in \bar{\mathcal{H}}$. This proves (8.44).

The condition **(i)** now implies

$$\mathcal{V}_\infty \cap L_+^\infty = \{0\}.$$

The separation theorem (Theorem 8.6) implies that **(ii)** holds.

(ii) $\Rightarrow$ (i)

Let $f \in L^1, P(f > 0) = 1$ be such that $\int fZdP = 0$ for all $Z \in \mathcal{H}$. For this f, there exists Young's function Ψ such that $a = \mathbb{E}\Psi(f) < \infty$. If $a \leq 1$, then $\|f\|_\Psi \leq 1$. On the other hand, if $a > 1$, then $\Psi(\frac{f}{a}) \leq \frac{1}{a}\Psi(f)$ and hence $\mathbb{E}\Psi(\frac{f}{a}) \leq 1$, and so $\|f\|_\Psi \leq a$. In either case, we have

$$\|f\|_\Psi < \infty. \tag{8.45}$$

Let Φ be the conjugate function of Ψ. Let $\Phi_m(x) = \Phi(mx)$ for $m \geq 1$. Let $Z \in \mathcal{U} \cap L_+^\infty$. For each $m \geq 1$, let $Z_{m,n} \in \mathcal{H}$ such that $P(Z_{m,n} \geq Z - \frac{1}{n}) \longrightarrow 1$ and $\mathbb{E}\Phi_m(Z_{m,n}^-) \longrightarrow 0$. Choose n_m such that $n_m \geq m$ and

$$P(Z_{m,n_m} \geq Z - \frac{1}{n_m}) \geq 1 - \frac{1}{m^2}$$

$$\mathbb{E}(\Phi_m(Z_{m,n_m}^-)) \leq 1.$$

Writing $U_m = Z_{m,n_m}$, it follows that $P(\liminf U_m^+ \geq Z) = 1$ and $\mathbb{E}\Phi(mU_m^-) \leq 1$ so that $\|U_m^-\|_\Phi \leq \frac{1}{m}$. Now $\int fU_m dP = 0$ implies

$$\int fU_m^+ dP = \int fU_m^- dP.$$

Using Fatou's Lemma it follows that

$$\begin{aligned} \int fZdP &\leq \int f(\liminf U_m^+)dP \\ &\leq \liminf \int fU_m^+ dP \\ &= \liminf \int fU_m^- dP. \end{aligned} \tag{8.46}$$

Since

$$\int fU_m^- dP \leq \|f\|_\Psi \|U_m^-\|_\Phi$$

and $\|U_m^-\|_\Phi \longrightarrow 0$, it follows from (8.46) that $\int fZdP = 0$. Since $P(f > 0) = 1$, we conclude $P(Z = 0) = 1$. □

We are now in a position to deduce our main result. It is a direct consequence of Theorem 8.13 and Lemma 8.11. The last assertion follows from Theorem 5.3.

Theorem 8.14 *Suppose $\tilde{S}_t^1, \ldots, \tilde{S}_t^k$ are locally bounded. Then $\tilde{S}$ satisfies the condition NAACR if and only if $\tilde{S}$ admits an EMM. Further, if $\tilde{S}$ satisfies NAACR, then $\tilde{S}$ is a semimartingale.*

As we remarked earlier, our framework can allow infinitely many stocks as in Kreps, who considered an arbitrary set of commodities being traded in the market. We state the result below and it follows again from Theorem 8.13, Lemma 8.11 and Theorem 5.3.

Theorem 8.15 *Consider a market consisting of $\{S^\alpha : \alpha \in \Delta\}$, where each S^α is assumed to be locally bounded. Let $\mathcal{K}$ be the class of contingent claims attainable via simple strategies over a finitely many stocks over a finite horizon. Then $\mathcal{K}$ satisfies NAACR if and only if $\{S^\alpha : \alpha \in \Delta\}$ admits an EMM. In particular, if $\mathcal{K}$ satisfies NAACR, then each S_t^α is a semimartingale.*

Remark It should be noted that NAACR implies that the underlying stocks are semimartingales. It was precisely in the context of characterization of semimartingales that these ideas led Yan to formulate his version of the separation theorem.

8.7 Fractional Brownian motion ($1/2 < H < 1$) and existence of arbitrage opportunities

Definition Normalized fractional Brownian motion $B_H(t), t \in [0, T], 1/2 < H < 1$, is a Gaussian process with zero mean and covariance function

$$EB_H(t)B_H(s) = (1/2)\{|t|^{2H} + |s|^{2H} - |t-s|^{2H}\}.$$

An introduction to the important properties of this process can be found in Samorodnitsky and Taqqu (1994) and also in Mandelbrot and Van Ness (1968). The property most important for us is the fact that $E\{B_H(t) - B_H(s)\}^2 = |t - s|^{2H}, 0 < s < t < T$.

In analogy with geometric Brownian motion, geometric fractional Brownian motion (GFBM) is a solution to the following stochastic differential equation:

$$dS_t = \mu S_t dt + \sigma S_t dB_H(t), \mu \in R, \sigma > 0, S_0 = 1, t \in [0, T].$$

This equation has been studied by Cutland et al. (1995) who use nonstandard analysis in their proof. We use L^1 convergence of Riemann sums. The main result we need is

Lemma 8.16 $S_t = \exp(\mu t + \sigma B_H(t))$ *which satisfies*

$$S(t) = 1 + \int_0^t \mu S(t)dt + \int_0^t \sigma S(t)dB_H(t), t > 0. \tag{8.47}$$

Proof For simplicity, we first assume $\mu = 0$ and $\sigma = 1$. Let $0 = t_0 < t_1 < \cdots < t_n < t_{n+1} = t$ be a partition of $[0, t], t > 0$. Let $\Delta B_H(t_k) = B_H(t_{k+1}) - B_H(t_k), k = 0, 1, \ldots, n$. Then using the Gaussian property of $\Delta B_H(t_k)$ it can be shown that

$$E|e^{B_H(t)} - 1 - \sum_k e^{B_H(t_k)} \Delta B_H(t_k)| \leq const. \sum_k (t_{k+1} - t_k)^{2H} \to 0 \tag{8.48}$$

as $\sup_k (t_{k+1} - t_k) \to 0$.

From (8.48) we have the following relation:

$$e^{B_H(t)} = 1 + \int_0^t e^{B_H(s)} dB_H(s), \tag{8.49}$$

where the integral is defined as the L^1 limit of Riemann sums. We write (8.49) as the differential

$$de^{B_H(t)} = e^{B_H(t)} dB_H(t)$$

and, more generally, as

$$de^{\sigma B_H(t)} = \sigma e^{\sigma B_H(t)} dB_H(t). \tag{8.50}$$

Now consider the differential of the process

$$S_t = \exp(\mu t + \sigma B_H(t)).$$

We then have

$$dS_t = \mu S_t dt + \sigma S_t dB_H(t). \tag{8.51}$$

Thus $S(t) = \exp(\mu t + \sigma B_H(t)), t \in [0, T]$, satisfies the following relation:

$$S(t) = 1 + \int_0^t \mu S(t)dt + \int_0^t \sigma S(t)dB_H(t). \tag{8.52}$$

□

Equation (8.52) has also been proved in (Lin, 1995), where the convergence of the Riemann sums is taken in probability.

An example of an *arbitrage* opportunity for geometric fractional Brownian motion (GFBM) has been constructed by Rogers (1997). In this example, using information from the entire past of fractional Brownian motion, one chooses stopping times to invest judiciously in such a way that starting from a unit wealth, one can make a positive gain with a probability one. Rogers also indicates how similar results can be proved for a larger class of processes, obtained by modifying the kernel of the fractional Brownian motion process.

Our example, however, does not use information from the past. We construct an *arbitrage* opportunity over $[0, T]$, first for geometric fractional Brownian motion and then for geometric versions of a class of Gaussian processes. The class of Gaussian processes we are referring to forms a subclass of the Gaussian processes considered by Gladyshev (Gladyshev, 1961). A special class of these processes was considered by Kolmogorov earlier and called Wiener spirals. For the sake of simplicity, we shall develop the arguments for fractional Brownian motion and, in the end, mention how the results can be extended to this subclass of Gladyshev processes.

Let us now take GFBM as our stock price process

$$S(t) = e^{\mu t + \sigma B_H(t)}, \tag{8.53}$$

which satisfies the equation

$$S(t) = 1 + \mu \int_0^t S(s)ds + \sigma \int_0^t S(s)dB_H(s). \tag{8.54}$$

Notice that (8.54) implies, for $\mu = 0$,

$$\sigma \int_0^t e^{\sigma B_H(s)} dB_H(s) = e^{\sigma B_H(t)} - 1, \tag{8.55}$$

a relation that will be needed. The bond price process, as before, is assumed to be

$$B(t) = e^{rt} \tag{8.56}$$

or

$$dB(t) = rB(t)dt. \tag{8.57}$$

Now we give our example of an *arbitrage* opportunity for GFBM. Consider the particular portfolio

$$\pi(t) = 2e^{\mu t + \sigma B_H(t)}(e^{\sigma B_H(t)} - 1). \tag{8.58}$$

Theorem 8.17 *π, as defined by (8.58), is an arbitrage opportunity for a geometric fractional Brownian motion.*

Proof With π defined in (8.58), the discounted value process becomes

$$\tilde{V}_t(\pi) = \int_0^t 2e^{\mu s + \sigma B_H(s)}(e^{\sigma B_H(s)} - 1)e^{-rs}\left\{\sigma dB_H(s) + (\mu - r)ds\right\}. \tag{8.59}$$

If we assume that $\mu = r$, (8.59) equals

$$\begin{aligned}
\tilde{V}_t(\pi) &= \int_0^t 2e^{\sigma B_H(s)}(e^{\sigma B_H(s)} - 1)\sigma dB_H(s) \\
&= 2\sigma \int_0^t e^{2\sigma B_H(s)} dB_H(s) - 2\sigma \int_0^t e^{\sigma B_H(s)} dB_H(s) \\
&= e^{2\sigma B_H(t)} - 1 - 2(e^{\sigma B_H(t)} - 1) \text{ (from (8.55))} \\
&= (e^{\sigma B_H(t)} - 1)^2. \qquad (8.60)
\end{aligned}$$

It can now be verified from (8.60) that π is an *arbitrage* opportunity for GFBM. To complete the proof, it remains to show that π is a self-financing strategy.

Notice that $(e^{\sigma B_H(t)} - 1)^2$ is the discounted value of the wealth accumulated up to time t and that the actual value is $V_t(\pi) = e^{rt}(e^{\sigma B_H(t)} - 1)^2$. In that case the number of shares of stock held is

$$\phi_t = \pi(t)/S(t),$$

and the number of shares of bond held is

$$\psi_t = (V_t(\pi) - \pi(t))/B(t).$$

With these we have $\phi_t S(t) + \psi_t B(t) = V_t(\pi)$. Since $\phi_0 S(0) + \psi_0 B(0) = 0$, to show that (ϕ_t, ψ_t) is self-financing, we need to show that

$$\int_0^t \phi_u dS(u) + \int_0^t \psi_u dB(u) = V_t(\pi) = e^{rt}(e^{\sigma B_H(t)} - 1)^2. \qquad (8.61)$$

We have

$$\begin{aligned}
&\int_0^t \phi_u dS(u) + \int_0^t \psi_u dB(u) \\
&= \int_0^u \pi(u)\{\sigma dB_H(u) + \mu du\} + \int_0^t \{V_u(\pi) - \pi(u)\} r du \\
&= \int_0^t \pi(u)\sigma dB_H(u) + \int_0^t V_u(\pi) r du \text{ (since } \mu = r). \qquad (8.62)
\end{aligned}$$

Plugging in the expressions for π and $V_t(\pi)$ in the above equation and then using equation (8.47), it is easy to see that (8.62) equals $e^{rt}(e^{\sigma B_H(t)} - 1)^2 = V_t(\pi)$, thus proving (8.61). □

In our example we assumed $\mu = r$ for notational convenience. In the general case, we can take

$$\pi(t) = 2e^{\mu t + \sigma B_H(t)}(e^{\mu t - rt + \sigma B_H(t)} - 1), \qquad (8.63)$$

and use (8.54) with $\mu - r$ in place of μ, yielding $\tilde{V}_t(\pi) = (e^{\mu t - rt + \sigma B_H(t)} - 1)^2$.

8.8 Extension to geometric Gladyshev processes

Fractional Brownian motion is a special case of the Gaussian processes $\xi(t)$ considered in the paper by Gladyshev (Gladyshev, 1961). The results of the previous section will now be extended to a subclass of these processes, which we call Gladyshev processes.

We first describe the Gladyshev processes. For simplicity we assume that the mean of the process is zero and the time interval is [0, 1]. The following conditions on the correlation function $r(t, s), 0 \le s, t \le 1$ are the main features of these Gaussian processes:

(a) $r(t, s)$ is continuous for $0 \le s, t \le 1$,

(b) for $t \neq s$ there exists $\partial^2 r(t, s)/\partial t \partial s$ and

$$|\partial^2 r(t, s)/\partial t \partial s| < L/|t - s|^{\gamma}, 0 < \gamma < 2,$$

(c) the expression

$$\frac{r(t, t) - 2r(t, t - h) + r(t - h, t - h)}{h^{2-\gamma}}$$

converges uniformly to some function $g(t)$ on $0 \le t \le 1$ as $h \to 0$.

Note that unlike fractional Brownian motion, Gladyshev processes do not generally have stationary increments. Gladyshev in Theorem 2 of (Gladyshev, 1961) assumes the condition $\int_0^1 g(t)dt > 0$. Instead of [0, 1] we can consider any other interval $[a, b] \subset [0, 1]$ and then Gladyshev's results will also hold for $[a, b]$.

We shall further assume that $\int_a^b g(s)ds > 0$, for all $[a, b] \subset [0, 1]$. This implies $g(t) > 0$ a.s. on [0, 1]. Including this we assume the following condition, which defines our subclass of Gladyshev processes as

$$0 < \gamma < 1, \; g(t) > 0 \text{ for a.e. } t. \tag{8.64}$$

Like fractional Brownian motion these processes are not semimartingales. Since the covariance functions are not explicitly known, the techniques of Section 4.9 (especially Theorem 3) of Liptser and Shiryayev (1989) cannot be directly used to prove this.

Lemma 8.18 *Under conditions (8.64), ξ is not a semimartingale.*

Proof The proof consists of two parts, first showing that the sample paths are a.s. nondifferentiable (hence, of unbounded variation) and then showing that the quadratic variation process of ξ is zero.

The proof of the first part follows an argument due to Mandelbrot and Van Ness (1968). Take any point $t_0 \in [0, 1]$. For a sequence of numbers $t_n \downarrow t_0$ and any $\epsilon > 0$, define the sets

$$A(t_n, \omega) = \left\{ \sup_{t_0 < x \le t_n} \left| \frac{\xi(t_n) - \xi(t_0)}{t_n - t_0} \right| > \epsilon \right\}.$$

If we show that

$$P\Big\{\lim_{n\to\infty} A(t_n,\omega)\Big\} = 1,$$

then the paths will be a.s. nondifferentiable at t_0 and similarly at other points. Note that the sets $A(t_n,\omega)$ are nonincreasing and hence

$$\begin{aligned}
& P\Big\{\lim_{n\to\infty} A(t_n,\omega)\Big\} \\
= \; & \lim_{n\to\infty} P(A(t_n,\omega)) \\
\geq \; & \lim_{n\to\infty} P\Big\{\Big|\frac{\xi(t_n)-\xi(t_0)}{t_n-t_0}\Big| > \epsilon\Big\} \\
= \; & P\left\{\frac{|\xi(t_n)-\xi(t_0)|}{\sqrt{E(\xi(t_n)-\xi(t_0))^2}} > \frac{|t_n-t_0|}{\sqrt{E(\xi(t_n)-\xi(t_0))^2}}\epsilon\right\} && (8.65)\\
\to \; & 1, && (8.66)
\end{aligned}$$

since in (8.65) the first term in the parentheses is the absolute value of a standard normal and the term on the right goes to zero, because from assumption (c) on the covariance and by the assumptions (8.64)

$$\begin{aligned}
\frac{|t_n-t_0|}{\sqrt{E(\xi(t_n)-\xi(t_0))^2}} \quad & \sim \quad \frac{|t_n-t_0|}{|t_n-t_0|^{1-\gamma/2}\sqrt{g(t_o)}} \\
& \to \quad 0. && (8.67)
\end{aligned}$$

For the other part take a partition $0 = t_0 < t_1 < \cdots < t_{n+1} = t$, of $[0,t]$ and consider the expression

$$\sum_k \Big\{\Delta\xi_{t_k}\Big\}^2.$$

If we take an expectation of the above, then we get

$$\begin{aligned}
& E\Big\{\sum_k \Big\{\Delta\xi_{t_k}\Big\}^2\Big\} \\
= \; & \sum_k \Big\{r(t_{k+1},t_{k+1}) - 2r(t_{k+1},t_k) + r(t_k,t_k)\Big\} \\
\leq \; & \sum_k \Big\{(\Delta t_k)^{2-\gamma} g(t_k) + \epsilon(\Delta t_k)^{2-\gamma}\Big\}, \; \epsilon \text{ independent of } k, && (8.68)
\end{aligned}$$

which goes to zero as $\sup \Delta t_k \to 0$ since $0 < \gamma < 1$ and $\int_0^t g(s)ds < \infty$. Thus, the quadratic variation process of ξ is zero and ξ cannot be a semimartingale because it has been shown to have paths of unbounded variation a.s. □

The case of ξ with stationary Gaussian increments has been discussed separately by Gladyshev and, in this case, $g(t)$ is a positive constant, so that our assumption on $g(t)$ in (8.64) is satisfied.

Now consider the geometric Gladyshev process $e^{\xi(t)}, 0 \leq t \leq 1$. To show that Lemma 8.16 holds for $e^{\xi(t)}$, we need to show that

$$\sum_k \left\{ E[e^{\Delta\xi(t_k)} - 1 - \Delta\xi(t_k)]^2 \right\}^{1/2} \to 0 \tag{8.69}$$

as $\sup \Delta t_k \to 0$. Denoting $r(t_{k+1}, t_{k+1}) - 2r(t_{k+1}, t_k) + r(t_k, t_k)$ by ρ_k, we note that $\rho_k = E\{\xi(t_{k+1}) - \xi(t_k)\}^2$. Thus in the argument of Lemma 8.16, we have to use ρ_k instead of $(\Delta t_k)^{2H}$. But from assumption (c) we get uniformly in k that

$$\rho_k \leq (\Delta t_k)^{2-\gamma} g(t_k) + \epsilon(\Delta t_k)^{2-\gamma} (\ \epsilon \text{ independent of } k\ ). \tag{8.70}$$

Again using the Gaussian property it can be shown that

$$\begin{aligned} & \sum_k \left\{ E[e^{\Delta\xi(t_k)} - 1 - \Delta\xi(t_k)]^2 \right\}^{1/2} \\ \leq \;& const. \sum_k \rho_k \\ \leq \;& const. \sum_k \left\{ (\Delta t_k)^{2-\gamma} g(t_k) + \epsilon(\Delta t_k)^{2-\gamma} \right\}. \end{aligned} \tag{8.71}$$

Since $0 < \gamma < 1$ and $\int_0^1 g(t)dt < \infty$, (8.69) follows, i.e., $e^{\xi(t)}$ satisfies

$$e^{\xi(t)} = 1 + \int_0^t e^{\xi(s)} d\xi(s), 0 \leq t \leq 1. \tag{8.72}$$

Similarly, we can show that $S(t) = e^{\mu t + \sigma\xi(t)}$ satisfies

$$S(t) = 1 + \mu \int_0^t S(s)ds + \sigma \int_0^t S(s)d\xi(s). \tag{8.73}$$

Then, following the proof of Theorem 8.17, we can show that

$$\pi(t) = 2e^{\mu t + \sigma\xi(t)}(e^{\sigma\xi(t)} - 1) \tag{8.74}$$

is an *arbitrage* opportunity for $e^{\mu t + \sigma\xi(t)}$ under the assumption $\mu = r$ and finally, under the general case in (8.63).

9
Complete Markets

In the discrete case, we saw that the notion of completeness plays an important role in option pricing if the underlying market is complete, and the price of both European and American call and put options are uniquely determined. The same is the case in continuous time, as we will see later. We begin by defining completeness of a market. Section 2 is a digression into stochastic calculus. Next we show that in the examples considered earlier, namely that of geometric Brownian motion and that of diffusions, the markets are complete. We will also show that the completeness of the underlying market is characterized by the uniqueness of the equivalent martingale measure.

9.1 Definition

Throughout this chapter we will consider a market consisting of a bond (S^0) and of k-stocks $(S^1, S^2, \ldots, S^k)$, where $S_t^1, S_t^2, \ldots, S_t^k$ are continuous semimartingales and S_t^0 is a continuous increasing process. We assume that the S^i are defined on a probability space $(\Omega, \mathcal{F}, P)$ and that the S_0^i are constants for each i. Let $(\mathcal{F}_t^S)$ denote the smallest filtration with respect to which the S_t^i, $1 \leq i \leq k$ are adapted. It follows that $\mathcal{F}_0^S$ is the σ-field generated by the P-null sets.

Let $\tilde{S}_t^i = S_t^i(S_t^0)^{-1}$, $1 \leq i \leq k$ be the discounted stock price process. We assume that $\mathcal{M}(P)$, the class of P-equivalent (local) martingale measures, is nonempty and we will fix an element of $\mathcal{M}(P)$ and denote it by Q. We recall definitions of relevant notions given in Chapter 7. A self-financing strategy in this market is denoted by $\theta = (x, \pi)$, where $x \geq 0$ and $\pi = (\pi^1, \ldots, \pi^k)$ is an $\mathbb{R}^k$-

valued $(\mathcal{F}_t^S)$-predictable process such that (7.19) holds. The (discounted) value of the investor invoking the strategy θ at time t is given by

$$\tilde{V}_t(\theta) = x + \sum_{i=1}^{k} \int_0^t \pi_u^i d\tilde{S}_u^i. \tag{9.1}$$

A contingent claim in this market is a non-negative $(\mathcal{F}_T^S)$-measurable Q integrable random variable. A contingent claim Z_T is said to be *attainable* if there exists a strategy θ such that the discounted value process $\tilde{V}_t(\theta)$ is a Q-martingale and the value process V_T satisfies

$$V_T(\theta) = Z_T \quad \text{a.s.} \tag{9.2}$$

The strategy θ satisfying (9.2) above is said to be a *hedging strategy* for the contingent claim Z_T.

Definition The market consisting of k-stocks $(S^1, S^2, \ldots, S^k)$ and bond (S^0) is said to be complete if every contingent claim is attainable.

We begin with a simple observation about the completeness of the market.

Lemma 9.1 *The market is complete if and only if all Q-martingales (M_t), with respect to the filtration $(\mathcal{F}_t^S)$, admit a representation*

$$M_t = E^Q(M_T) + \sum_{i=1}^{k} \int_0^t f^i d\tilde{S}^i \tag{9.3}$$

for an $\mathbb{R}^k$-valued predictable process f such that

$$\sum_{i=1}^{k} \int_0^t |f_s^i|^2 d\langle \tilde{S}^i, \tilde{S}^i \rangle_s < \infty \quad a.s.\ Q.$$

Proof First assume that (9.3) holds. If X is a contingent claim, writing $M_t = E^Q(X|\mathcal{F}_t)$, we have from (9.3), that $M_t = E^Q(X) + \sum_{i=1}^k \int_0^t f^i \, d\tilde{S}^i$, where the f^i are predictable and $\sum_{i=1}^k \int_0^T (f_u^i)^2 du < \infty$ a.s.

Let $\theta = (x, f', \ldots, f^k)$ where $x = E^Q(X)$ and define

$$f_t^0 = x + \sum_{i=1}^{k} \int_0^t f^i \, d\tilde{S}^i - \sum_{i=1}^{k} f_t^i \tilde{S}_t^i.$$

Then we have from the above that

$$M_t = f_t^0 + \sum_{i=1}^{k} f_t^i \tilde{S}_t^i = \tilde{V}_t(\theta).$$

It follows that

$$X = M_T = \tilde{V}_T(\theta),$$

which proves that X is attainable. Hence, the market is complete.

Next, suppose the market is complete. Let Y_t be a positive martingale and $Z_T = Y_T S_T^0$. Since Z_T is attainable, $Z_T = V_T(\theta)$ for some hedging strategy $\theta = (x, \pi)$. Hence $Y_T = \tilde{V}_T(\theta)$. Since $\tilde{V}_t(\theta)$ and Y_t are martingales, we have $\tilde{V}_t(\theta) = Y_t$. From (9.1) it follows that Y_t has the representation (9.3).

Finally, let M_t be a martingale. Applying the above result to the positive martingales $Y_t^+ = E(M_T^+|\mathcal{F}), Y_t^- = E(M_T^-|\mathcal{F})$, we obtain the required representation for M_t. □

The importance of the notion of completeness in the context of mathematical finance and option pricing theory is because of the fact (proved below) that in a complete market, the price of every contingent claim is determined uniquely by the market and the principle of no arbitrage.

Theorem 9.2 *Suppose that $Q \in \mathcal{M}(P)$. Suppose that a contingent claim Z_T is attainable using hedging strategy*

$$\theta = (x, \pi).$$

In other words,

$$\begin{aligned} Z_T &= V_T(\theta) \\ &= x + \sum_{i=1}^{k} \int_0^T \pi_u^i \, d\tilde{S}_u^i \end{aligned} \tag{9.4}$$

with $\tilde{V}_t$ being a Q-martingale. Then in order for the augmented market (consisting of the bond, the k-stocks and the contingent claim Z_T) to satisfy the NA property, the price at $t = 0$ of the claim Z_T must be x. Further, the price x is given by

$$x = \mathbb{E}_Q[Z_T(S_T^0)^{-1}]. \tag{9.5}$$

Proof Since Z_T is attainable, an investor may start with an initial investment x, follow the trading strategy θ and be exactly in the same position as someone who bought the contingent claim Z_T—no matter how the stock prices behave. Now if the price p at which the claim Z_T is trading at time $t = 0$ (there are enough buyers and sellers at this price p) is larger than x, then a broker can sell one claim Z_T at price p and follow the trading strategy $\bar{\theta} = (p, \pi)$. Note that at $t = 0$, the broker has not invested any money on his own. At time T, the liabilities of the investor are Z_T and the assets are $V_T(\bar{\theta})$. Noting that

$$\begin{aligned} \tilde{V}_T(\bar{\theta}) &= (p - x)S_T^0 + V_T\theta) \\ &= (p - x)S_T^0 + Z_T, \end{aligned}$$

it follows that the net (discounted) assets of the broker from the strategy described above is $(p - x)S_T^0$ and thus the broker has an arbitrage opportunity if the price

p is strictly greater than x. On the other hand, if the price p is strictly less than x, an investor can buy a claim at price p and follow an investment strategy

$$\bar{\theta} = (-p, -\pi).$$

This time the assets of the investor are Z_T and the liabilities are $\tilde{V}_T(\bar{\theta})$. Here

$$\begin{aligned} V_T(\bar{\theta}) &= S_T^0\Big(-p + \sum_{i=1}^k \int_0^t (-\pi_u^i) d\tilde{S}_u^i\Big) \\ &= S_T^0\Big(-p + x - x - \sum_{i=1}^k \int_0^t (\pi_u^i) d\tilde{S}_u^i\Big) \\ &= S_T^0\Big((x-p) - \tilde{V}_T(\theta)\Big) \\ &= S_T^0\Big((x-p) - Z_T\Big). \end{aligned}$$

It follows that the net (discounted) assets of the broker from the strategy described above is $(x - p)$ and thus any investor has an arbitrage opportunity if the price p is strictly less than x.

Thus if we require that the augmented market consisting of the bond, the k-stocks and the contingent claim Z_T be free of arbitrage opportunities, then the price of the contingent claim Z_T must be x. □

Now if the market is complete, all claims, including all European call options and European put options, admit a unique price. Later we will see that the same is the case with American options.

9.2 Representation of martingales

We saw in the previous section that completeness of the market is related to the martingale representation property discussed at length in Chapter 2. Here we need an integral representation of a martingale, which may not be a square integrable martingale; whereas in Chapter 3, we only discussed the representation of square integrable martingales. We will show that when the underlying processes are continuous local martingales, the representation property then holds for square integrable martingales if and only if it holds for all martingales. For the case of a Brownian motion, this is a result due to Clark (Clark, 1970) which we will prove first. The general case in finite dimension is treated later. Here is a technical result which we need for our result.

Lemma 9.3 *Let $(N_t)_{0 \le t \le T}$ be a continuous local martingale on $(\Omega^1, \mathcal{F}^1, P^1)$ such that $N_0 = 0$. Then given $\eta > 0$ and $\delta > 0$, there exists $\varepsilon > 0$ such that*

$$P^1(\langle N, N \rangle_T \ge \eta) \le P^1(\sup_{0 \le t \le T} |N_t| > \varepsilon) + \delta.$$

Proof Let $(W_t)_{0\leq t<\infty}$ be a Brownian motion on $(\Omega^2, \mathcal{F}^2, P^2)$ and let

$$(\Omega^0, \mathcal{F}^0, P^0) = (\Omega^1, \mathcal{F}^1, P^1) \otimes (\Omega^2, \mathcal{F}^2, P^2).$$

As usual, we regard processes N and W as defined on the product space. Let

$$Z_t = N_{t\wedge T} + W_{(t-T)\vee 0}.$$

Then it follows that Z_t is a martingale such that $\langle Z, Z\rangle_t$ increases to ∞ as t increases to ∞. For $0 \leq u < \infty$ let

$$\sigma_u = \inf\{t \geq 0 : \langle Z, Z\rangle_t \geq u\}.$$

Then σ_u is a stopping time and $\langle Z, Z\rangle_{\sigma_u} = u$. Hence $Y_u = Z_{\sigma_u}$ is a martingale and $\langle Y, Y\rangle_u = \langle Z, Z\rangle_{\sigma_u} = u$. Thus Y is a Brownian motion by Theorem 2.9. Now using $N_t = Y_t, 0 \leq t \leq T$, one has

$$\begin{aligned}
P^1(\langle N, N\rangle_T \geq \eta) &= P^0(\langle Z, Z\rangle_T \geq \eta) && (9.6)\\
&= P^0(\sigma_\eta \leq T) && (9.7)\\
&\leq P^0(\sup_{0\leq t\leq\sigma_\eta} |Z_t| > \varepsilon,\ \sigma_\eta \leq T) && (9.8)\\
&\quad + P^0(\sup_{0\leq t\leq\sigma_\eta} |Z_t| \leq \varepsilon) &&\\
&\leq P^0(\sup_{0\leq t\leq T} |N_t| > \varepsilon) + P^0(\sup_{0\leq t\leq\sigma_\eta} |Z_t| \leq \varepsilon) && (9.9)\\
&= P^1(\sup_{0\leq t\leq T} |N_t| > \varepsilon) + P^0(\sup_{0\leq u\leq\eta} |Y_u| \leq \varepsilon) && (9.10)\\
&= P^1(\sup_{0\leq t\leq T} |N_t| > \varepsilon) + P^0(|Y_\eta| \leq \varepsilon) && (9.11)
\end{aligned}$$

Now given η and δ, it is clear (since the distribution of Y_η is Gaussian with mean 0 and variance η) that we can choose $\varepsilon > 0$ such that $P^0(|Y_\eta| \leq \varepsilon) \leq \delta$. This completes the proof. □

Here is the extension of Theorem 3.2.

Theorem 9.4 *Suppose $(W_t)_{0\leq t\leq T}$ is a Wiener process on $(\Omega^0, \mathcal{F}^0, Q^0)$. If $(M_t, \mathcal{F}_t^W)_{0\leq t\leq T}$ is a martingale, then there exists a $(\mathcal{F}_t^W)$-predictable process f such that*

$$\int_0^t |f_s|^2 ds < \infty \ a.s. \qquad (9.12)$$

and

$$M_t = E^{Q^0}(M_0) + \int_0^t f dW. \quad 0 \leq t \leq T. \qquad (9.13)$$

Proof It suffices to consider martingales (M_t) such that $E^{Q^0}(M_T) = 0$. Let

$$M_T^n = (M_T \wedge n) \vee (-n) - E^{Q^0}((M_T \wedge n) \vee (-n))$$

and let

$$M_t^n = E^{Q^0}(M_T^n | \mathcal{F}_t^W).$$

It is clear that M_T^n converges to M_T pointwise as well as in $L^1(Q^0)$, and hence for all $\varepsilon > 0$,

$$\lim_{n \longrightarrow \infty} Q^0(\sup_{0 \le t \le T} |M_t^n - M_r| \ge \varepsilon) = 0. \tag{9.14}$$

Clearly M_t^n is bounded and hence is a square integrable martingale. Thus there exists an $(\mathcal{F}_t^W)$-predictable $f^{(n)}$ such that

$$M_t^n = \int_0^t f^{(n)} dW \tag{9.15}$$

(see Theorem 3.2). For $n, m \ge 1$, it follows that

$$\langle M^n - M^m, M^n - M^m \rangle_t = \int_0^t |f^{(n)} - f^{(m)}|^2 ds$$

and so, using (9.14) and Lemma 9.3, we get that, given $\eta > 0$,

$$\lim_{(n,m) \longrightarrow \infty} Q^0(\int_0^T |f^{(n)} - f^{(m)}|^2 ds > \eta) = 0. \tag{9.16}$$

Thus we can get a subsequence $g^{(k)} = f^{(n_k)}$ satisfying

$$Q^0(\int_0^T |g^{(k)} - g^{(k+1)}|^2 ds > 2^{-k}) \le 2^{-k}.$$

By the Borel–Cantelli lemma, it follows that

$$\sum_{k=1}^{\infty} \Big(\int_0^T |g^{(k)} - g^{(k+1)}|^2 ds \Big)^{\frac{1}{2}} < \infty \ a.s.\ Q^0.$$

Hence

$$\begin{aligned} \Big(\int_0^T \Big(\sum_{k=1}^{\infty} |g^{(k)} - g^{(k+1)}| \Big)^2 ds \Big)^{\frac{1}{2}} &\le \sum_{k=1}^{\infty} \Big(\int_0^T |g^{(k)} - g^{(k+1)}|^2 ds \Big)^{\frac{1}{2}} \\ &< \infty \ a.s.\ Q^0. \end{aligned} \tag{9.17}$$

$$\begin{aligned} &\lim_{j \longrightarrow \infty} \sup_{k \ge j, l \ge j} \int_0^T |g^{(k)} - g^{(l)}|^2 ds \\ \le\ &\lim_{j \longrightarrow \infty} \Big[\sum_{k=j}^{\infty} \Big(\int_0^T |g^{(k)} - g^{(k+1)}|^2 ds \Big)^{\frac{1}{2}} \Big]^2 \\ =\ &0 \ a.s.\ Q^0. \end{aligned} \tag{9.18}$$

Define

$$g_s(\omega) = \limsup_{k \longrightarrow \infty} g_s^{(k)}(\omega), \quad \omega \in \Omega^0.$$

It follows from (9.19) that

$$\int_0^T |g^{(k)} - g|^2 ds \longrightarrow 0 \text{ in } Q^0 \text{ probability}, \tag{9.19}$$

and so by Theorem 1.10

$$\lim_{n \longrightarrow \infty} Q^0(\sup_{0 \le t \le T} |M_t^{n_k} - \int_0^t g dW ds| > \eta) = 0. \tag{9.20}$$

Thus $M_t = \int_0^t g dW$. □

We will now give an extension of this result to the case of k-continuous local martingales such that all square integrable martingales are representable as stochastic integrals iff all martingales are representable. Many steps in the proof are similar to the one given above, and in such cases, details will be omitted.

Theorem 9.5 *Let* $U^1, U^2, \ldots, U^k$ *be continuous local martingales on* $(\Omega^0, \mathcal{F}^0, Q^0)$. *Let* $\mathcal{F}_t^U$ *be the filtration generated by* $U = (U^1, U^2, \ldots, U^k)$. *Suppose that every bounded martingale* $(M_t, \mathcal{F}_t^U)$ *on* $(\Omega^0, \mathcal{F}^0, Q^0)$ *admits a representation*

$$M_t = E^Q(M_0) + \sum_{i=1}^k \int_0^t f^i dU^i. \quad 0 \le t \le T, \tag{9.21}$$

where $f = (f^1, \ldots, f^k)$ *is a* $(\mathcal{F}_t^U)$*-predictable process satisfying*

$$E^{Q^0}\left[\sum_{i=1}^k \int_0^T |f_s^i|^2 d\langle U^i, U^i\rangle_s\right] < \infty. \tag{9.22}$$

Then every martingale $(M_t, \mathcal{F}_t^U)$ *on* $(\Omega^0, \mathcal{F}^0, Q^0)$ *also admits the representation (9.21) with* f *predictable, satisfying*

$$\sum_{i=1}^k \int_0^T |f_s^i|^2 d\langle U^i, U^i\rangle_s < \infty \quad a.s. Q^0. \tag{9.23}$$

Proof Let (M_t) be a mean zero martingale. As in the Brownian motion case, it suffices to consider such martingales. Let

$$M_T^n = (M_T \wedge n) \vee (-n) - E^{Q^0}((M_T \wedge n) \vee (-n))$$

and

$$M_t^n = E^{Q^0}(M_T^n | \mathcal{F}_t^U).$$

It is clear that M_T^n converges to M_T pointwise as well as in $L^1(Q^0)$ and hence for all $\varepsilon > 0$,

$$\lim_{n\longrightarrow\infty} Q^0(\sup_{0\le t\le T} |M_t^n - M_r| \ge \varepsilon) = 0. \tag{9.24}$$

Proceeding exactly as in the proof of the previous result (Theorem 9.4), we can extract a subsequence $\{n_m\}$ such that $N_t^m = M_t^{n_m}$ satisfies

$$Q^0(\langle N^{m+1} - N^m, N^{m+1} - N^m\rangle_T \ge 2^{-m}) \le 2^{-m}.$$

Then by the Borel–Cantelli lemma it follows that

$$\sum_{m=1}^{\infty} \left(\langle N^{m+1} - N^m, N^{m+1} - N^m\rangle_T\right)^{\frac{1}{2}} < \infty \ a.s.\ Q^0. \tag{9.25}$$

Clearly, N_t^m is a bounded martingale and hence admits a representation

$$N_t^m = \sum_{i=1}^{k} \int_0^t g^{m,i} dU^i. \quad 0 \le t \le T. \tag{9.26}$$

Let $\mathcal{Y}$ denote the class of $\mathbb{R}^k$-valued predictable processes $h = (h^1, h^2, \ldots, h^k)$ such that

$$\sum_{i=1}^{k} \int_0^T |h_s^i|^2 d\langle U^i, U^i\rangle_s < \infty \ a.s.\ Q^0.$$

For $f, h \in \mathcal{Y}$, let $C(f, h)$ be the random variable defined by

$$C(f, h) = \sum_{i,j=1}^{k} \int_0^T f^i h^j d\langle U^i, U^j\rangle_s.$$

Then C is a bilinear form and

$$C(f, h) = \langle M^1, M^2\rangle_T,$$

where $M_t^1 = \sum_{i=1}^k \int_0^t f^i dU^i$ and $M_t^2 = \sum_{i=1}^k \int_0^t h^i dU^i$. Thus, by the Kunita–Watanabe inequality (1.52),

$$|C(f, h)| \le \sqrt{C(f, f)}\sqrt{C(h, h)}.$$

As a consequence, we have

$$\begin{aligned} |C(f+h, f+h)| &= |C(f, f) + 2C(f, h) + C(h, h)| \\ &\le C(f, f) + 2\sqrt{C(f, f)}\sqrt{C(h, h)} + C(h, h) \\ &= (\sqrt{C(f, f)} + \sqrt{C(h, h)})^2, \end{aligned}$$

which gives us

$$\sqrt{C(f+h, f+h)} \leq \sqrt{C(f, f)} + \sqrt{C(h, h)}. \tag{9.27}$$

Note that $\langle N^m, N^m \rangle_T = C(g^{(m)}, g^{(m)})$ where $g^{(m)} = (g^{m,1}, \ldots, g^{m,k})$. Then (9.25) and (9.27) imply

$$\begin{aligned}\lim_{j \longrightarrow \infty} \sup_{m \geq j, l \geq j} & \sqrt{C(g^{(m)} - g^{(l)}), C(g^{(m)} - g^{(l)})} \\ \leq & \lim_{j \longrightarrow \infty} \sum_{m=j}^{\infty} \sqrt{C(g^{(m+1)} - g^{(m)}, g^{(m+1)} - g^{(m)})} \\ = & \ 0 \ a.s. \ Q^0. \end{aligned} \tag{9.28}$$

Let us define

$$g_s^i(\omega) = \limsup_{m \longrightarrow \infty} g_s^{m,i}(\omega) \quad \omega \in \Omega^0.$$

It follows from (9.28) that

$$C(g^{(m)} - g, g^{(m)} - g) \longrightarrow 0 \ a.s. \ Q^0 \tag{9.29}$$

and $g \in \mathcal{Y}$. Defining $N_t = \sum_{i=1}^k \int_0^t g^i dU^i$, we have

$$\langle N^m - N, N^m - N \rangle_T = C(g^{(m)} - g, g^{(m)} - g).$$

From Lemma 9.3 it follows that

$$Q^0(\sup_{0 \leq t \leq T} |N_t^m - N_t| > \varepsilon) \longrightarrow 0$$

for every $\varepsilon > 0$. This imples $N_t = M_t$, completing the proof. □

9.3 Examples of complete markets

We will show in the two examples of stock models considered in Section 7.4 that the market is complete.

9.3.1 *Geometric Brownian motion (GBM)*

Consider Example 1 in Section 7.4. Let (W_t) be a Brownian motion (defined on a probability space $(\Omega, \mathcal{F}, P)$) and consider a market consisting of a bond with rate of interest r and a single stock whose price (S_t) is given by

$$dS_t = \sigma S_t \, dW_t + \mu S_t \, dt,$$

where σ, μ are constants. Here

$$S_t = S_0 \exp\{\sigma W_t + (\mu - \frac{1}{2}\sigma^2)t\},$$

and thus

$$\tilde{S}_t = S_0 \exp\{-rt\} S_t \tag{9.30}$$

$$= S_0 \exp\{\sigma W_t + (\mu - r - \frac{1}{2}\sigma^2)t\}. \tag{9.31}$$

Let

$$R_T = \exp\{aW_T - \frac{1}{2}a^2 T\}$$

where $a = \frac{r-\mu}{\sigma}$. Write $\hat{W}_t = W_t - at$ and $R_T = e^{aW_T - \frac{1}{2}a^2 T}$. Then by Girsanov's theorem, Q, given by $dQ = R_T\, dP$, is a probability measure equivalent to P. We then have

$$\tilde{S}_t = e^{\sigma \hat{W}_t - \frac{1}{2}\sigma^2 t} \quad d\tilde{S}_t = \sigma \tilde{S}_t\, d\hat{W}_t.$$

Hence $\mathcal{F}_t^S = \mathcal{F}_t^{\tilde{S}} = \mathcal{F}_t^{\hat{W}}$. Then on $(\Omega, \mathcal{F}, Q)$, $\tilde{W}$ is a Brownian motion and, taking $S_0 = 1$,

$$\tilde{S}_t = \exp\{\sigma \hat{W} - \frac{1}{2}\sigma^2 t\}$$

$$d\tilde{S}_t = \sigma \tilde{S}_t d\hat{W}.$$

It follows that

$$\mathcal{F}_t^S = \mathcal{F}_t^{\hat{W}}. \tag{9.32}$$

Note that Q is an EMM for P. If X is a contingent claim (an $\mathcal{F}_T^S$-measurable integrable random variable), then consider the martingale

$$M_t = E^Q(X|\mathcal{F}_t^S) = E^Q(X|\mathcal{F}_t^{\hat{W}}).$$

By Theorem 9.4 it follows that M_t can be written as

$$M_t = E^Q(X) + \int_0^t f\, d\tilde{W}$$

for a predictable process f such that $\int_0^T f_s^2 ds < \infty$ *a.s.*. Writing $\phi_s = f_s(\tilde{S}_t^{-1})$, it follows that

$$M_t = E^Q(X) + \int_0^t \phi_s d\tilde{S},$$

and so the claim $X = M_T$ is attainable. This shows that the market consisting of the geometric Brownian motion and a bond is complete.

Consider a European call option in this market with striking price K and terminal time T. Then completeness of the market tells us (see Theorem 9.2) that the price of this contingent claim must be

$$x = \mathbb{E}_Q[(S_T - K)^+](\exp\{-rT\}).$$

This can be rewritten as

$$x = \mathbb{E}_Q[(\tilde{S}_T - Ke^{-rT})^+]$$

and, recalling that the distribution of $\tilde{S}_T(S_0)^{-1}$ under Q is the same as that of $\exp\{\sigma\xi - \frac{1}{2}\}$, where ξ has standard normal distribution, we get

$$x = \mathbb{E}\left[(\exp\left\{\sigma\xi - \frac{1}{2}\right\} - Ke^{-rT})^+\right]. \tag{9.33}$$

This integral can be evaluated in terms of the distribution function Φ of the standard normal distribution. This leads to the famous Black–Scholes formula. We will present the derivation due to Black–Scholes in the next chapter.

$$x = S_0^1\Phi(a) - K\exp -rT\Phi(b), \tag{9.34}$$

where

$$a = \frac{\log(S_0) - \log(K) + (r + \frac{1}{2}\sigma^2)T}{\sigma\sqrt{T}} \tag{9.35}$$

$$b = \frac{\log(S_0) - \log(K) + (r - \frac{1}{2}\sigma^2)T}{\sigma\sqrt{T}}. \tag{9.36}$$

9.3.2 *Diffusion model for stock prices*

Let the stock prices S_t^i $(i = 1, \ldots, k)$ be given as the unique solutions of

$$dS_t^i = S_t^i\left\{\sum_{j=1}^k \sigma_{ij}(S_t, t)dW_t^j + b_i(S_t, t)dt\right\}, \quad i = 1, \ldots, k. \tag{9.37}$$

We assume that the matrix $a(t, x) = \sigma(t, x)\sigma^*(t, x)$ is positive definite for all (t, x). It is further assumed that σ_{ij} and b_i satisfy the conditions which ensure the existence of a unique solution to the above system of equations.

Define

$$R_T = \exp\left\{\sum_{i=1}^k \int_0^T f_i(S_u, u)dW_u^i - \frac{1}{2}\sum_{i=1}^k \int_0^T f_i^2(S_u, u)du\right\},$$

where

$$f_i(S_u, u) = -\sum_{j=1}^k \sigma_{ij}^{-1}b_j(X_u, u).$$

Further assume that $\mathbb{E}\, R_T = 1$ and define $dQ = R_T dP$. Then by Girsanov's Theorem, under Q, $\tilde{W}_t$, given by $\tilde{W}_t^i = W_t^i - \int_0^t f_i(S_u, u)du$, is a Brownian motion. We have $d\tilde{W}_t = dW_t + \sigma^{-1}b(S_t, t)\, dt$ and from (9.37),

$$dS_t^i = S_t^i \sum_{j=1}^{k} \sigma_{ij} d\tilde{W}_t^j. \tag{9.38}$$

Thus, S_t^i $(i = 1, \ldots, k)$ is a martingale under Q and it follows that Q is an EMM for P. It also follows from (9.38) that $\mathcal{F}_t^S = \mathcal{F}_t^{\tilde{W}}$.

Let X be a contingent claim and let $M_t = E^Q(X|\mathcal{F}_t^S) = E^Q(X|\mathcal{F}_t^{\tilde{W}})$. From Theorem 9.4, we have

$$M_t = E^Q(X) + \sum_{i=1}^{k} \int_0^t g_i \, d\tilde{W}_t^i,$$

where g_i's are predictable and $\int_0^T |g_i|^2 dt < \infty$ a.s. From (9.38),

$$d\tilde{W}_t^i = \sum_{j=1}^{k} (\sigma^{-1})_{ij} \frac{dS_t^i}{S_t^i}, \quad \sigma^{-1} \text{ being the inverse matrix.}$$

Hence

$$M_t = E^Q(X) + \sum_{j=1}^{k} \int_0^t \phi_u^j \, dS_u^j,$$

where $\phi_u^j = \sum_{i=1}^k g_i (\sigma^{-1})_{ij}$. Then

$$X = M_T = V_T(\theta) := x + \sum_{j=1}^{k} \int_0^T \phi_u^j \, dS_u^j,$$

where $\theta = (x, \phi^1, \ldots, \phi^k)$, $x = E^Q(X)$ is the hedging strategy. Hence X is attainable.

9.4 Equivalent martingale measures

We saw that in the discrete case, the market consisting of the stocks $(S^1, S^2, \ldots, S^k)$ is complete if and only if $\mathcal{M}(P)$ is a singleton. We will prove this for the model considered at the beginning of this chapter.

We begin by noting that for a market to be complete, it suffices to require that all bounded claims be attainable.

Theorem 9.6 *The following are equivalent:*

(i) All (integrable) contingent claims are attainable.

(ii) All bounded contingent claims are attainable.

Proof Of course (i) implies (ii). Suppose (ii) holds. Then it follows that all bounded martingales (M_t) admit a stochastic integral representation with respect to $(\tilde{S}^1, \tilde{S}^2, \dots, \tilde{S}^k)$ (consider the contingent claim $M_T + a$ where a is a bound for $|M_T|$). From Theorem 9.5, it follows that all Q-martingales admit a stochastic integral representation w.r.t. $(\tilde{S}^1, \tilde{S}^2, \dots, \tilde{S}^k)$. Given a contingent claim X, consider the martingale

$$M_t = E^Q(X|\mathcal{G}_t).$$

Using the integral representation of M, we can conclude that X is attainable. □

Here is the main result of this chapter. This gives a characterization of complete markets analogous to the characterization in the discrete case, shown in Theorem 6.3.

Theorem 9.7 *Suppose $Q \in \mathcal{M}(P)$. The market consisting of $(S^0, S^1, S^2, \dots, S^k)$ is then complete if and only if $\mathcal{M}(P) = \{Q\}$.*

Proof Suppose the market is complete. Let $Q^1 \in \mathcal{M}(P)$ be arbitrary. Let

$$Q^0 = \frac{1}{2}(Q + Q^1).$$

Then clearly, Q^0 is absolutely continuous w.r.t. Q. Let $R = \frac{dQ^0}{dQ}$ and let $R_t = \mathbb{E}^Q[R|\mathcal{F}_t^S]$. Then (R_t) is a martingale. Since the market is complete, there exists a $(\mathcal{F}_t^S)$-predictable process g ($\mathbb{R}^k$-valued) such that the $(\mathcal{F}_t^S)$-martingale (R_t) can be represented as

$$R_t = 1 + \int_0^t \sum_{i=1}^k g_u^i d\tilde{S}_u^i$$

(since $\mathbb{E}^Q(R_T) = 1$). Let

$$\tau_m = inf\{t \geq 0 : \ |\tilde{S}_t| \geq m\} \wedge T.$$

Since $\tilde{S}^i$ are continuous processes, $S_{t\wedge\tau_m}^i$ is a bounded process for every m. Since $Q, Q^0 \in \mathcal{M}(P)$, $\tilde{S}_{t\wedge\tau_m}^i$ are martingales for each m under Q as well as under Q^0. Fix $s < t$ and $B \in \mathcal{F}_s^S$ and i. Then we have

$$\mathbb{E}^{Q^0}[1_B \tilde{S}_{t\wedge\tau_k}^i] = \mathbb{E}^{Q^0}[1_B \tilde{S}_{s\wedge\tau_k}^i] \tag{9.39}$$

and, as a consequence,

$$E^Q[1_B \tilde{S}_{t\wedge\tau_k}^i R_t] = E^Q[1_B \tilde{S}_{s\wedge\tau_k}^i R_s]. \tag{9.40}$$

Thus $\tilde{S}_{t\wedge\tau_k}^i R_t$ is a Q-martingale. It follows that

$$\langle \tilde{S}^i, R\rangle_t = 0 \ \ \forall t \ \ a.s. Q.$$

As a consequence,

$$\begin{aligned} \langle R, R\rangle_T &= \sum_{i=0}^{k} \int_0^T g_t^i d\langle \tilde{S}^i, R\rangle_t & (9.41) \\ &= 0 & (9.42) \end{aligned}$$

so that R_t^2 is a Q-martingale. It follows that

$$R = R_T = 1 \quad a.s.\ Q.$$

This shows that $Q^0 = Q$ and hence $Q^1 = Q$. Thus $\mathcal{M}(P)$ is a singleton.

For the other part, suppose that $\mathcal{M}(P)$ is a singleton. To show that the market is complete, it suffices to show that every bounded contingent claim is attainable, in view of Theorem 9.6. Let X be a bounded contingent claim, $|X| \le K$. Using Theorem 3.5 and writing $X_t = E^Q[X|\mathcal{F}_t^S]$, we have

$$X_t = E^Q(X) + \int_0^t \sum_{i=1}^{k} f_u^i d\tilde{S}_u^i + Z_t, \tag{9.43}$$

where Z_t is a martingale such that $E^Q(Z_t) = 0$ and

$$\langle \tilde{S}^i, Z\rangle_t = 0. \tag{9.44}$$

Let

$$\sigma_m = \inf\left\{t \ge 0 : \left|\int_0^t \sum_{i=1}^{k} f_u^i d\tilde{S}_u^i\right| \ge m\right\} \wedge T.$$

Then the continuity of $\tilde{S}^i$ and, as a consequence, the continuity of the stochastic integral appearing above imply that the martingale $Z_t^m = Z_{t\wedge\sigma_m}$ is bounded by $K^1 = 2K + m$. Further, for every $m \ge 1$

$$\langle \tilde{S}^i, Z^m\rangle_t = 0, \tag{9.45}$$

and hence $\tilde{S}_t^i Z_t^m$ is a Q-martingale. Let $R_t = \frac{1}{K^1}(K^1 + Z_t^m)$. Then it follows that $\mathbb{E}^Q(R_t) = 1$ and $\tilde{S}_t^i R_t$ is a Q-martingale. Arguments in (9.39) and (9.40) above yield that $\tilde{S}_t^i$ is a Q^0-local martingale where

$$Q^0(B) = \int_B R_T dQ.$$

Thus, $Q^0 \in \mathcal{M}(P)$ and hence $Q^0 = Q$ since $\mathcal{M}(P)$ is assumed to be a singleton. This implies $R = 1$ *a.s.* Q and so $Z^m = 0$ *a.s.* Q.

This holds for all m and it follows that

$$X_t = E^Q(X) + \int_0^t \sum_{i=1}^{k} f_u^i d\tilde{S}_u^i. \tag{9.46}$$

This completes the proof. □

9.5 Incomplete markets

In general, given a semimartingale X, $\mathcal{M}(P)$ may have more than one element. In this case, the market is called incomplete since there will exist claims that are not attainable.

In connection with their work on minimizing the risk involved in using an equivalent martingale measure, Föllmer and Schweitzer defined what they call a minimal EMM, which is a uniquely determined element $\hat{P}$ of $\mathcal{M}(P)$ (Föllmer and Schweizer, 1991). In addition to the risk minimality property, this enables one to choose a unique $\hat{P} \in \mathcal{M}(P)$. So while we do not discuss the risk minimality the ability to choose a unique EMM is important and hence we will consider this question here.

We start with the setup of Section 5.2 and consider a continuous P-semimartingale (X_t). We assume that $\mathcal{M}(P)$ is non-empty. If $X_t = X_0 + M_t + A_t$ is the canonical decomposition of X, then we saw from Section 5.2 (Theorem 5.6) that

$$A_t^j = \sum_{i=1}^d \int_0^t h_s^i d[M^i, M^j]_s, \tag{9.47}$$

where h^j are predictable processes satisfying

$$\int_0^T \sum_{i=1}^d |h_s^i|^2 d\left[M^i, M^i]_s\right] < \infty \ a.s. \tag{9.48}$$

Let ρ_t be defined by

$$\rho_t = \exp\left\{\sum_{j=1}^d - \int_0^t h_s^j dM_s^j - \frac{1}{2}\sum_{j=1}^d \int_0^t |h_s^j|^2 d[M^j, M^j]_s\right\}. \tag{9.49}$$

Suppose that $\mathbb{E}[\rho_T] = 1$ and define Q^* by $dQ^* = \rho_T dP$. Then, writing $X_t = (X_t^1, \ldots, X_t^d)$, we have

$$d(X_t^i \rho_t) = \rho_t \, dX_t^i + X_t^i \, d\rho_t + d[X^i, \rho]_t.$$

Noting that ρ_t satisfies

$$\rho_t = 1 - \sum_{j=1}^d \int_0^t \rho_s h_s^j dM_s^j \tag{9.50}$$

and using (9.47), we have

$$[X_i, \rho]_t = -\sum_{j=1}^d \int_0^t \rho_s h_s^j \, d[M^i, M^j]_s.$$

Hence

$$\begin{aligned}
d(X_t^i\rho_t) &= \rho_t dX_t^i - X_t^i\Big(\sum_{j=1}^d \rho_t h_t^j dM_t^j\Big) - \sum_{j=1}^d \rho_t h_t^j\, d[M^i, M^j]_t \\
&= \rho_t dM_t^i + \sum_{i=1}^d \rho_t h_t^i d[M^i, M^j]_t - X_t^i\Big(\sum_{j=1}^d \rho_t h_t^j dM_t^j\Big) \\
&\quad - \sum_{j=1}^d \rho_t h_t^j d[M^i, M^j]_t \\
&= \rho_t(1 - h_t^i X_t^i)\, dM_t^i - \sum_{j\neq i} \rho_t h_t^j X_t^j\, dM_t^j.
\end{aligned}$$

Hence $X_t^i\rho_t$ is a P-local martingale and thus X_t^i is a Q^*-local martingale for each i; that is, X_t is an $\mathbb{R}^d$-valued Q^*-local martingale, proving that $Q^* \in \mathcal{M}(P)$.

Let $\mathcal{H}$ be the class of random variables Z that can be written as $Z = c + \{\sum_{j=1}^d \int_0^T f_s^j dM_s^j$, where c is a constant and (f^j) are predictable processes that satisfy $\int_0^T \sum_{i=1}^d |f_s^i|^2 d[M^i, M^i]_s < \infty$. From (9.50) it follows that $\rho_T \in \mathcal{H}$. The next result shows that Q^* is the only measure in $\mathcal{M}(P)$ whose density belongs to $\mathcal{H}$.

Theorem 9.8 *Let X be a continuous P-semimartingale with canonical decomposition $X_t = X_0 + M_t + A_t$. Suppose $\mathcal{M}(P)$ is non-empty and $\mathbb{E}[\rho_T] = 1$, where ρ_t is defined by (9.48)–(9.49).*

Let $Q \in \mathcal{M}(P)$ and let $L_t = \mathbb{E}[\frac{dQ}{dP}|\mathcal{F}_t]$. Then L admits a decomposition

$$L_t = \rho_t R_t,$$

where R is a local martingale with the property $[R, M^j] = 0$ for $1 \leq j \leq d$.

Further, Q^*, defined by $\frac{dQ^*}{dP} = \rho_T$, is the only $Q \in \mathcal{M}(P)$ such that

$$\frac{dQ}{dP} \in \mathcal{H}. \tag{9.51}$$

Proof Let Q be any EMM ($Q \in \mathcal{M}(P)$) and let

$$L_t = \mathbb{E}_P[\frac{dQ}{dP}|\mathcal{F}_t]. \tag{9.52}$$

Let

$$V_t = \int_0^t (L_{s-})^{-1} dL_s. \tag{9.53}$$

As in Theorem 5.6, we can decompose V as

$$V_t = Y_t + Z_t,$$

where $Y \in \mathcal{H}$ is given by

$$Y_t = \sum_{j=1}^{d} \int_0^t g_s^j dM_s^j$$

and Z is a P-local martingale such that $M_t^j Z_t$ is a P-local martingale for $1 \leq i \leq d$. Thus $[M^j, Z]_t = 0$. As in the proof of Theorem 5.6, we can prove that

$$A_t^j = -\sum_{i=1}^{d} \int_0^t g_s^i d[M^i, M^j]_s \tag{9.54}$$

and hence, using (9.47), it follows that for $1 \leq j \leq d$,

$$B_t^j := \sum_{i=1}^{d} \int_0^t (g_s^i + h_s^i) d[M^i, M^j]_s = 0. \tag{9.55}$$

Let $N = \sum_{j=1}^d \int_0^t (g_s^j + h_s^j) dM_s^j$. Then

$$\begin{aligned}
[N, N]_t &= \sum_{j=1}^{d} \sum_{i=1}^{d} \int_0^t (g_s^j + h_s^j)(g_s^i + h_s^i) d[M^i, M^j]_s \\
&= \sum_{j=1}^{d} \int_0^t (g_s^j + h_s^j) dB_s^j \\
&= 0.
\end{aligned} \tag{9.56}$$

Thus, Y also can be written as $Y = -\sum_{j=1}^d \int_0^t h_s^j dM_s^j$ and hence

$$\rho_t = 1 + \int_0^t \rho_{s-} dY_s.$$

By Ito's formula, (used for $f(x) = x^{-1}$)

$$\begin{aligned}
d\rho_t^{-1} &= = (-1)\rho_t^{-2} d\rho_t + \rho_t^{-3} d[\rho, \rho]_t \\
&= (-1)\rho_t^{-2} \rho_t dY_t + \rho_t^{-3} \rho_t^2 d[Y, Y] \\
&= (-1)\rho_t^{-1} dY_t + \rho_t^{-1} d[Y, Y].
\end{aligned} \tag{9.57}$$

Let $R_t = L_t(\rho_t)^{-1}$. Using

$$L_t = 1 + \int_0^t L_{s-} dV_s \tag{9.58}$$

and (9.57) and Itô's formula (this time for $f(x_1, x_2) = x_1 x_2$), we get

$$\begin{aligned}
d(L_t \rho_t^{-1}) &= L_t d\rho_t^{-1} + \rho_t^{-1} dL_t + d[L, \rho^{-1}]_t \\
&= -L_{t-}\rho_t^{-1} dY_t + L_{t-}\rho_t^{-1} d[Y, Y]_t + L_{t-}\rho_t^{-1} dV_t \qquad (9.59) \\
&\quad - L_{t-}\rho_t^{-1} d[V, Y]_t \\
&= L_{t-}\rho_t^{-1} dZ_t + L_{t-}\rho_t^{-1} d[Y, Y]_t - L_{t-}\rho_t^{-1} d[V, Y]_t \\
&= L_{t-}\rho_t^{-1} dZ_t.
\end{aligned} \tag{9.60}$$

Here we have used $V = Y + Z$ and the fact that $[Y, Z]_t = 0$ so that $[V, Y]_t = [Y, Y]_t$. We thus conclude that $R_t = L_t \rho_t^{-1}$ satisfies

$$dR_t = R_{t-} dZ_t.$$

Since $[M^j, Z]_t = 0$ for all j, it follows that $[M^j, R]_t = 0$ as well.

Now if Q is such that $\frac{dQ}{dP} \in \mathcal{H}$, then U, defined via (9.52)–(9.53) satisfies $U_t \in \mathcal{H}$ and hence $Z_t \in \mathcal{H}$. Since $[Z, M^i]_t = 0$ for $1 \leq i \leq d$, it follows that $Z_t = 0$. This proves $R_t = 1$ and $L_t = \rho_t$. Thus $Q = Q^*$ which proves the last assertion. □

Remark The measure Q^* in the theorem above is called the *minimal martingale measure*. It can also be characterized by the following property: *If N is a P-local martingale such that $[X^i, N]_t = 0$ for $1 \leq i \leq d$, then N is also a Q^*-local martingale*. Here the property $[X^i, N]_t = 0$ is equivalent to $[M^i, N]_t = 0$, where M^i is the local martingale part of X^i. The last statement is, in turn, equivalent to $M_t^i N_t$ being a P-local martingale. Thus, the change from P to Q^* is minimal in the sense that while it changes the semimartingales X_t^i, $1 \leq i \leq d$, into local martingales, it does not alter the local martingale character of any local martingale that is *orthogonal* to M^i (i.e., $[M^i, N]_t = 0$), $1 \leq i \leq d$.

In an incomplete market, prices of claims that are not attainable cannot be uniquely determined via Theorem 9.2. However, the requirement that the augmented market, which consists of the bond, the k-stocks and a contingent claim, must be free of arbitrage opportunities does puts bounds on the price of the contingent claim at time $t = 0$.

Let us consider a contingent claim Z_T and let $\tilde{Z}_T = Z_T (S_T^0)^{-1}$ be its discounted value of the claim at time zero. Let us fix an EMM Q. Let $\mathcal{A}^+$ be the set of x such that there exists a trading strategy $\pi = (x, \pi^1, \ldots, \pi^k)$ such that $\tilde{V}(x, \theta)_t$ is a Q-martingale and

$$\tilde{V}(\pi)_T \geq \tilde{Z}_T \ a.s. \ P. \tag{9.61}$$

Let $\mathcal{A}^-$ be the set of y such that there exists a trading strategy $\bar{\pi} = (y, \bar{\pi}^1, \ldots, \bar{\pi}^k)$ and such that $\tilde{V}(\bar{\pi})_t$ is a Q-martingale and

$$\tilde{V}(\bar{\pi})_T \leq \tilde{Z}_T \ a.s. \ P. \tag{9.62}$$

Let

$$x^+ = \inf \mathcal{A}^+ \tag{9.63}$$

and

$$x^- = \sup \mathcal{A}^+. \tag{9.64}$$

With these notations we have the following result:

Theorem 9.9 *In order that the augmented market, consisting of the bond, the k-stocks and the contingent claim Z_T, be free of arbitrage opportunities, the price p of the contingent claim Z_T must satisfy*

$$x^- \leq p \leq x^+. \tag{9.65}$$

Proof Suppose that the contingent claim is traded at price p and that the augmented market satisfies NA. The proof is essentially contained in the proof of Theorem 9.2. The proof there shows that if $x \in \mathcal{A}^+$ is such that $x < p$, then there exists an arbitrage opportunity. Also, $x \in \mathcal{A}^-$ and $x > p$ lead to the existence of an arbitrage opportunity. Hence we have $x^+ \geq p$ and $x^- \leq p$. It remains to prove that the bounds are consistent—namely

$$x^- \leq x^+.$$

To see this, note that if π and $\bar{\pi}$ satisfy (9.61) and (9.62), respectively, and $\tilde{V}_t(\pi)$ and $\tilde{V}_t(\bar{\pi})$ are Q-martingales, then

$$\mathbb{E}[\tilde{V}_T(\pi)] = x$$

and

$$\mathbb{E}[\tilde{V}_T(\bar{\pi})] = y.$$

Together with (9.61) and (9.62), these observations give us

$$y \leq \mathbb{E}_Q[\tilde{Z}_T] \leq x.$$

This holds for all $x \in \mathcal{A}^+$ and all $y \in \mathcal{A}^-$ and hence

$$x^- \leq \mathbb{E}_Q[\tilde{Z}_T] \leq x^+ \tag{9.66}$$

This completes the proof. □

So if a claim is not attainable, its price must be within the bounds given above.

9.6 Completeness and underlying filtration

It is customary in many treatments of stochastic calculus to treat the filtration as a mere technicality. This is all right if the only interest is to define the stochastic integral. However, in the application to mathematical finance (as in the application to stochastic filtering theory and stochastic control theory), one must keep track of the observables in the model, and then all trading strategies (filter or estimate in stochastic filtering theory and control or action in stochastic control theory) have to be adapted to the filtration generated by the observations.

In the context of option pricing (and mathematical finance), it is very important to keep track of observable processes and ensure that the trading strategies under consideration are predictable for the filtration generated by the observable random variables. This would ensure that any strategy that we come up with can actually be implemented by an investor.

Just as we do not allow processes that anticipate the future to be used as trading strategies, we cannot allow trading strategies that are predictable for a filtration that is larger than the one generated by the observables. And since the notion of completeness, which plays a very important role in the theory, depends crucially on the class of trading strategies, it also depends on the choice of filtration. We elaborate on this theme because in the literature the filtration seems to have been treated as a mere technicality.

Suppose that the stock price processes $(S_t^1, \dots, S_t^k)$ are semimartingales on $(\Omega, \mathcal{F}, P)$ with respect to a filtration $(\mathcal{F}_t)$. Let the smallest filtration with respect to which $(S_t^1, \dots, S_t^k)$ are adapted be $(\mathcal{G}_t)$. It is possible that the market is complete if one allows all $(\mathcal{F}_t)$-predictable trading strategies but may not be complete if one only allows $(\mathcal{G}_t)$-predictable trading strategies. One can consider a situation where the observed information is more than the information furnished by the stock prices under consideration. However, in such situations, one has to stipulate this explicitly.

The following model has been immensely popular in the literature (Karatzas, 1997; Karatzas and Shreve, 1988; Levental and Skorohod, 1995; Duffie, 1992). It consists of a bond $S_t^0 = \exp\{rt\}$ and d-stocks given by the stochastic differential equation

$$dS_t^i = S_t^i\{\sum_{i=1}^{d} \sigma_{ij} dW_t^j + \mu_i dt\} \quad 1 \le i \le d, \tag{9.67}$$

where (σ_{ij}) is now a $d \times d$ non-singular matrix-valued $(\mathcal{F}_t)$ adapted process and (μ_i) is an $\mathbb{R}^d$ $(\mathcal{F}_t)$ adapted process. $(\mathcal{F}_t)$ is the smallest filtration with respect to which $(W_t^1, \dots, W_t^d)$ are adapted. σ is usually assumed to be positive definite, and suitable boundedness or integrability assumptions are imposed on σ and μ. In view of the integral representation theorem (Theorem 9.4), the market is complete *if we allow all $(\mathcal{F}_t)$-predictable processes as trading strategies.* However, it would not be reasonable to assume that $(W_t^1, \dots, W_t^d)$ are observed. Thus the only realistic case where the market would turn out to be complete would be when the filtration, generated by the driving Brownian motion $(W_t^1, \dots, W_t^d)$, is the same as the filtration generated by the stock price processes $(S_t^1, \dots, S_t^d)$. (The situation is analogous to the situation in stochastic control theory, where the controls are required to be functionals of the observed state (process) and functionals of the driving noise itself are not allowed as controls.) All the results on this generalized diffusion model are derived using the completeness of the market. Thus while the results derived are mathematically correct, applicability of these is doubtful at the level of generality at which they are considered. This includes the results on the price of the options—European and American in (Karatzas, 1997; Karatzas and Shreve, 1988) and the characterization of the absence of the arbitrage property (NA) in (Levental and Skorohod, 1995). If one assumes that the functional σ_{ij} is of Markovian type, i.e.,

$$\sigma_{ij}(u,\omega) = \bar{\sigma}_{ij}(u, S_u^1(\omega), S_u^2(\omega), \dots, S_u^d(\omega)),$$

and if $\bar{\sigma}_{ij}(u, x^1, x^2, \dots, x^d)$ is assumed to be positive definite for all $(u, x^1, x^2, \dots, x^d)$, then it follows that

$$\sigma\{(W_s^1, \dots, W_s^d) : 0 \le s \le t\} = \sigma\{(S_s^1, \dots, S_s^d) : 0 \le s \le t\}.$$

To conclude, we would like to stress that in modeling stock prices and available trading strategies, the underlying σ-fields should not be treated merely as a technicality that one has to use in order to apply results from stochastic calculus. A proper choice of the σ-fields that corresponds to the information available to an investor is essential if the results are to be relevant.

10
Black and Scholes Theory

In this chapter, we present the work of Black–Scholes and derive the Black–Scholes option pricing formula that follows their argument. While we have essentially derived the formula in the previous chapter, this approach has the added advantage that we can explicitly compute the hedging strategy. We also consider the diffusion model of the stock prices and obtain the price of the option via the Feynman–Kac formula.

10.1 Preliminaries

We consider a market consisting of one stock and one bond. The stock price process S is assumed to be a geometric Brownian motion (G.B.M.) such that

$$S_t = S_0 \exp\{\sigma W_t + (\mu - \frac{1}{2}\sigma^2)t \quad (0 \leq t \leq T)$$

or, in Itô's differential form,

$$dS_t = \mu S_t dt + \sigma S_t dW_t, \quad S_0 > 0.$$

Here, $\mu \in \mathbb{R}$ is the average rate of return (since $\mathbb{E}[S_t] = S_0 e^{\mu t}$) and σ is the volatility. Suppose that the price of bond at time t is given by $B_t = e^{rt}$, where r is being the interest rate and assumed to be a constant.

Earlier, we have seen that the market, consisting of the bond and the stock, does not admit arbitrage opportunities and is complete; thus the price of every claim, including the European call option, is determined uniquely. Here we will derive the hedging strategy for the option.

Recall that a trading strategy consists of a pair (π^0, π^1) of predictable processes, where π_t^0 is the number of bonds and π_t^1 is the number of stocks the investor holds at time t. The value of the portfolio corresponding to the strategy (π^0, π^1) at time t is represented by $V_t(\pi^0, \pi^1)$ and is given by

$$V_t(\pi^0, \pi^1) = \pi_t^0 B_t + \pi_t^1 S_t.$$

The gain $G_t(\pi^0, \pi^1)$, accrued to the investor via the strategy (π^0, π^1) up to time t, is given by

$$\int_0^t \pi_u^0 dB_u + \int_0^t \pi_u^1 dS_u.$$

The strategy (π^0, π^1) is said to be self-financing if no fresh investment is made at any time $t > 0$ and there is no consumption. In such a case, the value equals the gain plus the initial investment, so that for a self-financing strategy (π^0, π^1),

$$V_t(\pi^0, \pi^1) = V_0(\pi^0, \pi^1) + \int_0^t \pi_u^0 dB_u + \int_0^t \pi_u^1 dS_u. \tag{10.1}$$

From now on we only consider self-financing strategies.

Suppose that the call option (with terminal time T and striking price K) is itself traded in the market. It is reasonable to expect that the price p_t of the option at any time depends upon (i) the current price of the stock and (ii) the time until expiration $T - t$ (equivalently on t). Thus, the price p_t can be expressed as

$$p_t = P(t, S_t)$$

for a suitable function P. We require that the option is valued in such a manner that the augmented market, consisting of the bond B_t, the stock S_t and the option p_t, does not admit arbitrage oportunities. We want to see if a hedging strategy exists and obtain it explicitly. Recall that a hedging strategy is a strategy for which the value (at time T) equals the worth of the option.

Lemma 10.1 *Suppose (π^0, π^1) is a strategy such that*

$$V_T(\pi^0, \pi^1) = P(T, S_T) = (S_T - K)^+. \tag{10.2}$$

Then

$$V_0(\pi^0, \pi^1) = P(0, S_0). \tag{10.3}$$

and for all t, $0 \leq t \leq T$,

$$V_t(\pi^0, \pi^1) = P(t, S_t). \tag{10.4}$$

Proof For simplicity of presentation, we first prove the case $t = 0$. Suppose $V_0 := V_0(\pi^0, \pi^1) < P(0, S_0)$. Let $z = P(0, S_0) - V_0 > 0$. Define a new strategy $(\bar{\pi}^0, \bar{\pi}^1, \bar{\pi}^2)$ in the augmented market, consisting of B_t, S_t, p_t, given by

$$\bar{\pi}^0 = \pi^0 + z \quad \bar{\pi}^1 = \pi^1 \quad \bar{\pi}^2 = -1.$$

The value function of this strategy $\bar{V}_t = V_t(\bar{\pi}^0, \bar{\pi}^1, \bar{\pi}^2)$ is given by

$$\begin{aligned}
\bar{V}_t &= \bar{\pi}_t^0 B_t + \bar{\pi}_t^1 S_t + \bar{\pi}_t^2 p_t \\
&= zB_t + \pi_t^0 B_t + \pi_t^1 S_t - p_t \\
&= z + \int_0^t z dB_u + \pi_0^0 B_0 + \pi_0^1 S_0 + \int_0^t \pi_u^0 dB_u \\
&\quad + \int_0^t \pi_u^1 dS_u - \int_0^t dp_u - p_0 \\
&= (z + \pi_0^0 B_0 + \pi_0^1 S_0 - p_0) + \int_0^t \bar{\pi}_u^0 dB_u + \int_0^t \bar{\pi}_u^1 dS_u + \int_0^t \bar{\pi}_u^2 dp_u \\
&= \bar{V}_0 + \int_0^t \bar{\pi}_u^0 dB_u + \int_0^t \bar{\pi}_u^1 dS_u + \int_0^t \bar{\pi}_u^2 dp_u.
\end{aligned}$$

Here we have used the fact that (π^0, π^1) is a self-financing strategy. The calculations given above prove that $(\bar{\pi}^0, \bar{\pi}^1, \bar{\pi}^2)$ is a self-financing strategy as well. By the choice of z, we have that $\bar{V}_0 = 0$, and since $V_T = p_T$,

$$\bar{V}_T = z.$$

Thus $(\bar{\pi}^0, \bar{\pi}^1, \bar{\pi}^2)$ is an arbitrage opportunity. Hence $z = P(0, S_0) - V_0$ cannot be strictly positive. Similarly, we can show that if $P(0, S_0) < V_0$, one has an arbitrage. Thus $V_0 = P(0, S_0)$. This proves the first part.

We now come to the case $t > 0$. Let $Z_t = P(t, S_t) - V_t$ and let $C = \{Z_t > 0\}$ and let $D = \{Z_t < 0\}$. Define a new strategy $(\bar{\pi}^0, \bar{\pi}^1, \bar{\pi}^2)$ in the augmented market consisting of B_t, S_t, p_t, given by $\bar{\pi}_u^0 = 0, \bar{\pi}_u^1 = 0, \bar{\pi}_u^2 = 0$ for $0 \le u \le t$, and for $t < u \le T$,

$$\begin{aligned}
\bar{\pi}_u^0 &= (1_C - 1_D)(\pi_u^0 + Z_t) \\
\bar{\pi}_u^1 &= (1_C - 1_D)\pi_u^1 \\
\bar{\pi}_u^2 &= -(1_C - 1_D).
\end{aligned}$$

Proceeding as in the case where $t = 0$, we can show that the strategy $(\bar{\pi}^0, \bar{\pi}^1, \bar{\pi}^2)$ is a self-financing strategy with zero initial investment and

$$\begin{aligned}
V_T(\bar{\pi}^0, \bar{\pi}^1, \bar{\pi}^2) &= (1_C - 1_D)Z_t \\
&= |Z_t|.
\end{aligned}$$

If $P(|Z_t| = 0) < 1$, $(\bar{\pi}^0, \bar{\pi}^1, \bar{\pi}^2)$ would be an arbitrage opportunity. Hence

$$V_t(\pi^0, \pi^1) = P(t, S_t).$$

10.2 The Black–Scholes PDE

We now look for a strategy (π^0, π^1) for which (10.3) holds. In view of Lemma 10.1 proved above, we can and do assume (10.4) which is a necessary condition. For now, assume that

$$P(t,x) \in C^{1,2}([0,T) \times \mathbb{R}).$$

The idea is to write V_t and $p_t = P(t, S_t)$ as Itô differentials:

$$\begin{aligned} dV_t &= a_t dt + b_t dW_t, \\ dp_t &= \alpha_t dt + \beta_t dW_t \end{aligned}$$

and then to set $a_t = \alpha_t, b_t = \beta_t$. (Actually, we can only conclude that $\int_0^T |a_t - \alpha_t| dt = 0$ *a.s.* and $\int_0^T |b_t - \beta_t|^2 dt = 0$ *a.s.*. We ignore this for now, until we come to the next theorem.) Now

$$\begin{aligned} dV_t &= \pi_t^1 dS_t + \pi_t^0 dB_t \\ &= \pi_t^1(\mu S_t dt + \sigma S_t dW_t) + r\pi_t^0 B_t dt \\ &= (\mu\pi_t^1 S_t + r\pi_t^0 B_t)dt + \sigma\pi_t^1 S_t dW_t, \end{aligned} \tag{10.5}$$

using the fact that $dB_t = rB_t dt, r > 0$. By Ito's formula we get

$$\begin{aligned} dp_t &= \left\{ \tfrac{\partial P}{\partial t}(t,S_t) + \mu \tfrac{\partial P}{\partial x}(t,S_t)S_t + \tfrac{1}{2}\sigma^2 S_t^2 \tfrac{\partial^2 P}{\partial x^2}(t,S_t) \right\} dt \\ &\quad + \sigma \tfrac{\partial P}{\partial x}(t,S_t) S_t dW_t. \end{aligned} \tag{10.6}$$

Hence we get (remember that $S_t > 0$ *a.s.*)

$$\pi_t^1 = \frac{\partial P}{\partial x}(t, S_t) \tag{10.7}$$

and

$$\mu\pi_t^1 S_t + r\pi_t^0 B_t = \frac{\partial P}{\partial t}(t,S_t) + \mu \frac{\partial P}{\partial x}(t,S_t)S_t + \frac{1}{2}\sigma^2 S_t^2 \frac{\partial^2 P}{\partial x^2}(t,S_t). \tag{10.8}$$

Substituting for π^1 from (10.7) into (10.8), we can cancel the term involving μ to conclude that

$$r\pi_t^0 B_t = \frac{\partial P}{\partial t}(t,S_t) + \frac{1}{2}\sigma^2 S_t^2 \frac{\partial^2 P}{\partial x^2}(t,S_t). \tag{10.9}$$

Next, from the equations $\pi_t^1 S_t + \pi_t^0 B_t = V_t$ and $V_t = P(t, S_t)$ (see 10.4), we have

$$\pi_t^0 B_t = \{P(t,S_t) - \frac{\partial P}{\partial x}(t,S_t)S_t\}$$

or

$$\pi_t^0 = \frac{1}{B_t}\left\{P(t,S_t) - \frac{\partial P}{\partial x}(t,S_t)S_t\right\}. \tag{10.10}$$

Substituting from (10.10) into (10.9), we obtain the equation

$$\frac{\partial P}{\partial t} + \frac{1}{2}\sigma^2 S_t^2 \frac{\partial^2 P}{\partial x^2} + r S_t \frac{\partial P}{\partial x}(t, S_t) - rP(t, S_t) = 0, \quad 0 < t < T$$

with $P(T, S_T) = (S_T - K)^+$. Thus, the option price process $P(t, S_t)$ is the solution to the boundary value problem:

$$\begin{aligned} \frac{\partial P}{\partial t} + \frac{1}{2}\sigma^2 x^2 \frac{\partial^2 P}{\partial x^2} + rx\frac{\partial P}{\partial x} - rP(t,x) &= 0, \quad 0 < t < T \quad (10.11) \\ P(T,x) &= (x-K)^+. \quad (10.12) \end{aligned}$$

The partial differential equation (PDE) (10.11) is the Black and Scholes PDE and is a special case of the backward parabolic equation introduced in Section 4.3. We thus have identified the hedging strategy (π^0, π^1) modulo a technical assumption. We have also ignored null sets in the preceding discussion. In the next section, we will show that the boundary value problem (10.11)–(10.12) admits a unique solution $P(t, x)$, and then we will go on to prove that (π^1, π^0) defined by (10.10) and (10.7), respectively, constitutes a self-financing hedging strategy. We will also show that this strategy is an admissible strategy (see Chapter 7). Finally, to obtain (π^1, π^0) and the initial investment required for $V_0(\pi^1, \pi^0)$ explicitly, we need an explicit form for $P(t, x)$.

10.3 Explicit solution of the Black–Scholes PDE

We follow the original paper of Black and Scholes to obtain the solution of the Cauchy problem (10.11)–(10.12) in closed form and also show that it is unique. The idea is to reduce (10.11)–(10.12) to the heat equation with a new boundary condition and apply Theorem 2.11.

Introduce new independent variables τ, ζ by

$$\tau = \gamma(T-t), \quad \zeta = \alpha\left\{\log\frac{x}{K} + \beta(T-t)\right\}, \tag{10.13}$$

where α, β, γ are constants to be chosen later. Define the function $y(\tau, \zeta)$ by

$$P(t,x) = e^{-r(T-t)} y(\tau, \zeta). \tag{10.14}$$

Then

$$\begin{aligned} \frac{\partial P}{\partial t} &= re^{-r(T-t)}y + e^{-r(T-t)}\frac{\partial y}{\partial t}; \\ \frac{\partial y}{\partial t} &= \frac{\partial y}{\partial \zeta}\frac{\partial \zeta}{\partial t} - \frac{\partial y}{\partial \tau}\gamma = -\alpha\beta\frac{\partial y}{\partial \zeta} - \gamma\frac{\partial y}{\partial \tau}. \end{aligned}$$

Hence

$$\frac{\partial P}{\partial t} = re^{-r(T-t)}y - \alpha\beta e^{-r(T-t)}\frac{\partial y}{\partial \zeta} - \gamma e^{-r(T-t)}\frac{\partial y}{\partial \tau}. \tag{10.15}$$

Also,

$$\begin{aligned}\frac{\partial P}{\partial x} &= e^{-r(T-t)}\frac{\partial y}{\partial x} = e^{-r(T-t)}\frac{\partial y}{\partial \zeta}\frac{\partial \zeta}{\partial x}\\ &= e^{-r(T-t)}\frac{\partial y}{\partial \zeta}\frac{\alpha}{x}.\end{aligned}$$

Hence

$$x\frac{\partial P}{\partial x} = \alpha e^{-r(T-t)}\frac{\partial y}{\partial \zeta}, \tag{10.16}$$

noting that

$$\begin{aligned}\frac{\partial^2 P}{\partial x^2} &= e^{-r(T-t)}\frac{\partial}{\partial x}\left[\frac{\partial y}{\partial \zeta}\frac{\partial \zeta}{\partial x}\right]\\ &= e^{-r(T-t)}\left\{\frac{\partial}{\partial x}\left(\frac{\partial y}{\partial \zeta}\right)\frac{\partial \zeta}{\partial x} + \frac{\partial y}{\partial \zeta}\frac{\partial^2 \zeta}{\partial x^2}\right\}\\ &= e^{-r(T-t)}\left\{\frac{\partial^2 y}{\partial \zeta^2}(\frac{\partial \zeta}{\partial x})^2 + \frac{\partial y}{\partial \zeta}\frac{\partial^2 \zeta}{\partial x^2}\right\}.\end{aligned}$$

Substitute

$$\frac{\partial \zeta}{\partial x} = \frac{\alpha}{x}, \quad \frac{\partial^2 \zeta}{\partial x^2} = -\frac{\alpha}{x^2}$$

in the above to get

$$\frac{\partial^2 P}{\partial x^2} = e^{-r(T-t)}\left\{\frac{\alpha^2}{x^2}\frac{\partial^2 y}{\partial \zeta^2} - \frac{\alpha}{x^2}\frac{\partial y}{\partial \zeta}\right\};$$

so

$$x^2\frac{\partial^2 P}{\partial x^2} = e^{-r(T-t)}\left\{\alpha^2\frac{\partial^2 y}{\partial \zeta^2} - \alpha\frac{\partial y}{\partial \zeta}\right\}. \tag{10.17}$$

Then we have

$$\begin{aligned}&\frac{\partial P}{\partial t} + \frac{1}{2}\sigma^2 x^2\frac{\partial^2 P}{\partial x^2} + rx\frac{\partial P}{\partial x} - rP =\\ &\quad re^{-r(T-t)}y - \alpha\beta e^{-r(T-t)}\frac{\partial y}{\partial \zeta} - \gamma e^{-r(T-t)}\frac{\partial y}{\partial \tau}\\ &\quad + \frac{\sigma^2}{2}e^{-r(T-t)}\left\{\alpha^2\frac{\partial^2 y}{\partial \zeta^2} - \alpha\frac{\partial y}{\partial \zeta}\right\} + r\alpha e^{-r(T-t)}\frac{\partial y}{\partial \zeta} - re^{-r(T-t)}y.\end{aligned} \tag{10.18}$$

Cancelling the common factor $e^{-r(T-t)}$ and using that P satisfies the PDE (10.11), we conclude that y satisfies

$$-\alpha\beta\frac{\partial y}{\partial \zeta} - \gamma\frac{\partial y}{\partial \tau} + \frac{\sigma^2}{2}\left\{\alpha^2\frac{\partial^2 y}{\partial \zeta^2} - \alpha\frac{\partial y}{\partial \zeta}\right\} + r\alpha\frac{\partial y}{\partial \zeta} = 0. \tag{10.19}$$

The coefficient of $\frac{\partial y}{\partial \zeta}$ in (10.19) is

$$-\alpha\beta - \frac{\alpha\sigma^2}{2} + r\alpha = \alpha\left[r - \beta - \frac{\sigma^2}{2}\right].$$

Choosing $\beta = r - \frac{\sigma^2}{2}$, this equals zero, and (10.19) becomes

$$-\gamma\frac{\partial y}{\partial \tau} + \frac{\sigma^2\alpha^2}{2}\frac{\partial^2 y}{\partial \zeta^2} = 0.$$

Now choose $\gamma = \sigma^2\alpha^2$ where (α is arbitrary $\neq 0$). The Black–Scholes PDE (10.11) is thus transformed into the heat equation

$$\frac{\partial y}{\partial \tau} = \frac{1}{2}\frac{\partial^2 y}{\partial \zeta^2}. \tag{10.20}$$

We need to see how the boundary condition $P(T, x) = (x - K)^+$ (10.12) is transformed. We have, setting $\alpha = 1$ (recall, we have chosen $\beta = (r - \frac{\sigma^2}{2})$ and $\gamma = \sigma^2\alpha^2$),

$$P(T, x) = y(0, \log\frac{x}{K}). \tag{10.21}$$

Thus,

$$y(0, u) = (Ke^u - K)^+. \tag{10.22}$$

Thus, we have to solve the Cauchy problem (initial value problem) for the heat equation,

$$\frac{\partial y}{\partial \tau} = \frac{1}{2}\frac{\partial^2 y}{\partial u^2}, \tag{10.23}$$

with the boundary condition

$$y(0, u) = K(e^u - 1)^+ \equiv f(u). \tag{10.24}$$

From Theorem 2.11, we conclude that the Cauchy problem (10.23)–(10.24) admits a unique solution given by

$$y(\tau, \zeta) = \frac{1}{\sqrt{2\pi}}\int_{-\infty}^{\infty} e^{\frac{-(\zeta - x)^2}{2\tau}} f(\zeta)d\zeta, \tag{10.25}$$

where

$$f(x) = K(e^x - 1)^+.$$

Clearly, f satisfies the condition $0 \leq f(x) < C_1 e^{ax^2}$ for constants $0 < a < \frac{1}{4T}$ and C_1, suitably chosen. Thus

$$y(\tau, \zeta) = \frac{K}{\sqrt{2\pi\tau}}\int_0^{\infty} e^{-\frac{(u-\zeta)^2}{2T}}(e^u - 1)\, du.$$

Putting $z = \frac{u-\zeta}{\sqrt{\tau}}$, we get

$$\begin{aligned} y(\tau,\zeta) &= \frac{K}{\sqrt{2\pi}}\int_{-\zeta/\sqrt{\tau}}^{\infty} e^{-\frac{z^2}{2}}\left(e^{\zeta+z\sqrt{\tau}}-1\right)dz \\ &= \frac{K}{\sqrt{2\pi}}\int_{-\zeta/\sqrt{\tau}}^{\infty} e^{-\frac{z^2}{2}+z\sqrt{\tau}+\zeta}dz - \frac{K}{\sqrt{2\pi}}\int_{-\zeta/\sqrt{\tau}}^{\infty} e^{-\frac{z^2}{2}}dz. \end{aligned} \tag{10.26}$$

The first integral in (10.27) equals

$$\begin{aligned} \frac{K}{\sqrt{2\pi}}\int_{-\zeta/\sqrt{\tau}}^{\infty} e^{-\frac{1}{2}(z-\sqrt{\tau})^2+\frac{1}{2}\tau+\zeta}dz &= Ke^{\zeta+\frac{1}{2}\tau}\frac{1}{\sqrt{2\pi}}\int_{-(\zeta/\sqrt{\tau}+\sqrt{\tau})}^{\infty} e^{-\frac{1}{2}v^2}dv \\ &= Ke^{\zeta+\frac{1}{2}\tau}\Phi\left(\frac{\zeta}{\sqrt{\tau}}+\sqrt{\tau}\right). \end{aligned}$$

(Here, Φ is the distribution function of the standard normal distribution.) The second integral in (10.27) equals $K\Phi(\zeta/\sqrt{\tau})$. Hence

$$y(\tau,\zeta) = Ke^{\zeta+\frac{1}{2}\tau}\Phi\left(\frac{\zeta}{\sqrt{\tau}}+\sqrt{\tau}\right) - K\Phi\left(\frac{\zeta}{\sqrt{\tau}}\right). \tag{10.27}$$

Let us now substitute in terms of the original variables t, x. We have chosen $\alpha = 1$, $\beta = (r - \frac{\sigma^2}{2})$ and $\gamma = \sigma^2\alpha^2$. Thus

$$\begin{aligned} e^{\zeta+\frac{1}{2}\tau} &= e^{\log\frac{x}{K}+(r-\frac{\sigma^2}{2})(T-t)+\frac{\sigma^2}{2}(T-t)}, \\ &= \frac{x}{K}e^{r(T-t)}; \end{aligned}$$

and

$$\begin{aligned} \frac{\zeta}{\sqrt{\tau}}+\sqrt{\tau} &= \frac{\log\frac{x}{K}+(r-\frac{\sigma^2}{2})(T-t)}{\sigma\sqrt{T-t}} + \sigma\sqrt{T-t} \\ &= \frac{\log\frac{x}{K}+(r-\frac{\sigma^2}{2})(T-t)+\sigma^2(T-t)}{\sigma\sqrt{T-t}} \\ &= \frac{\log\frac{x}{K}+(r+\frac{1}{2}\sigma^2)(T-t)}{\sigma\sqrt{T-t}} \\ &= g(x, T-t), \end{aligned}$$

where

$$g(x,t) = \frac{\log\frac{x}{K}+(r+\frac{\sigma^2}{2})t}{\sigma\sqrt{t}}. \tag{10.28}$$

Next,

$$\begin{aligned} \frac{\zeta}{\sqrt{\tau}} &= g(x, T-t) - \sigma\sqrt{T-t} \\ &= h(x, T-t), \end{aligned}$$

where

$$h(x,t) = g(x,t) - \sigma\sqrt{t}. \tag{10.29}$$

From (10.14) and (10.27), we get

$$P(t,x) = e^{-r(T-t)}y(\tau,\zeta),$$

where

$$y(\tau,\zeta) = e^{r(T-t)}x\Phi[g(x,T-t)] - k\Phi[h(x,T-t)].$$

Hence

$$P(t,x) = x\Phi[g(x,T-t)] - Ke^{-r(T-t)}\Phi[h(x,T-t)] \tag{10.30}$$

is a solution to the Cauchy problem for the Black–Scholes equation. Thus we have proved the following:

Theorem 10.2 *The Black–Scholes PDE (10.11) with boundary condition (10.12) admits a unique solution P as defined by (10.30), where g, h are defined by (10.28),(10.29) and Φ is the distribution function of the standard normal distribution.*

10.4 The Black–Scholes formula

We have now proved that the Black–Scholes PDE (10.11) with boundary condition (10.12) admits a unique solution P. We will now prove that the strategy (π^0, π^1), defined by (10.7) and (10.10), is a self-financing trading strategy and is a hedging strategy for the call option. It is necessary to do this all over again because while deriving the PDE for P, we had ignored null sets in one of the steps.

The value function $V_t = V_t(\pi^0, \pi^1)$ is given by

$$\begin{aligned} V_t &= \pi_t^0 B_t + \pi_t^1 S_t \\ &= \{P(t,S_t) - \frac{\partial P}{\partial x}(t,S_t)S_t\} + \frac{\partial P}{\partial x}(t,S_t)S_t \\ &= P(t,S_t). \end{aligned} \tag{10.31}$$

The gain $G_t = G_t(\pi^0, \pi^1)$ is given by

$$G_t = \int_0^t \pi_u^0 dB_u + \int_0^t \pi_u^1 dS_u.$$

Using the fact the P satisfies the PDE (10.11) and Ito's formula, we get

$$\begin{aligned} G_t &= r\int_0^t \{P(s,S_s) - \frac{\partial P}{\partial x}(s,S_s)S_s\}ds + \int_0^t \frac{\partial P}{\partial x} dS_s \\ &= \int_0^t \frac{\partial P}{\partial x} dS_s + \int_0^t \left\{ \frac{\partial P}{\partial s} + \frac{1}{2}\sigma^2 S_s^2 \frac{\partial^2 P}{\partial x^2} \right\} ds. \\ &= P(t,S_t) - P(0,S_0). \end{aligned} \tag{10.32}$$

The equations (10.31) and (10.32) imply that $V_t = V_0 + G_t$, and hence the strategy (π^0, π^1) is self-financing. (10.31) for $t = T$ also implies that (π^0, π^1) is a hedging strategy. Thus we have proved the following fundamental result of Black and Scholes which they derived in 1973.

Theorem 10.3 *Consider the market consisting of one stock S and one bond B specified at the beginning of Section 10.1. Then the price p of the European call option with terminal time T and striking price K on the stock is given by*

$$p = P(0, S_0), \tag{10.33}$$

where P is the unique solution to the PDE (10.11) with boundary condition (10.12) and is given by (10.30).

Proof We have observed earlier that there exists a hedging strategy for the contingent claim $(S_T - K)^+$ which requires an initial investment $V_0 = P(0, S_0)$. Hence if the price p differs from $P(0, S_0)$, it would lead to an arbitrage opportunity. This completes the proof. □

The solution $P(t, x)$ of the Black–Scholes PDE can be expressed via the Feynman–Kac formula (Theorem 4.6) and applied to the PDE directly without transforming it into the heat equation.

From the definitions of g and h given above, it is easy to verify that

$$\sup_{0 \le t \le T} |P(t, x)| \le |x| + K. \tag{10.34}$$

It is easy to see from the expression for P that

$$P \in C([0, T] \times \mathbb{R}) \cap C^{1,2}((0, T) \times \mathbb{R}).$$

Thus $P(t, x)$ satisfies the conditions of Theorem 4.6. The operator A_t, in the present case, is given by $A_t = \frac{\partial}{\partial t} + \frac{1}{2}\sigma^2 x^2 \frac{\partial^2}{\partial x^2} + rx\frac{\partial}{\partial x}$. Hence the coefficients satisfy conditions of Theorem 4.6. The function $c(t, x) \equiv r > 0$, in this case, and $f(x) = (x - K)^+$. Thus Theorem 4.6 applies to the Black and Scholes Cauchy problem and we have

Theorem 10.4

$$P(t, x) = E_{P_{t,x}}\left[(X_T - K)^+ e^{-r(T-t)}\right], \tag{10.35}$$

where $P_{t,x}$ is the law of the solution $X^{t,x}$ of the SDE

$$X_s^{t,x} = x + \int_t^s rX_v^{t,x} dv + \sigma \int_t^s X_v^{t,x} dW_v. \tag{10.36}$$

10.5 Diffusion model

Let us consider Example 2 in Chapter 7. We consider a market consisting of k stocks, whose prices $S^1, S^2, \ldots, S_t^k$ are given by

$$dS_t^i = S_t^i \left\{ \sum_{j=1}^{d} \sigma^{ij}(t, S_t) dW_t^j + b^i(t, S_t) dt \right\}, \quad i = 1, \ldots, k,$$

where $S = (S^1, \ldots, S^k)$ and

$$a^{ij}(t, x) := \sum_m \sigma^{im}(t, x)\sigma^{jm}(t, x)$$

is assumed to be positive definite for each (t, x); $\sigma^{ij}(t, x)$, $b^i(t, x)$ are bounded Lipschitz continuous functions and $(W^1, W^2, \ldots, W^d)$ are independent Brownian motions. Under these conditions the SDE admits a unique solution (see Theorem 4.3). Suppose the bond price is given by $S_t^0 = \exp\{rt\}$. It was noted in Section 9.3 that this market is complete and hence by Theorem 9.2, the price p of the call option on S^1, with terminal time T and striking price K, is given by

$$p = \mathbb{E}_Q[(S_T^1 - K)^+ \exp\{-rT\}] \tag{10.37}$$

where Q is the EMM (See Section 9.3). Let us note here that

$$dS_t^i = S_t^i \left\{ \sum_{j=1}^{d} \sigma^{ij}(t, S_t) d\tilde{W}_t^j + r dt \right\} \quad i = 1, \ldots, k,$$

where $(\tilde{W}^1, \ldots, \tilde{W}^k)$ is a k-dimensional standard Brownian motion under Q.

Here it is not possible to obtain an explicit expression for p since the distribution of S^1 under Q cannot be written down explicitly. The integral in (10.37) can be evaluated by Monte–Carlo simulation. Alternatively, as in the case of the Black–Scholes model, we can consider the price $P(t, x)$ at time t, if the price of S_t then is x, and obtain a PDE for $P(t, x)$.

Consider the PDE

$$\frac{\partial}{\partial t}\hat{P}(t, x) + \frac{1}{2}\sum_{i,j=1}^{k} a^{ij}(t, x) x_i x_j \frac{\partial^2}{\partial x_i \partial x_j}\hat{P}(t, x) + \sum_{i=1}^{k} r x_i \frac{\partial}{\partial x_i}\hat{P}(t, x) = 0 \tag{10.38}$$

with boundary condition

$$\hat{P}(T, x) = (x_1 - K)^+. \tag{10.39}$$

From the Feynman–Kac formula (Theorem 4.6), it follows that if the Cauchy problem (10.38)–(10.39) admits a solution having at most polynomial growth, then

$$\hat{P}(0, S_0) = \mathbb{E}_Q[(S_T^1 - K)^+].$$

Defining

$$P(t,x) = \hat{P}(t,x)\exp\{-r(T-t)\},$$

we conclude that $P(t,x)$ would satisfy the PDE

$$\frac{\partial}{\partial t}P(t,x) + \frac{1}{2}\sum_{i,j=1}^{k} a^{ij}(t,x)x_i x_j \frac{\partial^2}{\partial x_i \partial x_j}P(t,x) \qquad (10.40)$$

$$+\sum_{i=1}^{k} r x_i \frac{\partial}{\partial x_i}P(t,x) - rP(t,x) = 0$$

with boundary condition

$$P(T,x) = (x_1 - K)^+. \qquad (10.41)$$

These remarks lead us to the following result:

Theorem 10.5 *Suppose that for* $1 \le i, j \le k$ *that*

$$\sigma^{ij}, \frac{\partial \sigma^{ij}}{\partial x_i}, \frac{\partial^2}{\partial x_i \partial x_j}\sigma^{ij}$$

are bounded Hölder continuous functions. Then we have the following:

1. *The Cauchy problem (10.40)–(10.41) admits a unique solution* P *in the class of* $C([0,T] \times \mathbb{R}^k) \cap C^{1,2}((0,T) \times \mathbb{R}^k$ *functions* v *having at most linear growth).*

2. *The price* p *of the European call option on* S^1 *with terminal time* T *and striking price* K *is given by*

$$p = P(0, S_0). \qquad (10.42)$$

Proof As in the one-dimensional case with constant coefficients considered earlier in the chapter, let us introduce variables $\tau, \zeta = (\zeta_1, \ldots, \zeta_k)$ and a function $y(\tau, \zeta)$ as follows:

$$\tau = (T-t) \quad \zeta_i = \log(x_i) \ \ 1 \le i \le k \qquad (10.43)$$

$$y(\tau,\zeta) = P(t,x)e^{-r(T-t)}. \qquad (10.44)$$

Also, let $A^{ij}(\tau,\zeta) = a^{ij}(t,x)$ where (τ,ζ) and (t,x) are related via (10.43). The Cauchy problem (10.40)–(10.41) for P can be recast as the following Cauchy problem for y:

$$\frac{\partial}{\partial \tau}y(\tau,\zeta) = \frac{1}{2}\sum_{i,j=1}^{k} A^{ij}(\tau,\zeta)\frac{\partial^2}{\partial \zeta_i \partial \zeta_j}y(\tau,\zeta) + \sum_{i=1}^{k}(r - A^{ii}(\tau,\zeta))\frac{\partial}{\partial \zeta_i}y(\tau,\zeta),$$

(10.45)

$$y(0, \zeta) = (e_1^{\zeta} - K)^+. \tag{10.46}$$

The assumptions on a^{ij} imply that the Cauchy problem (10.45)–(10.46) admits a unique solution in the class of functions having at most exponential growth. It follows that the Cauchy problem (10.41)–(10.41) admits a unique solution in the class of functions having linear growth.

The Feynman–Kac formula (Theorem 4.6) would now imply that

$$P(0, S_0) = \mathbb{E}_Q[(S_T^1 - K)^+ e^{-rT}].$$

This completes the proof, in view of the observation made earlier that the price of the option under consideration is given by (10.37). □

Remark As in the Black–Scholes model, it can be verified that $\theta = (p, \pi^1, \dots, \pi^k)$, where

$$\pi_t^i = \frac{\partial}{\partial x_i} P(t, S_t)$$

is a self-financing strategy such that the value function $V_t(\theta)$ satisfies

$$V_t(\theta) = P(t, S_t),$$

and thus θ is the hedging strategy for the European call option under consideration.

11

Discrete Approximations

In this chapter we first derive a discrete approximation to the Black and Scholes PDE as well as to the Feynman–Kac formula for its solution. We follow, in part, the work of Merton and Samuelson as presented in the book by (Merton, 1990), and, also we follow the treatment of the binomial option pricing formula introduced in (Duffie, 1992) for the purpose of working out numerical approximations.

11.1 The binomial model

Let us consider a model for a stock (S_n) in discrete time, $n = 1, \dots, N$ given by

$$S_{n+1} = \xi_{n+1} S_n, \quad 1 \leq n \leq N$$

where $\xi_1, \xi_2, \dots, \xi_N$ are independent random variables and taking two values, U and D, with

$$P(\xi_1 = U) = p, \; P(\xi_1 = D) = 1 - p$$

and $0 < p < 1$. Thus, the distribution of S_n is given by

$$P(S_n = S_0 U^k D^{n-k}) = {}^n C_k \, p^k (1-p)^{n-k},$$

hence the name *binomial model.*

Suppose that there is a bond whose price at time n is

$$B_n = R^n$$

so that R is the rate of return ($R - 1$ is the rate of interest) per unit time. Assume that U, D, R satisfy

$$D < R < U.$$

It is easy to see that the market consisting of the bond B_n and stock S_n satisfies the no arbitrage property. Further, by Theorem 6.3, the market is complete and admits a unique equivalent martingale measure. Thus, for the European call option on S_n with terminal time N and striking price K, there exists a hedging strategy. We will compute the hedging strategy explicitly below. In order to treat call and put options in one stroke, we consider a contingent claim that entitles the holder to receive at time N an amount $H(S_N)$ (when the stock price is S_N). Suppose that the hedging strategy (also required to be self-financing) $\theta = (\pi^0, \pi^1)$ is given by

$$\pi^1_{m+1} = N(S_m, m) \qquad \pi^0_{m+1} = M(S_m, m).$$

The value $V_{m+1}(\theta)$ of this strategy at time $m + 1$ is given by

$$V_{m+1}(\theta) = N(S_m, m)S_{m+1} + M(S_m, m)R^{m+1}. \tag{11.1}$$

In order that θ be a hedging strategy, we must have

$$V_N(\theta) = H(S_N). \tag{11.2}$$

Since the hedging strategy θ is required to be self-financing, we also have for $0 \leq m < N$, that the value of the portfolio at time m must equal the amount required in order to implement the strategy at that time, i.e.,

$$V_m(\theta) = N(S_m, m)S_m + M(S_m, m)R^m. \tag{11.3}$$

Let us write $B(S, j) = M(S, j)R^j$ and

$$F(S, j) = N(S, j)S + B(S, j). \tag{11.4}$$

Then (11.3) can be rewritten as

$$V_m(\theta) = F(S_m, m). \tag{11.5}$$

We thus have

$$N(S_{m-1}, m - 1)S_m + B(S_{m-1}, m - 1)R = F(S_m, m). \tag{11.6}$$

The identity (11.6) should hold for all possible values of S_{m-1}, S_m. In our model, if $S_{m-1} = S$, then $S_m = US$ or $S_m = DS$. Thus, we must have (writing $j = m - 1$)

$$N(S, j)US + B(S, j)R \;=\; F(US, j + 1) \tag{11.7}$$

$$N(S, j)DS + B(s, j)R \;=\; F(DS, j + 1). \tag{11.8}$$

Solving these simultaneous equations, we get

$$N(S, j) = \frac{F(US, j+1) - F(DS, j+1)}{(U-D)S} \tag{11.9}$$

$$B(S, j) = \frac{UF(DS, j+1) - DF(US, j+1)}{(U-D)R}. \tag{11.10}$$

Substituting the values of $N(S, j)$ and $B(S, j)$ from (11.9)–(11.10) into (11.4), we get

$$F(S, j) = \frac{1}{R}\Big(F(US, j+1)\frac{R-D}{U-D} + F(DS, j+1)\frac{U-R}{U-D}\Big).$$

Writing $q = \frac{R-D}{U-D}$, we thus have

$$F(S, j) = \frac{1}{R}\Big(qF(US, j+1) + (1-q)F(DS, j+1)\Big). \tag{11.11}$$

From (11.2) and (11.5) we have

$$F(S, N) = H(S). \tag{11.12}$$

The relations (11.11)–(11.12) determine $F(S, j)$ recursively and then the hedging strategy for $N(S, j)$ and $M(S, j)$ is determined by (11.7)–(11.8).

11.2 A binomial Feynman–Kac formula

Recall that the stock model is defined in terms of Bernoulli random variables $\xi_1, \dots, \xi_N$. Define a new probability measure Q such that $\xi_1, \dots, \xi_N$ are independent under Q and

$$Q(\xi_j = U) = q\ ,\ Q(\xi_j = D) = (1-q) \tag{11.13}$$

It can be seen that Q is the unique equivalent martingale measure. It should be noted that p does not play any role in the analysis. In fact, the analysis remains valid as long as

$$P(S_n = S_0 U^k D^{n-k} \text{ for some } k) = 1,\ 1 \le n \le N.$$

This had been observed in Chapter 6 as well. The formula (11.11) can now be written as

$$F(S, j) = \frac{1}{R}\mathbb{E}_Q[F(\xi_{j+1}S, j+1)], \tag{11.14}$$

and hence it gives

$$F(S, j) = \frac{1}{R^{N-j}}\mathbb{E}_Q[F(\xi_{j+1}\xi_{j+2}\dots\xi_N S, N)]. \tag{11.15}$$

Recalling that $S_N = S_0 \xi_{j+1} \xi_{j+2} \cdots \xi_N$ and (11.12), we get

$$F(S_0, 0) = \frac{1}{R^N} \mathbb{E}_Q[H(S_N)], \tag{11.16}$$

For $H(x) = (x - K)^+$ and $R = e^{rN}$, (11.16) becomes

$$F(S_0, 0) = \mathbb{E}_Q\big[e^{-rT} (S_T - K)^+\big]. \tag{11.17}$$

Letting $X_j = \log \xi_j$ ($X_j = u = \log U$ with probability q and $= d = \log D$ with probability $1 - q$), we have

$$F(S_0, 0) = \mathbb{E}_Q\big\{e^{-rT} \big(e^{\sum_{j=1}^n X_j} - K\big)^+\big\}. \tag{11.18}$$

Formulas (11.16), (11.17) or (11.18) may be called the binomial version of the Feynman–Kac formula.

11.3 Approximation of the Black–Scholes PDE

The idea behind the approximation is similar to that of passing from the difference equation of the symmetric random walk to the heat equation. It will be recalled in this connection that in the previous chapter the Black–Scholes PDE was solved by suitably transforming it into the heat equation. Instead of the time $n \in \{1, 2, \ldots, N\}$, we consider time $t \in \{0, h, 2h, \ldots, Nh = T\}$. Later, we will take the limit as $N \longrightarrow \infty$ which is the same as $h \longrightarrow 0$. The stock model over this discrete set of points is given by

$$S_{jh}^N = S_0 \xi_1 \xi_2 \ldots \xi_j \tag{11.19}$$

where $\xi_1, \ldots, \xi_N$ are i.i.d. random variables with

$$P(\xi_1 = U) = p_h \quad P(\xi_1 = D) = 1 - p_h, \tag{11.20}$$

where $U = 1 + \sigma h^{\frac{1}{2}}$, $D = 1 - \sigma h^{\frac{1}{2}}$, and

$$p_h = \frac{1}{2} + \frac{1}{2}\frac{\mu}{\sigma} h^{\frac{1}{2}}. \tag{11.21}$$

The return on the bond in an interval $(t, t + h)$ is assumed to be $R = e^{rh}$. It is assumed that σ, μ, r are constants and $h < \frac{1}{\sigma^2}$. Let $Y_t = \sigma W_t + (\mu - \frac{1}{2}\sigma^2)$ and let $S_t = S_0 \exp\{Y_t\}$, (W_t) being a standard Brownian motion. Note that S_t is the Black–Scholes stock price model considered in the previous section. As $h \longrightarrow 0$ (equivalently $N \longrightarrow \infty$), the model approximates the Black–Scholes model in the following sense—the process (S_t^N) converges in distribution to the process

(S_t). (See (Billingsley, 1968), Chapter 1 for a discussion of convergence in distribution. Here, we will outline the arguments leaving the details to the reader.) Fix N and define (S_t^N) as a $D[0, T]$-valued process as follows:

$$S_t^N = S_{jh}^N \quad jh \leq t < (j+1)h. \tag{11.22}$$

Let $Y_t^N = \log(S_t^N)$ and $\zeta_j = \log(\xi_j)$. Then

$$Y_t^N = \sum_{i=1}^{[Nt]} \zeta_i.$$

It is easy to verify that

$$\mathbb{E}_P[Y_t^N] \longrightarrow (\mu - \frac{1}{2}\sigma^2)t \;\; as\; N \longrightarrow \infty \tag{11.23}$$

and

$$\mathbb{E}_P\Big[\big(Y_t^N - \mathbb{E}_P(Y_t^N)\big)^2\Big] \longrightarrow \sigma^2 \mu t \;\; as\; N \longrightarrow \infty. \tag{11.24}$$

The Lindeberg–Feller central limit theorem (See (Billingsley, 1968), Theorem 7.2.) implies that for each t, $Y_t^N \longrightarrow Y_t$, in the sense of convergence in distribution. We need to use the relations (11.23)–(11.24) and the fact that $\zeta_1, \zeta_2, \ldots,$ are i.i.d.

Indeed, one can deduce that for $t_1, t_2, \ldots, t_m \in [0, T]$,

$$(Y_{t_1}^N, Y_{t_2}^N, \ldots, Y_{t_m}^N) \longrightarrow (Y_{t_1}, Y_{t_2}, \ldots, Y_{t_m}) \text{ in distribution.} \tag{11.25}$$

From the definition of (Y_t^N) it can be verified that for $0 \leq t_1 \leq t \leq t_2 \leq T$,

$$\mathbb{E}_P[|Y_t^N - Y_{t_1}^N|^2 |Y_{t_2}^N - Y_t^N|^2] \leq C|t_2 - t_1|^2 \tag{11.26}$$

for some constant $C < \infty$. The relations (11.25)–(11.26) imply that the process Y^N converges in distribution to the process Y, as a $D[0, T]$ valued process (see (Billingsley, 1968), Theorem 15.6), and hence

$$(S_t^N) \text{ converges in distribution to } (S_t).$$

We will see that the hedging strategy and the value function for this model lead us to the Black–Scholes PDE.

As in the previous section, let $F(S, t)$ be the function that determines the value of the hedging portfolio in this model. Let q_h denote the probability of the event $\xi_1 = U$ under the equivalent probability measure Q. Then, in this case, q_h is given by

$$q_h = \frac{e^{rh} - (1 - \sigma h^{\frac{1}{2}})}{2\sigma h^{\frac{1}{2}}}. \tag{11.27}$$

Noting that

$$q_h = \frac{1}{2}\big(1 + \frac{r}{\sigma}h^{\frac{1}{2}}\big) + o(h), \tag{11.28}$$

the equation (11.11) here gives

$$\begin{aligned} e^{rh} F(S,t) &= \big\{\frac{1}{2}(1 + \frac{r}{\sigma}h^{\frac{1}{2}}) + o(h)\big\} F\big[S(1+\sigma h^{\frac{1}{2}}),\ t+h\big] \\ &+ \big\{\frac{1}{2}(1 - \frac{r}{\sigma}h^{\frac{1}{2}}) + o(h)\big\} F\big[S(1-\sigma h^{\frac{1}{2}}),\ t+h\big]. \end{aligned} \tag{11.29}$$

Write $k = S\sigma h^{\frac{1}{2}}$ and $G(S+k,\ t+h) = F[S(1+\sigma h^{\frac{1}{2}}),\ t+h]$. Assuming sufficient smoothness properties for F, a finite Taylor expansion gives

$$\begin{aligned} G(S+k,\ t+h) &= G(S,t) + \big(k\frac{\partial G}{\partial S} + h\frac{\partial G}{\partial t}\big) \\ &+ \frac{1}{2}\big(k^2\frac{\partial^2 G}{\partial S^2} + h^2\frac{\partial^2 G}{\partial t^2}\big) + o(\sqrt{k^2+h^2}). \end{aligned}$$

Since $h^2\,\frac{\partial^2 G}{\partial t^2} = o(h)$, we get

$$\begin{aligned} F[S(1+\sigma h^{\frac{1}{2}}),\ t+h] &= F(S,t) + h\frac{\partial F}{\partial t}(S,t) + S\sigma h^{\frac{1}{2}}\frac{\partial F}{\partial S}(S,t) \\ &+ \frac{1}{2}S^2\sigma^2 h\frac{\partial^2 F}{\partial S^2}(S,t) + o(h). \end{aligned} \tag{11.30}$$

Similarly,

$$\begin{aligned} F[S(1-\sigma h^{\frac{1}{2}}),\ t+h] &= F(S,t) + h\frac{\partial F}{\partial t}(S,t) - S\sigma h^{\frac{1}{2}}\frac{\partial F}{\partial S}(S,t) \\ &+ \frac{1}{2}S^2\sigma^2 h\frac{\partial^2 F}{\partial S^2}(S,t) + o(h). \end{aligned} \tag{11.31}$$

From (11.29), (11.30) and (11.31),

$$\begin{aligned} \{1+rh+o(h)\}F(S,t) &= \Big\{\big\{\frac{1}{2}(1+\frac{r}{\sigma}h^{\frac{1}{2}}) + o(h)\big\}F(S,t) + h\frac{\partial F}{\partial t} \\ &+ S\sigma h^{\frac{1}{2}}\frac{\partial F}{\partial S} + \frac{1}{2}S^2\sigma^2 h\frac{\partial^2 F}{\partial S^2} + o(h)\Big\} \\ &+ \Big\{\frac{1}{2}(1-\frac{r}{\sigma}h^{\frac{1}{2}}) + o(h)\Big\}\Big\{F(S,t) + h\frac{\partial F}{\partial t} \\ &- S\sigma h^{\frac{1}{2}}\frac{\partial F}{\partial S} + \frac{1}{2}S^2\sigma^2 h\frac{\partial^2 F}{\partial S^2} + o(h)\Big\}. \end{aligned}$$

Dividing both sides by h, we get

$$\begin{aligned} rF(S,t) &= \left(\frac{1}{2}\frac{\partial F}{\partial t} + \frac{1}{2}rS\frac{\partial F}{\partial S} + \frac{1}{4}S^2\sigma^2\frac{\partial^2 F}{\partial S^2}\right) \\ &\quad + \left(\frac{1}{2}\frac{\partial F}{\partial t} + \frac{1}{2}rS\frac{\partial F}{\partial S} + \frac{1}{4}S^2\sigma^2\frac{\partial^2 F}{\partial S^2}\right) + o(h). \end{aligned}$$

Making $h \longrightarrow 0$, we get the continuous time equation

$$\frac{1}{2}\sigma^2 S^2 \frac{\partial^2 F(S,t)}{\partial S^2} + rS\frac{\partial F}{\partial S}(S,t) + \frac{\partial F(S,t)}{\partial t} - rF(S,t) = 0. \tag{11.32}$$

The equation (11.11) with T as the expiration date gives $F(S, T) = H(S)$. Thus in the continuous time limiting case if $H(S_T) = (S_T - K)^+$ (a European call option with K being the exercise price and T being the terminal time) we have the following: The value of the option at time t is $F(S_t, t)$ where $F(x,t)$ $(t > 0)$ is the solution of the boundary value problem,

$$\frac{\partial F}{\partial t}(x,t) + \frac{1}{2}\sigma^2 x^2 \frac{\partial^2 F}{\partial x^2} + rx\frac{\partial F}{\partial x} - rF(x,t) = 0,\ 0 < t < T, \tag{11.33}$$

with boundary condition

$$F(x, T) = (x - K)^+. \tag{11.34}$$

Let us note also that the hedging strategy in the discrete case was given by (11.9), which here becomes

$$N(S_t, t) = \frac{F(S_t(1 + \sigma h^{\frac{1}{2}}), t + h) - F(S_t(1 - \sigma h^{\frac{1}{2}}), t + h)}{2\sigma h^{\frac{1}{2}} S_t}.$$

Taking the limit as $h \longrightarrow 0$, we conclude that the hedging strategy in the continuous case is given by

$$\pi_t^1 = \frac{\partial F}{\partial x}(S_t, t).$$

Thus the hedging stategy and the value function for the discrete approximation to the Black–Scholes model converge to the respective quantities for the Black–Scholes model.

11.4 Approximation to the Black–Scholes formula

In the previous section, we considered an approximation to the Black–Scholes model of stock prices and showed that the Black–Scholes PDE can be obtained via the approximation. Here, we will show that the Black–Scholes formula, which is the Feynman–Kac formula for the solution of the PDE, can be obtained as a limit of the formula (11.17) for the option price in the discrete case.

We will continue to use the set up from the previous section. The formula (11.17), giving the price of the option in discrete time, becomes (writing $F_N(T)$ as the price of the option on S^N, striking price K, terminal time T)

$$F_N(T) = \mathbb{E}_Q[e^{-rT}(S_T^N - K)^+], \tag{11.35}$$

where Q is the equivalent martingale measure. Here Q is such that $\xi_1, \ldots, \xi_N$ are i.i.d. random variables with

$$Q(\xi_1 = U) = q_h \quad Q(\xi_1 = D) = 1 - q_h, \tag{11.36}$$

where $U = 1 + \sigma h^{\frac{1}{2}}$, $D = 1 - \sigma h^{\frac{1}{2}}$ and

$$q_h = \frac{1}{2} + \frac{1}{2}\frac{r}{\sigma}h^{\frac{1}{2}} + o(h). \tag{11.37}$$

Recall that

$$S_{jh}^N = S_0 \xi_1 \xi_2 \ldots \xi_j. \tag{11.38}$$

As in the previous section, the Lindeberg–Feller central limit theorem implies that the random variables S_T^N, under the probability measure Q, converge in distribution to $S_T = \exp\{\sigma W_T + (r - \frac{1}{2}\sigma^2)T\}$, where (W_t) is a standard Brownian motion. We will prove below that $F_N(T)$ converges to the option price for the continuous time case $F(T)$, which in turn is given by the Feynman–Kac formula

$$\mathbb{E}[e^{\{-rT\}}(S_T - K)^+].$$

(This simplifies to the Black–Scholes formula.)

Since we have already proved that S_T^N converges in distribution to S_T, the convergence of $F_N(T)$ to $F(T)$ would follow if we show that $(S_T^N - K)^+$ are uniformly integrable (See Theorem 5.4 in (Billingsley 1968).) Uniform integrability of $(S_T^N - K)^+$ would follow, if we prove uniform integrability of S_T^N. A sufficient condition for the latter is

$$\mathbb{E}_Q[S_T^N] \longrightarrow \mathbb{E}[S_T].$$

Since $\mathbb{E}[S_T] = \exp\{rT\}$, the required result, namely the convergence of $F_N(T)$ to $F(T)$, would follow once we prove that

$$\mathbb{E}_Q[S_T^N] \longrightarrow \exp\{rT\}. \tag{11.39}$$

Using (11.36)–(11.38), it follows that

$$\begin{aligned}
\mathbb{E}_Q[S_T^N] &= \left\{(1 + \sigma h^{\frac{1}{2}})q_h + (1 - \sigma h^{\frac{1}{2}})(1 - q_h)\right\}^N & (11.40)\\
&= \left\{1 + 2\sigma h^{\frac{1}{2}} q_h - q_h\right\}^N & (11.41)\\
&= \left\{1 + (2\sigma h^{\frac{1}{2}})(\frac{1}{2}\frac{r}{\sigma}h^{\frac{1}{2}} + o(h))\right\}^N & (11.42)\\
&= \left\{1 + rh + o(h))\right\}^N. & (11.43)
\end{aligned}$$

Recalling that $Nh = T$, it follows that as $N \longrightarrow \infty$, $F_N(T) \longrightarrow \exp\{rT\}$, thus completing the proof.

12
The American Options

In this chapter, we consider the American call option in a continuous time model of stock prices. The development is similar to that in discrete time and follows our general approach of deriving upper and lower bounds based on the NA principle. We will show that in a complete market, the two bounds coincide.

We will consider a general American type security and derive results on the call option as special a case. A significant result is that the American call option has the same price as the corresponding European option.

The treatment is given with full detail, without invoking results on optimal stopping for the sake of completeness. We will again restrict ourselves to the case when the stock prices are continuous (semimartingales).

12.1 Model

We consider a market consisting of k stocks, whose prices are given by $S_t^1, S_t^2, \ldots, S_t^k$, respectively. We assume that $S^1, S^2, \ldots, S^k$ are continuous semimartingales. The bond price S_t^0 at time t is assumed to satisfy

$$0 < S_u^0 \leq S_t^0 \quad 0 \leq u \leq t.$$

We will assume that S_0^i is a constant, $1 \leq i \leq k$. Let

$$\tilde{S}_t^j = S_t^j (S_t^0)^{-1}$$

be the discounted price. Recall that the class of (self-financing) investment strategies are given by

$$\theta = (y, \pi^1, \dots, \pi^k), \tag{12.1}$$

such that $\pi^1, \dots, \pi^k$ are $\mathcal{F}_t^S$-predictable and the corresponding value process

$$V_t(\theta) := S_t^0(y + \sum_{i=1}^{k} \pi_t^i \tilde{S}_t^i), \quad (t > 0) \tag{12.2}$$

satisfies

$$V_t(\theta) := S_t^0(y + \sum_{i=1}^{k} \int_0^t \pi_u^i d\tilde{S}_u^i). \tag{12.3}$$

We will assume that $Q \in \mathcal{E}(P)$, i.e., there exists a probability measure Q under which $(\tilde{S}_1^j, \mathcal{F}_t^S)$ are martingales for $1 \leq j \leq k$.

As seen earlier, this implies that NA holds. The American call and put options are similar to the European options considered earlier, the difference being that they an be exercised by the holder (buyer) at any time of his choice up to time T. Clearly, the holder can use any information he has until time t, to decide whether to exercise his option at that time or to continue. He is not allowed to anticipate the future and thus it follows that the time he chooses to collect his payoff must be a stopping time with respect to the observation σ-fields.

We are thus lead to the following definition:

Definition An American call option with terminal time T and striking price K on the ith security entitles the holder of this claim to exercise his option at any stopping time τ, $\tau \leq T$, and collect a reward of

$$(S_\tau^i - K)^+.$$

The corresponding European option would entitle the owner to collect a reward of $(S_T^i - K)^+$. Since the holder of the American option is allowed to choose $\tau = T$, he is always better off than someone holding the corresponding European option and hence must pay a higher price.

Indeed, if the price of the American option is lower (by an amount x) than the corresponding European option, an investor can sell the European option and buy the American option and invest the difference in bonds. At time T, if the European option is exercised, he can exercise his American option as well. In either case, he would have made a profit of

$$xS_T^0$$

and this would thus be an arbitrage opportunity.

The case for the American put option (defined below) and the corresponding European put option is similar.

Definition An American put option with terminal time T and striking price K on the ith security entitles the holder of this claim to exercise his option at any stopping time τ, $\tau \leq T$ and collect a reward of

$$(K - S_\tau^i)^+.$$

Given below is the American analogue of a general contingent claim h. This includes both the call and put options as particular cases. Thus, we will discuss the question of pricing for a general American contingent claim and deduce the results for the call and put options as special cases.

Definition An American contingent claim with terminal time T consists of a $\mathcal{F}_t^S$-adapted process (h_t) (written as $ACC((h_t), T)$). The holder of this claim can choose a stopping time τ, $\tau \leq T$ and collect a reward of h_τ.

If $h_t = (S_t^i - K)^+$, the $ACC((h_t), T)$ is the American call option on the ith security, and if $h_t = (K - S_t^i)^+$, the $ACC((h_t), T)$ is the American put option on the ith security. Corresponding to an $ACC((h_t), T)$, we can consider the European contingent claim h_T and, as argued above for the case of call options, the principle of no arbitrage implies that the price of $ACC((h_t), T)$ must be at least as much as the price of the European contingent claim h_T.

We will assume that (h_t) is a continuous process satisfying

$$\mathbb{E}_Q\left(\sup_{0\leq t\leq T} |h_t|\right) < \infty. \tag{12.4}$$

12.2 Upper and lower bounds

We will follow our approach as in the earlier chapters and here derive bounds for the price of an American contingent claim. We would like to stress again that the bounds are derived just on the basis of the no arbitrage principle.

As in the discrete case, here is a set of upper bounds for the price of the American contingent claim.

Let $B^+ = B^+((h_t), T)$ consist of all $y \in [0, \infty)$ such that there exists a self-financing strategy

$$\theta = (y, \pi^1, \ldots, \pi^k), \tag{12.5}$$

such that $\pi^1, \ldots, \pi^k$ are $\mathcal{F}_t^S$-predictable and the discounted value process

$$\tilde{V}_t(\theta) = y + \sum_{i=1}^{k} \pi_t^i \tilde{S}_t^i,$$

is a Q-martingale and satisfies

$$\tilde{V}_t(\theta) \geq h_t (S_t^0)^{-1} \quad \forall t \text{ a.s.} \tag{12.6}$$

Lemma 12.1 *Let $y \in B^+$. Suppose the price p of the $ACC((h_t), T)$ is more than y. Then the market, consisting of the bond, k-stocks along with the American option, will admit an arbitrage opportunity.*

Proof Let θ be a strategy given by (12.5) satisfying (12.6). Consider the strategy of selling the American contingent claim at price p and investing it on stocks following the strategy

$$\hat{\theta} = (p, \pi^1, \ldots, \pi^k).$$

The value process $V_t(\hat{\theta})$ is given by

$$\begin{aligned} V_t(\hat{\theta}) &= S_t^0(p + \sum_{i=1}^k \pi_t^i \tilde{S}_t^i) \\ &= S_t^0(y + \sum_{i=1}^k \pi_t^i S_t^i) + S_t^0(p - y) \\ &= V_t(\theta) + S_t^0(p - y) \\ &\geq h_t + S_t^0(p - y). \end{aligned}$$

Thus if the buyer of the American contingent claim chooses to collect his payoff at time τ, the investor can also liquidate (sell) his holdings of the stocks and his net assets at time τ will be

$$V_\tau(\hat{\theta}) - h_\tau,$$

which is at least $S_\tau^0(p - y)$. This is an arbitrage opportunity. □

As a consequence of the result given above, it follows that if we impose the condition of no arbitrage, then

$$y^+ = \inf B^+$$

is an upper bound for the price of the American contingent claim.

Let $B^- = B^-((h_t))$ consist of all $y \in [0, \infty)$ such that there exists a self-financing strategy $\theta = (y, \pi^1, \ldots, \pi^k)$ and a stopping time τ, $(\tau \leq T)$ such that $\tilde{V}_t(\theta)$, the discounted value process for the strategy θ, is a Q martingale and

$$\tilde{V}_\tau(\theta) \leq h_\tau (S_\tau^0)^{-1} \text{ a.s.}. \tag{12.7}$$

Here is the result in which the set B^- is the set of lower bounds for the price of the American contingent claim.

Lemma 12.2 *Let $y \in B^-$. Suppose the price p of the $ACC((h_t), T)$ is less than y. Then the market consisting of the bond, k-stocks along with the American contingent claim, will admit an arbitrage opportunity.*

Proof Let θ be a strategy given by (12.5), satisfying (12.7). Consider the strategy of buying the American option at the price p and following the strategy

$\hat{\theta} = (-p, -\pi^1, \dots, -\pi^k)$. Note that to implement this strategy, no investment is required from the investor. The value process $V_t(\hat{\theta})$ is given by

$$\begin{aligned} V_t(\hat{\theta}) &= S_t^0(-p - \sum_{i=1}^k \pi_t^i \tilde{S}_t^i) \\ &= -V_t(\theta) + S_t^0(y - p). \end{aligned}$$

The investor should exercise his option at time τ, and his net assets are

$$\begin{aligned} h_\tau + V_\tau(\hat{\theta}) &= h_\tau - V_\tau(\theta) + S_\tau^0(y - p) \\ &\geq S_\tau^0(y - p) \end{aligned}$$

by choice of the strategy and of τ. This is an arbitrage opportunity since y is larger than p. □

Let us define

$$y^- = \sup\ B^-.$$

It follows from the previous lemma that y^- is a lower bound for the price of the American contingent claim.

Lemma 12.3 *The upper and lower bounds are consistent, i.e.,*

$$y^- \leq y^+.$$

Proof Let us write

$$\tilde{h}_t = h_t \cdot (S_t^0)^{-1}.$$

Then if $y \in B^+$, one has

$$V_t(\theta) \geq h_t \text{ a.s.}$$

for some strategy $\theta = (y, \pi^1, \dots, \pi^k)$ for which $\tilde{V}_t(\theta)$ is a Q-martingale. Hence

$$\tilde{V}_t(\theta) \geq \tilde{h}_t \text{ a.s.} \tag{12.8}$$

Since $\tilde{V}_t(\theta)$ is a Q-martingale, for any stopping time τ, one has (recalling that $\tilde{V}_0 = y$)

$$y = \mathbb{E}_Q(\tilde{V}_\tau(\theta)) \geq \mathbb{E}_Q(\tilde{h}_\tau).$$

It follows that

$$y \geq \sup_{\tau \leq T} \mathbb{E}_Q[\tilde{h}_\tau], \tag{12.9}$$

(where the supremum is taken over all stopping times) which in turn implies that

$$y^+ \geq \sup_{\tau \leq T} \mathbb{E}_Q[\tilde{h}_\tau]. \tag{12.10}$$

Now for $y \in B^-$, let $\theta = (y, \pi^1, \ldots, \pi^k)$ be such that $\tilde{V}_t(\theta)$ is a Q-martingale and has a stopping tme σ such that

$$\tilde{V}_\sigma \leq \tilde{h}_\sigma.$$

Using $\tilde{V}_0 = y$, it follows that

$$\begin{aligned} y &= \mathbb{E}_Q[\tilde{V}_\sigma] \\ &\leq \mathbb{E}_Q[\tilde{h}_\sigma] \\ &\leq \sup_{\tau \leq T} \mathbb{E}_Q[\tilde{h}_\tau]. \end{aligned}$$

This now implies that

$$y^- \leq \sup_{\tau \leq T} \mathbb{E}_Q[\tilde{h}_\tau]. \tag{12.11}$$

The required result now follows from (12.10) and (12.11). □

We have proved that for any EMM Q, the upper bound y^+ and the lower bound y^- of the $ACC((h_t), T)$ satisfy

$$y^- \leq \sup_{\tau \leq T} \mathbb{E}_Q[\tilde{h}_\tau] \leq y^+. \tag{12.12}$$

We will show that in the case where the market consisting of $(S^0, S^1, \ldots, S^k)$ is complete, the upper and lower bounds coincide, just as in the case of the European claims.

12.3 American claims in complete markets

We now consider an American contingent claim (h_t) in a complete market $(S^0, S^1, \ldots, S^k)$. Let Q denote the EMM. Recall that it is unique since the market is complete. Here is our main result.

Theorem 12.4 *Let*

$$y_0 = \sup_{\tau \leq T} \mathbb{E}_Q(\tilde{h}_\tau). \tag{12.13}$$

There exists a strategy $\theta = (y_0, \pi^1, \ldots, \pi^k)$ *such that* $\tilde{V}_t(\theta)$ *is a* Q*-martingale,*

$$\tilde{V}_t(\theta) \geq \tilde{h}_t \ \ \forall t \quad a.s. \tag{12.14}$$

and there exists a stopping time σ^* *such that*

$$\tilde{V}_{\sigma^*}(\theta) = \tilde{h}_{\sigma^*} \quad a.s. \tag{12.15}$$

As a consequence,

$$y^+ = y^- = y_0. \tag{12.16}$$

Proof The proof is divided into two steps.

Step 1: There exists a stopping time σ^* such that

$$y_0 = \mathbb{E}_Q\left(\tilde{h}_{\sigma^*}\right).$$

We begin with an observation. For $0 < a < y_0, 0 < b < y_0$, let σ_a, σ_b be stopping times such that

$$\mathbb{E}_Q(\tilde{h}_{\sigma_a}) \geq y_0 - a$$

and

$$\mathbb{E}_Q(\tilde{h}_{\sigma_b}) \geq y_0 - b.$$

Then

$$\mathbb{E}_Q(\tilde{h}_{\sigma_a \wedge \sigma_b}) \geq y_0 - (a + b).$$

To see this, note that

$$\tilde{h}_{\sigma_a} + \tilde{h}_{\sigma_b} = \tilde{h}_{\sigma_a \wedge \sigma_b} + \tilde{h}_{\sigma_a \vee \sigma_b}$$

and hence

$$\begin{aligned} E^a(\tilde{h}_{\sigma_a \wedge \sigma_b}) &= \mathbb{E}_Q(\tilde{h}_{\sigma_a}) + \mathbb{E}_Q(\tilde{h}_{\sigma_b}) - \mathbb{E}_Q(\tilde{h}_{\sigma_a \vee \sigma_b}) \\ &\geq (y_0 - a) + (y_0 - b) - y_0 \\ &= y_0 - (a + b). \end{aligned}$$

Here, we have used the fact that $\sigma_a \vee \sigma_b$ is a stopping time and hence

$$\mathbb{E}_Q(\tilde{h}_{\sigma_a} \vee \tilde{h}_{\sigma_b}) \leq y_0.$$

Now, for $n \geq 1$, let σ_n be a stopping time such that

$$\mathbb{E}_Q(\tilde{h}_{\sigma_n}) \geq y_0 - \frac{1}{2^n}$$

and for $m \geq 1$, define

$$\begin{aligned} \sigma_{n,1} &= \sigma_n \wedge \sigma_{n+1} \\ \text{and} \quad \sigma_{n,m+1} &= \sigma_{n,m} \wedge \sigma_{(n+m+1)}. \end{aligned}$$

Then it follows that

$$\begin{aligned} \mathbb{E}_Q\left(\tilde{h}_{\sigma_{n,m}}\right) &\geq y_0 - \left(\frac{1}{2^n} + \frac{1}{2^{n+1}} + \ldots + \frac{1}{2^{n+m}}\right) \\ &\geq y_0 - \frac{1}{2^{n-1}}. \end{aligned} \tag{12.17}$$

Let $\sigma_n^* = \lim_m \sigma_{n,m}$ and $\sigma^* = \lim_n \sigma_n^*$. These limits exist as $\sigma_{n,m} \geq \sigma_{n,m+1}$ and $\sigma_n^* \leq \sigma_{n+1}^*$. Also, σ_n^* and σ^* are $(\mathcal{F}_t^S)$-stopping times. Since $\tilde{h}_t$ is a continuous process and

$$\mathbb{E}_Q\left(\sup_{t \leq T} |\tilde{h}_t|\right) < \infty,$$

we also have

$$\lim_{m\to\infty} \mathbb{E}_Q(\tilde{h}_{n,m}) = \mathbb{E}_Q(\tilde{h}_{\sigma_n^*}).$$

Using (12.17), it follows that

$$\mathbb{E}_Q(\tilde{h}_{\sigma^*}) \geq y_0.$$

Since σ^* is a stopping time, $y_0 \geq \mathbb{E}_Q(\tilde{h}_{\sigma^*})$. This completes step 1.

Step 2: To show the existence of $\theta = (y_0, \pi^1, \dots, \pi^k)$ satisfying the required properties:

For $0 \leq t \leq T$ define

$$Y_t = \text{essential supremum}\{\mathbb{E}_Q(\tilde{h}_\tau | \mathcal{F}_t) \quad t \leq \tau \leq T\}$$

where the essential supremum is taken over all stopping times τ. It is clear that Y_t is a super martingale and hence that it admits a modification that has right continuous paths with left limits.

Since

$$Y_t \leq \mathbb{E}_Q\left(\sup_{0\leq u\leq T} |h_u| \ |\mathcal{F}_t\right)$$

it follows that the Doob–Meyer decomposition result can be used on (Y_t) to get a martingale (M_t) with $M_0 = 0$ and an increasing process D_t such that $D_t \geq 0$ and

$$Y_t = y_0 + M_t - D_t.$$

Since $Y_t \geq \tilde{h}_t$, we have

$$y_0 + M_t \geq \tilde{h}_t \ \ \forall t \text{ a.s.}$$

Now, using the completeness of the market, we can get $\pi^1, \dots, \pi^k$ such that

$$M_t = \sum_{j=1}^{k} \int_0^t \pi_u^j d\tilde{S}_u^j. \tag{12.18}$$

Then

$$\begin{aligned} \tilde{V}_t(\theta) &= y_0 + M_t \\ &\geq \tilde{h}_t \ \ \forall t \text{ a.s.} \end{aligned}$$

and of course, $\tilde{V}_t$ is a martingale and so $y_0 \in B^+$.

On the other hand, for the stopping time σ^* constructed in step 1,

$$\tilde{V}_{\sigma^*}(\theta) \geq \tilde{h}_{\sigma^*} \ \text{ a.s.} \tag{12.19}$$

and

$$y_0 = \mathbb{E}_Q(\tilde{V}_{\sigma^*}(\theta)) = \mathbb{E}_Q(\tilde{h}_{\sigma^*}).$$

These two relations together imply that

$$\tilde{V}_{\sigma^*}(\theta) = \tilde{h}_{\sigma^*} \text{ a.s.} \tag{12.20}$$

Thus

$$y_0 \in B^-. \tag{12.21}$$

We have thus proved that $y_0 \in B^+$ and $y_0 \in B^-$. Hence (12.16) holds. This completes the proof. □

We have thus proved that in a complete market, the price of an American contingent claim (h_t) equals

$$p = \sup_{0 \leq \tau \leq T} \mathbb{E}_Q(\tilde{h}_\tau)$$

where Q is the EMM. As a consequence, it follows that if $(h_t, \mathcal{F}_t)$ is a Q-submartingale, then

$$p = \mathbb{E}_Q(\tilde{h}_T).$$

This is the case when considering the American call option

$$h_t = (S_t^1 - K)^+$$

with $S_t^0 = \exp(rt)$. Hence we have the following result (as in the discrete case).

Theorem 12.5 *In a complete market, the price of an American call option with terminal time T and striking price K is the same as the price of the European call option with terminal time T and striking price K. Thus the Black–Scholes formula for the price of a European call option also gives the price of the corresponding American call option.*

For a discussion of the American put option, we refer the reader to the book of Musiela and Rutkowski (Musiela and Rutkowski, 1997).

13

Asset Pricing with Stochastic Volatility

13.1 Introduction

We consider a market consisting of a stock S_t and a bond B_t governed by the following equations:

$$dS_t = a(t, S_t)S_t dt + \sigma_t S_t dW_t \tag{13.1}$$

and

$$dB_t = r_t B_t dt, \qquad B_0 = 1 \tag{13.2}$$

where W_t is a Brownian motion, S_0 is a given random variable independent of W, r_t is a bounded, non-negative, progressively measurable interest rate process.

We assume that the volatility $\sigma_t = \rho(\tilde{\sigma}_t)$, where $\rho \in C_b^2(\mathbb{R})$ is a strictly increasing function and $\tilde{\sigma}_t$ is a stochastic process satisfying the stochastic differential equation (SDE):

$$d\tilde{\sigma}_t = \tilde{\alpha}(t, \tilde{\sigma}_t)dt + \tilde{\beta}(t, \tilde{\sigma}_t)dW'_t \tag{13.3}$$

and W'_t is another Brownian motion independent of W_t. $\tilde{\sigma}_0$ is a given random variable independent of W and W'. By Itô's formula, we have

$$d\sigma_t = \alpha(t, \sigma_t)\sigma_t dt + \beta(t, \sigma_t)\sigma_t dW'_t \tag{13.4}$$

where, ρ' denoting the derivative of ρ,

$$\alpha(t, \sigma) = \left(\rho'(\rho^{-1}(\sigma))\tilde{\alpha}(t, \rho^{-1}(\sigma)) + \frac{1}{2}\rho''(\rho^{-1}(\sigma))\tilde{\beta}(t, \rho^{-1}(\sigma))^2\right) \Big/ \sigma$$

and

$$\beta(t, \sigma) = \rho'(\rho^{-1}(\sigma))\tilde{\beta}(t, \rho^{-1}(\sigma))/\sigma.$$

When the short rate process r_t is constant, it is well-known that, with respect to an equivalent martingale measure (EMM), the distribution of $\log S_T$ conditional on $\{\sigma_t : t \in [0, T]\}$ is normal and the option price can be calculated by taking expectation over the Black–Scholes formula as pointed out by (Duffie, 1992, page 182). However, assuming a general short rate process and a general contingent claim, we will investigate several interesting problems as noted below.

First, we shall prove that the market (S_t, B_t) is incomplete and price any contingent claim by its minimal EMM (cf. (Musiela and Rutkowski, 1997)).

Secondly, we assume that the volatility process fluctuates at a slower pace than the stock price process. In this case, a series expansion of the solution and a recursive formula for calculating the terms of the series are derived.

Thirdly, we assume that σ_t is not directly observable. Instead, it is observed subject to random noise

$$dY_t = h(\sigma_t)dt + dW_t'', \tag{13.5}$$

where W_t'' is another Brownian motion being independent of W_t and W_t' and h satisfies the following condition:

$$\int_0^T \mathbb{E}h(\sigma_t)^2 dt < \infty.$$

Let $\mathcal{F}_t^Y = \sigma(Y_s, s \le t)$ and $\hat{\sigma}_t = \mathbb{E}(\sigma_t|\mathcal{F}_t^Y)$. Consider the following *effective stock price model*:

$$d\hat{S}_t = a(t, \hat{S}_t)\hat{S}_t dt + \hat{\sigma}_t \hat{S}_t dW_t. \tag{13.6}$$

It will be shown in Proposition 13.8 that (13.6) has a unique solution. Again, we shall price any contingent claim by minimal EMM.

Finally, we shall consider the case when S_t is observable. In this case, the market is complete and the contingent claim will be priced by arbitrage instead of the minimal EMM. Since S_t is not necessarily a Markov process, a PDE of the form of (13.14) is not available. We shall use an idea similar to that in Section 3 to derive a PDE in an enlarged state space.

13.2 Incompleteness of the market

We consider the market model (13.1, 13.4) with numeraire given by (13.2).

The following assumptions will remain in force throughout the paper.

Assumptions (B):

(B1) There exist constants σ^m, $\sigma^M \in (0, \infty)$ such that

$$\sigma^m \le \rho(x) \le \sigma^M, \qquad \forall\, x \in \mathbb{R}.$$

(B2) $a(t, s)$ is a bounded function.
(B3) r_t is a bounded process.
(B4) $\tilde{\alpha}(t, \tilde{\sigma})$, $\tilde{\beta}(t, \tilde{\sigma})$ are bounded functions and there exists a constant ℓ^m such that

$$\tilde{\beta}(t, \tilde{\sigma}) \geq \ell^m > 0, \qquad \forall t \in [0, T], \ \ \tilde{\sigma} \in \mathbb{R}.$$

Assumptions (C): $\forall M > 0$, $\exists K_M > 0$ such that for any $t \geq 0$, $|x| \leq M$ and $|y| \leq M$, we have
(C1)

$$|a(t, x) - a(t, y)| \leq K_M |x - y|.$$

(C2)

$$|\tilde{\alpha}(t, x) - \tilde{\alpha}(t, y)| \leq K_M |x - y|$$

and

$$|\tilde{\beta}(t, x) - \tilde{\beta}(t, y)| \leq K_M |x - y|.$$

Proposition 13.1 *i) Under Assumptions (B4) and (C2), the SDE (13.3) has a unique solution.*
ii) Under Assumptions (B2) and (C1), the SDE (13.1) has a unique solution.

Proof i) follows from the well-known result on SDE (cf. (Ikeda and Watanabe, 1981, Chapter IV)). To prove ii), we define $\tilde{S}_t = \ln S_t$. (13.1) is equivalent to

$$d\tilde{S}_t = \left(\tilde{a}(t, \tilde{S}_t) - \frac{1}{2}\sigma_t^2 \right) dt + \sigma_t dW_t, \tag{13.7}$$

where $\tilde{a}(t, \tilde{s}) = a(t, e^{\tilde{s}})$. Under Assumptions (B2) and (C1), it is easy to verify the boundedness and local Lipschitz property for the coefficients of (13.7). Therefore, (13.7), and hence (13.1) has a unique solution (cf. (Ikeda and Watanabe, 1981, Chapter IV; Kallianpur, 1980, Chapter 5)). □

Let $Z_t = S_t / B_t$. Then

$$dZ_t = \frac{S_t}{B_t}(a(t, S_t) - r_t)dt + \frac{\sigma_t S_t}{B_t} dW_t. \tag{13.8}$$

Let

$$\eta_t = \exp\left(\int_0^t \gamma_u dW_u - \frac{1}{2} \int_0^t |\gamma_u|^2 du \right)$$

where $\gamma_t = \frac{r_t - a(t, S_t)}{\sigma_t}$.

Let $\mathbb{P}'$ be the probability measure given by

$$\frac{d\mathbb{P}'}{d\mathbb{P}} = \eta_T, \qquad \mathbb{P} - a.s. \tag{13.9}$$

By Girsanov's formula (Kreps, 1981, Theorem 7.1.3),

$$\tilde{W}_t \equiv W_t - \int_0^t \gamma_u du$$

is a $\mathbb{P}'$-Wiener process. Hence

$$Z_t = Z_0 + \int_0^t \frac{\sigma_u S_u}{B_u} d\tilde{W}_u$$

is a $\mathbb{P}'$-martingale.

Proposition 13.2 *The market consisting of the stock S_t and the bond B_t is incomplete.*

Proof Let

$$\eta_t' = \exp\left(W_t' - \frac{1}{2}t\right).$$

Under assumptions (B), it is easy to see that η_t and η_t' are $\mathbb{P}$-square-integrable martingales. Let

$$\frac{d\tilde{\mathbb{P}}}{d\mathbb{P}} = \eta_T \eta_T', \qquad \mathbb{P} - a.s.$$

Then $\tilde{\mathbb{P}}$ is a probability measure such that Z_t is also a $\tilde{\mathbb{P}}$-martingale. Therefore, we have two distinct EMMs. Hence the market is incomplete. □

Next, we consider the price of a European contingent claim X settled at time T. If X is attainable (cf. (Musiela and Rutkowski, 1997, page 72 for definition)), the price of X at time t is given by any EMM, e.g. $\tilde{\mathbb{P}}$ or $\mathbb{P}'$ (cf. (Musiela and Rutkowski, 1997, page 235, Proposition 10.1.3)):

$$v_t = B_t \mathbb{E}^{\tilde{\mathbb{P}}}(X B_T^{-1} \big| \mathcal{F}_t). \tag{13.10}$$

v_t given by (13.10) is also called the arbitrage price. If X is not attainable, we shall use the risk-minimizing hedge price.

Lemma 13.3 *$\mathbb{P}'$ is the minimal martingale measure associated with $\mathbb{P}$ (cf. (Musiela and Rutkowski, 1997, page 254 for definition)).*

Proof Suppose W and W' are $(\mathcal{F}_t)$-adapted; If N_t is an L^2-$\mathbb{P}$ local martingale, then by the Kunita–Watanabe representation,

$$N_t = N_0 + \int_0^t \ell_u dW_u + \int_0^t \ell'_u dW'_u + Z_t$$

where

$$\langle W, Z \rangle = \langle W', Z \rangle = 0.$$

Let N be strongly orthogonal to $\int_0^t \sigma_u dW_u$. Then we have

$$0 = \left\langle N, \int_0^{\cdot} \sigma_u dW_u \right\rangle_t = \int_0^t \ell_u \sigma_u du.$$

Hence $\ell_t = 0$ a.e. $t \in [0, T]$ a.s. Therefore

$$\begin{aligned} d(N_t \eta_t) &= N_t d\eta_t + \eta_t dN_t + d\langle N, \eta \rangle_t \\ &= N_t d\eta_t + \eta_t dN_t + \gamma_t \ell_t dt \\ &= N_t d\eta_t + \eta_t dN_t. \end{aligned}$$

$N_t \eta_t$ is a local $\mathbb{P}$-martingale. Hence, N_t is a local $\mathbb{P}'$-martingale. The conclusion of the lemma then follows from the definition of the minimal martingale measure. □

Proposition 13.4 *Let X be an European contingent claim settled at time T (not necessarily attainable). Then the risk-minimizing hedge price is*

$$v_t = B_t \mathbb{E}^{\mathbb{P}'}\left(X B_T^{-1} \middle| \mathcal{F}_t\right). \tag{13.11}$$

Proof By Lemma 13.3, $\mathbb{P}'$ is the unique minimal EMM. If the interest rate $r_t \equiv 0$, it follows from Theorem 3.14 in (Föllmer and Schweizer, 1991) that the risk-minimizing hedge price is $\mathbb{E}^{\mathbb{P}'}\left(X \middle| \mathcal{F}_t\right)$. In general, we can discount the contingent claim and obtain, as indicated by (10.30) in (Musiela and Rutkowski, 1997) as well as by (2.5) in (Hofmann et al., 1992), that v_t is given by (13.11). □

To derive a partial differential equation (PDE) for v_t, we assume that $r_t = r(t, S_t)$ and $X = g(S_T)$. In this case,

$$B_t B_T^{-1} = \exp\left(-\int_t^T r(u, S_u) du\right).$$

As (S_t, σ_t) is a Markov process, we have

$$v_t = \mathbb{E}^{\mathbb{P}'}\left(g(S_T) \exp\left(-\int_t^T r(u, S_u) du\right) \middle| S_t, \sigma_t\right) \equiv V(t, S_t, \sigma_t).$$

Note that, under $\mathbb{P}'$, (S_t, σ_t) satisfies the following SDE

$$\begin{cases} dS_t = r(t, S_t)S_t dt + \sigma_t S_t d\tilde{W}_t \\ d\sigma_t = \alpha(t, \sigma_t)\sigma_t dt + \beta(t, \sigma_t)\sigma_t dW'_t. \end{cases}$$

As

$$\frac{d\mathbb{P}'}{d\mathbb{P}} = \exp\left(\int_0^t (\gamma_u, 0) \cdot (dW_u, dW'_u) - \frac{1}{2}\int_0^t |(\gamma_u, 0)|^2 du, \right)$$

it follows from Girsanov's formula (Kreps, 1981, Theorem 7.1.3) that $\tilde{W}_t$ and W'_t are independent Brownian motions under $\mathbb{P}'$. Therefore the generator of (S_t, σ_t) is

$$\begin{aligned} \mathcal{A}_t f(s, \sigma) &= \frac{1}{2}\sigma^2 s^2 \frac{\partial^2 f(s, \sigma)}{\partial s^2} + \frac{1}{2}\sigma^2 \beta(t, \sigma)^2 \frac{\partial^2 f(s, \sigma)}{\partial \sigma^2} \\ &\quad + r(t, s)s\frac{\partial f(s, \sigma)}{\partial s} + \sigma\alpha(t, \sigma)\frac{\partial f(s, \sigma)}{\partial \sigma}. \end{aligned}$$

By (13.7), we have

$$d\tilde{S}_t = \tilde{a}(t, \tilde{S}_t, \tilde{\sigma}_t)dt + \rho(\tilde{\sigma}_t)dW_t \tag{13.12}$$

where

$$\tilde{a}(t, \tilde{S}, \tilde{\sigma}) = a\left(t, e^{\tilde{S}}\right) - \frac{1}{2}\rho(\tilde{\sigma})^2.$$

We make the following *Assumptions (FK)*:
(FK1) $\tilde{\alpha}$, $\tilde{\beta}$, a and r are uniformly Hölder continuous.
(FK2) g has at most polynomial growth.

Note that $(\tilde{S}_t, \tilde{\sigma}_t)$ is a Markov process with generator

$$\begin{aligned} \tilde{\mathcal{A}}_t f(\tilde{s}, \tilde{\sigma}) &= \frac{1}{2}\rho(\tilde{\sigma})^2 \frac{\partial^2 f(\tilde{s}, \tilde{\sigma})}{\partial \tilde{s}^2} + \frac{1}{2}\tilde{\beta}(t, \tilde{\sigma})^2 \frac{\partial^2 f(\tilde{s}, \tilde{\sigma})}{\partial \tilde{\sigma}^2} \\ &\quad + \tilde{a}(t, \tilde{s}, \tilde{\sigma})\frac{\partial f(\tilde{s}, \tilde{\sigma})}{\partial \tilde{s}} + \tilde{\alpha}(t, \tilde{\sigma})\frac{\partial f(\tilde{s}, \tilde{\sigma})}{\partial \tilde{\sigma}}. \end{aligned}$$

Further

$$\begin{aligned} &\tilde{V}(t, \tilde{S}_t, \tilde{\sigma}_t) \equiv V(t, S_t, \sigma_t) \\ = \ &\mathbb{E}^{\mathbb{P}'}\left(g\left(e^{\tilde{S}_T}\right)\exp\left(-\int_t^T r\left(u, e^{\tilde{S}_u}\right)du\right)\bigg|\tilde{S}_t, \tilde{\sigma}_t\right). \end{aligned}$$

We shall denote $r\left(u, e^{\tilde{s}}\right)$ by $\tilde{r}(u, \tilde{s})$. By Remark 5.7.8 in (Karatzas and Shreve, 1988), the following Cauchy problem:

$$\begin{cases} -\frac{\partial \tilde{V}}{\partial t} + \tilde{r}\tilde{V} = \tilde{\mathcal{A}}_t \tilde{V} \\ \tilde{V}(T, \tilde{s}, \tilde{\sigma}) = g\left(e^{\tilde{s}}\right) \end{cases} \tag{13.13}$$

has a solution satisfying the exponential growth condition:

$$\max_{0\le t\le T} |\tilde{V}(t,\tilde{s},\tilde{\sigma})| \le M e^{\mu(\tilde{s}^2+\tilde{\sigma}^2)}.$$

It is clear that Assumptions (B4) and (FK) imply the assumptions of Theorem 5.7.6 (Feynman–Kac formula) in (Karatzas and Shreve, 1988). Therefore, the solution of (13.13) admits the stochastic representation

$$\tilde{V}(t,\tilde{S}_t,\tilde{\sigma}_t) = \mathbb{E}^{\mathbb{P}'}\left(g\left(e^{\tilde{S}_T}\right) \exp\left(-\int_t^T r\left(u, e^{\tilde{S}_u}\right) du\right) \Big| \tilde{S}_t, \tilde{\sigma}_t \right).$$

Theorem 13.5 *$V(t,s,\sigma)$ is the unique solution of the following PDE:*

$$\frac{\partial V}{\partial t} + \frac{1}{2}\sigma^2 s^2 \frac{\partial^2 V}{\partial s^2} + \frac{1}{2}\sigma^2 \beta(t,\sigma)^2 \frac{\partial^2 V}{\partial \sigma^2} + r(t,s)s\frac{\partial V}{\partial s} + \sigma\alpha(t,\sigma)\frac{\partial V}{\partial \sigma} = r(t,s)V \tag{13.14}$$

with the terminal condition

$$V(T,s,\sigma) = g(s).$$

Proof It is clear that the map $\tilde{T} : (\sigma,s) \mapsto (\tilde{\sigma},\tilde{s}) = (\rho^{-1}(\sigma), \ln s)$ is a one-to-one correspondence between the solutions of (13.13) and those of (13.14). □

Remark i) If we take $\alpha = \beta = 0$ in (13.14) (that is, σ=constant), we obtain the well-known Black–Scholes PDE (cf. (Duffie, 1992, (23) on page 86)), as a special case.
ii) (13.14) coincides with the PDE obtained by Garman with the market price for volatility risk there, $\lambda(t,\sigma) = 0$ (cf. (Musiela and Rutkowski, 1997, line 8, page 156)).

Example 13.6 *Consider*

$$\begin{cases} dS_t = \mu S_t dt + \sigma_t S_t dW_t \\ \sigma_t \equiv \sigma_0 \qquad\qquad (\alpha = \beta = 0) \\ B_t \equiv 1 \end{cases}$$

where $\sigma_0 \in L^2(\Omega, \mathcal{F}.\mathbb{P})$ is independent of W_t. Let $X = (S_T - K)^+$. Evaluate v_t.

Solution: Note that $\gamma_t = -\frac{\mu}{\sigma_t}$. Hence

$$\frac{d\mathbb{P}'}{d\mathbb{P}} = \exp\left(-\frac{\mu}{\sigma_0} W_T - \frac{\mu^2}{2\sigma_0^2} T\right).$$

Under $\mathbb{P}'$, we have a Wiener process $\tilde{W}$ such that

$$dS_t = \sigma_0 S_t d\tilde{W}_t.$$

Then

$$S_T = S_t \exp\left(\sigma_0(\tilde{W}_T - \tilde{W}_t) - \frac{\sigma_0^2}{2}(T-t)\right).$$

Then

$$\begin{aligned} v_t & \\ &= \mathbb{E}^{\mathbb{P}'}\left(\left(S_t \exp(\sigma_0(\tilde{W}_T - \tilde{W}_t) - \frac{\sigma_0^2}{2}(T-t)) - K\right)^+ \Big| \mathcal{F}_t\right) \qquad (13.15) \\ &= S_t \Phi\left(\frac{\ln\frac{S_t}{K} + \frac{\sigma_0^2}{2}(T-t)}{\sigma_0\sqrt{T-t}}\right) - K\Phi\left(\frac{\ln\frac{S_t}{K} - \frac{\sigma_0^2}{2}(T-t)}{\sigma_0\sqrt{T-t}}\right). \qquad (13.16) \end{aligned}$$

The validity of (13.16) follows from the fact that σ_0 is $\mathcal{F}_0$-measurable so that (13.15) yields (13.16).

Remark If W and W' are correlated, we may consider the following equivalent volatility model:

$$d\sigma_t = \alpha(t, \sigma_t)\sigma_t dt + \beta(t, \sigma_t)\sigma_t(c dW_t + \sqrt{1-c^2} d\tilde{W}_t')$$

where $\tilde{W}_t'$ is another Brownian motion being independent of W_t and $c = \mathbb{E} W_1 W_1'$.

In this case, $\mathbb{P}'$ defined by (13.9) is still the minimal EMM and the pricing formula of X is still given by (13.11). However, the generator of (S_t, σ_t) (under $\mathbb{P}'$) is given by

$$\begin{aligned} \mathcal{A}_t f(s,\sigma) &= \frac{1}{2}\sigma^2 s^2 \frac{\partial^2 f(s,\sigma)}{\partial s^2} + c\sigma^2 s\beta(t,\sigma)\frac{\partial^2 f(s,\sigma)}{\partial s \partial \sigma} \\ &+ \frac{1}{2}\sigma^2 \beta(t,\sigma)^2 \frac{\partial^2 f(s,\sigma)}{\partial \sigma^2} + r(t,s)s\frac{\partial f(s,\sigma)}{\partial s} \\ &+ (\sigma\alpha(t,\sigma) + c\beta(t,\sigma)(r(t,s) - a(t,s)))\frac{\partial f(s,\sigma)}{\partial \sigma}. \end{aligned}$$

The PDE (13.14) needs to be adjusted accordingly.

13.3 Asymptotic analysis for models with two scales

In this section, we adopt the approach of (Fouque et al., 1988). We consider a series representation for the price function $V(t, s, \sigma)$ when the volatility process fluctuates at a slower pace than the stock price process. We assume that the stock price and volatility processes are time homogeneous, i.e., the coefficients are independent of the time. Let $\epsilon > 0$ and let S_t^ϵ and σ_t^ϵ be governed by the following SDEs:

$$dS_t^\epsilon = a(S_t^\epsilon)S_t^\epsilon dt + \sigma_t^\epsilon S_t^\epsilon dW_t, \tag{13.17}$$

$$d\sigma_t^\epsilon = \frac{1}{\epsilon}\alpha(\sigma_t^\epsilon)\sigma_t^\epsilon dt + \frac{1}{\sqrt{\epsilon}}\beta(\sigma_t^\epsilon)\sigma_t^\epsilon dW_t'. \tag{13.18}$$

Note that (13.18) is obtained from (13.4) by defining $\sigma_t^\epsilon = \sigma_{t/\epsilon}$ and noting that $W'_{t/\epsilon} \approx \frac{1}{\sqrt{\epsilon}}W'_t$. By Theorem 13.5, the price function $V^\epsilon(t,s,\sigma)$ is the unique solution of the following PDE:

$$\mathcal{L}(\sigma^2)V^\epsilon + \frac{1}{\epsilon}\mathcal{A}V^\epsilon = 0 \tag{13.19}$$

with the terminal condition

$$V^\epsilon(T,s,\sigma) = g(s)$$

where $\mathcal{L}(\lambda)$ and $\mathcal{A}$ are differential operators on variables t, s and on variable σ respectively, and are given by

$$\mathcal{L}(\lambda)f_1(t,s) = \frac{\partial f_1}{\partial t} + \frac{1}{2}\lambda s^2\frac{\partial^2 f_1}{\partial s^2} + r(t,s)s\frac{\partial f_1}{\partial s} - r(t,s)f_1$$

and

$$\mathcal{A}f_2(\sigma) = \frac{1}{2}\sigma^2\beta(\sigma)^2\frac{\partial^2 f_2}{\partial \sigma^2} + \sigma\alpha(\sigma)\frac{\partial f_2}{\partial \sigma}.$$

Next, we shall solve the PDE (13.19) by a series expansion:

$$V^\epsilon(t,s,\sigma) = \sum_{n=0}^{\infty} V_n(t,s,\sigma)\epsilon^n. \tag{13.20}$$

Let μ be the invariant measure of the diffusion process on $[\sigma_m, \sigma_M]$ with generator $\mathcal{A}$.

Theorem 13.7 *i) V_0 does not depend on the volatility σ and is governed by the following Black–Scholes PDE*

$$\mathcal{L}(\bar{\sigma^2})V_0 = 0$$

where $\bar{\sigma^2} = \int \sigma^2\mu(d\sigma)$.
ii) $\{V_n\}$ is obtained recursively:

$$\mathcal{A}V_{n+1} = -\mathcal{L}(\sigma^2)V_n, \qquad \forall n \geq 0.$$

Proof i) Applying the expansion (13.20) to the PDE (13.19), we have

$$\frac{1}{\epsilon}\mathcal{A}V_0 + \sum_{n=0}^{\infty}\left(\mathcal{L}(\sigma^2)V_n + \mathcal{A}V_{n+1}\right)\epsilon^n = 0.$$

Then

$$\mathcal{A}V_0 = 0 \tag{13.21}$$

and

$$\mathcal{L}(\sigma^2)V_n + \mathcal{A}V_{n+1} = 0, \qquad \forall n \geq 0. \tag{13.22}$$

Let $p(t, \sigma, A)$ be the transition function of the Markov process on $[\sigma_m, \sigma_M]$ with generator $\mathcal{A}$. Regarding V_0 as a function of σ with $t,\ s$ fixed, it follows from (13.21) that

$$\int V_0(\sigma_1)p(t, \sigma, d\sigma_1) = V_0(\sigma), \qquad \forall \sigma \in [\sigma_m, \sigma_M].$$

Since V_0 is a continuous function on $[\sigma_m, \sigma_M]$, its maximum is obtained, say, at σ_0. Then

$$V_0(\sigma_0) = \int V_0(\sigma_1)p(t, \sigma_0, d\sigma_1) \leq \int V_0(\sigma_0)p(t, \sigma_0, d\sigma_1) = V_0(\sigma_0).$$

Since the diffusion is non-degenerate, it is easy to see that $V_0(\sigma_0) = V_0(\sigma_1)$, for all $\sigma_1 \in [\sigma_m, \sigma_M]$, i.e., V_0 does not depend on the volatility σ.

Taking $n = 0$ in (13.22), we have

$$\mathcal{L}(\sigma^2)V_0 + \mathcal{A}V_1 = 0. \tag{13.23}$$

Since V_0 does not depend on σ,

$$\int \mathcal{L}(\sigma^2)V_0\mu(d\sigma) = \mathcal{L}(\bar{\sigma^2})V_0.$$

Since μ is the invariant measure of the diffusion process with generator $\mathcal{A}$,

$$\int \mathcal{A}V_1\mu(d\sigma) = 0.$$

Taking integrals with respect to μ on both sides of (13.23), we get

$$\mathcal{L}(\bar{\sigma^2})V_0 = 0.$$

ii) follows from (13.22) directly. □

13.4 Filtering of the stochastic volatility

Let $\pi_t f = \mathbb{E}(f(\sigma_t)\big|\mathcal{F}_t^Y)$. Let $\nu_t = Y_t - \int_0^t \pi_s h ds$ be the innovation process. Then π_t satisfies the following FKK equation (cf. (Kallianpur, 1980)):

$$\pi_t f = \pi_0 f + \int_0^t \pi_s A_s f ds + \int_0^t (\pi_s(fh) - \pi_s f \pi_s h) d\nu_s \tag{13.24}$$

where from (13.4) we have

$$A_t f(x) = \frac{x^2}{2}\beta(t,x)^2 f''(x) + x\alpha(t,x) f'(x).$$

Let $f(x) = I(x) \equiv x$. Then

$$\begin{aligned} \hat{\sigma}_t &= \hat{\sigma}_0 + \int_0^t \pi_s(\cdot\alpha(s,\cdot))ds + \int_0^t (\pi_s(Ih) - \hat{\sigma}_s \pi_s h) d\nu_s \\ &\equiv \hat{\sigma}_0 + \int_0^t \hat{\alpha}_s ds + \int_0^t \hat{\beta}_s d\nu_s. \end{aligned}$$

By the Kallianpur–Striebel formula,

$$\pi_t f = \frac{\tilde{\pi}_t f}{\tilde{\pi}_t 1}, \qquad \forall f \in C_b(\mathbb{R}), t \in [0, T]$$

where $\tilde{\pi}_t$, the unnormalized conditional measure, satisfies the Zakai equation on $\mathcal{M}(\mathbb{R})$:

$$\tilde{\pi}_t f = \tilde{\pi}_0 f + \int_0^t \tilde{\pi}_s(A_s f) ds + \int_0^t \tilde{\pi}_s(hf) dY_s, \qquad \forall f \in C_b^2(\mathbb{R}). \tag{13.25}$$

Similar to Proposition 13.1, we have

Proposition 13.8 *Under Assumptions (B2) and (C1), the SDE (13.6) has a unique solution.*

Now, we proceed as in Section 2. Let $\hat{Z}_t = \hat{S}_t / B_t$. Then

$$d\hat{Z}_t = \frac{\hat{S}_t}{B_t}(a(t,\hat{S}_t) - r_t)dt + \frac{\hat{\sigma}_t \hat{S}_t}{B_t} dW_t. \tag{13.26}$$

Let

$$\hat{\eta}_t = \exp\left(\int_0^t \hat{\gamma}_u dW_u - \frac{1}{2}\int_0^t |\hat{\gamma}_u|^2 du\right)$$

where $\hat{\gamma}_t = \frac{r_t - a(t,\hat{S}_t)}{\hat{\sigma}_t}$. By Assumptions (B), it is easy to see that $\hat{\eta}_t$ is a $\mathbb{P}$-square-integrable martingale. Let

$$\frac{d\hat{\mathbb{P}}'}{d\mathbb{P}} = \hat{\eta}_T, \qquad \mathbb{P} - a.s.$$

Then $\hat{\mathbb{P}}'$ is an EMM for the effective model. As in Lemma 13.3, $\hat{\mathbb{P}}'$ is the minimal martingale measure associated with $\mathbb{P}$. If X is a European contingent claim settled at time T, then the risk-minimizing hedge price is

$$\hat{v}_t = B_t \mathbb{E}^{\hat{\mathbb{P}}'}\left(X B_T^{-1} \middle| \mathcal{F}_t\right). \tag{13.27}$$

Since $(\hat{S}_t, \hat{\sigma}_t)$ is not a Markov process, a PDE of the form of (13.14) cannot be obtained.

Lemma 13.9 *$(\hat{S}_t, \tilde{\pi}_t)$ is a Markov process taking values on $\mathbb{R} \times \mathcal{M}(\mathbb{R})$.*

Proof It follows from Theorem 4.1 in (Kurtz and Ocone, 1988) that $\tilde{\pi}_t$ is the unique solution to (13.25) (this theorem was generalized by (Bhatt et al., 1995). It then follows that the equations (13.25) and (13.6) have a unique solution $(\hat{S}_t, \tilde{\pi}_t)$ and hence, $(\hat{S}_t, \tilde{\pi}_t)$ is a Markov process. □

Let $\mathcal{B}_t$ be the generator for $(\hat{S}_t, \tilde{\pi}_t)$. Consider the following PDE for $(t, \hat{s}, \tilde{\pi}) \in [0, T] \times \mathbb{R} \times \mathcal{M}(\mathbb{R})$:

$$\begin{cases} \frac{\partial}{\partial t}\hat{V}(t, \hat{s}, \tilde{\pi}) + \mathcal{B}_t \hat{V}(t, \hat{s}, \tilde{\pi}) = r(t, \hat{s})\hat{V}(t, \hat{s}, \tilde{\pi}) \\ \hat{V}(T, \hat{s}, \tilde{\pi}) = g(\hat{s}) \end{cases} \tag{13.28}$$

The proposition given below follows from arguments similar to those in the proof of Theorem 2.4.1 in (Kallianpur and Karandikar, 1988) with minor modifications.

Proposition 13.10 *Suppose that (13.28) has a solution $\hat{V}$ such that*

$$|\hat{V}(t, \hat{s}, \tilde{\pi})| \le K(1 + \hat{s}^n + (\tilde{\pi}1)^m), \qquad \forall (t, \hat{s}, \tilde{\pi}) \in [0, T] \times \mathbb{R}_+ \times \mathcal{M}(\mathbb{R})$$

where K, n and m are constants. Then $\hat{v}_t = \hat{V}(t, \hat{S}_t, \tilde{\pi}_t)$.

Proof Note that

$$\begin{aligned} M_t \equiv \hat{V}(t_0 + t, \hat{S}_{t_0+t}, \tilde{\pi}_{t_0+t}) &- \hat{V}(t_0, \hat{S}_{t_0}, \tilde{\pi}_{t_0}) \\ &- \int_{t_0}^{t_0+t} \left(\frac{\partial}{\partial u} + \mathcal{B}_u\right) \hat{V}(u, \hat{S}_u, \tilde{\pi}_u) du \end{aligned}$$

is a local martingale. Let

$$N_t = \exp\left(-\int_{t_0}^{t_0+t} r(u, \hat{S}_u) du\right) - 1.$$

Then

$$M'_t \equiv M_t N_t - \int_0^t M_s dN_s + M_t$$

is a local martingale. Note that

$$M'_t = \hat{V}(t_0+t, \hat{S}_{t_0+t}, \tilde{\pi}_{t_0+t}) \exp\left(-\int_{t_0}^{t_0+t} r(u, \hat{S}_u)du\right) - \hat{V}(t_0, \hat{S}_{t_0}, \tilde{\pi}_{t_0}),$$

and hence,

$$|M'_t| \leq K(2 + |\hat{S}_{t_0+t}|^n + |\hat{S}_{t_0}|^n + (\tilde{\pi}_{t_0+t}1)^m + (\tilde{\pi}_{t_0}1)^m). \tag{13.29}$$

It is then easy to see that $\mathbb{E}\sup_{0\leq t\leq T-t_0}|M'_t| < \infty$ and hence, M'_t is a martingale. So $\mathbb{E}^{\mathbb{P}'}M'_{T-t_0} = M'_0$. Therefore

$$\mathbb{E}^{\mathbb{P}'}\hat{V}(T, \hat{S}_T, \tilde{\pi}_T) \exp\left(-\int_{t_0}^{T} r(u, \hat{S}_u)du\right) = \hat{V}(t_0, \hat{S}_{t_0}, \tilde{\pi}_{t_0}),$$

i.e.

$$\hat{V}(t, \hat{S}_t, \tilde{\pi}_t) = \mathbb{E}^{\mathbb{P}'}g(\hat{S}_T) \exp\left(-\int_{t}^{T} r(u, \hat{S}_u)du\right).$$

□

Note that $\tilde{\pi}_t$ is adapted to $\mathcal{F}_t^Y$. Hence

$$\tilde{\pi}_t = \tilde{\pi}(t, Y) = \tilde{\pi}(t, Y_s, s \leq t). \tag{13.30}$$

Theorem 13.11 *Let $\hat{V}$ be the solution to (13.28) and let $\tilde{\pi}$ be given by (13.30). Then*

$$\hat{v}_t = \hat{V}(t, \hat{S}_t, \tilde{\pi}(t, Y)).$$

Example 13.12 *Let S_t, σ_t, B_t and X be given by Example 13.6. Let*

$$Y_t = \sigma_0 t + W'_t,$$

where σ_0 is a random variable such that there are constants $a < b$

$$\mathbb{P}(\sigma_0 = a) = \mathbb{P}(\sigma_0 = b) = \frac{1}{2}.$$

Let $\hat{\sigma}_t = \mathbb{E}(\sigma_0|\mathcal{F}_t^Y)$ and

$$d\hat{S}_t = \mu\hat{S}_t dt + \hat{\sigma}_t\hat{S}_t dW_t.$$

Obtain the price of the call option X based on the observation process Y_t and the stock price process $\hat{S}_t$.

Solution: First, we calculate $\hat{\sigma}_t$. Note that

$$\hat{\sigma}_t = a\mathbb{P}(\sigma_0 = a|\mathcal{F}_t^Y) + b\mathbb{P}(\sigma_0 = b|\mathcal{F}_t^Y) \equiv aF(Y) + bG(Y).$$

For any $A \in C([0,t])$, we have

$$\begin{aligned}\mathbb{P}(\sigma_0 = a, Y \in A) &= \mathbb{P}(\sigma_0 = a)\mathbb{P}(a \cdot + \omega \in A) \\ &= \frac{1}{2}\int_A \mathcal{E}_a(\omega)d\mathbb{P}(\omega)\end{aligned}$$

where $\mathcal{E}_a(\omega) = \exp\left(a\omega_t - \frac{1}{2}a^2 t\right)$.

On the other hand, we have

$$\begin{aligned}\mathbb{P}(\sigma_0 = a, Y \in A) &= \int_A F(Y)d\mathbb{P} \\ &= \int\int_A F(\omega)\mathcal{E}_{\omega_0}(\omega)d\mathbb{P}(\omega)d\mathbb{P}_0(\omega_0) \\ &= \int_A F(\omega)\frac{1}{2}(\mathcal{E}_a(\omega) + \mathcal{E}_b(\omega))d\mathbb{P}(\omega).\end{aligned}$$

Hence,

$$F(\omega) = \frac{\mathcal{E}_a(\omega)}{\mathcal{E}_a(\omega) + \mathcal{E}_b(\omega)}.$$

Similarly,

$$G(\omega) = \frac{\mathcal{E}_b(\omega)}{\mathcal{E}_a(\omega) + \mathcal{E}_b(\omega)}.$$

Therefore,

$$\hat{\sigma}_t = \frac{a\mathcal{E}_a(\omega) + b\mathcal{E}_b(\omega)}{\mathcal{E}_a(\omega) + \mathcal{E}_b(\omega)}.$$

Let $\hat{\gamma}_t = -\frac{\mu}{\hat{\sigma}_t}$ and

$$\frac{d\mathbb{P}'}{d\mathbb{P}} = \exp\left(\int_0^t \hat{\gamma}_u dW_u - \frac{1}{2}\int_0^t |\hat{\gamma}_u|^2 du\right).$$

Since

$$\hat{S}_T = \hat{S}_t \exp\left(\int_t^T \hat{\sigma}_s d\tilde{W}_s - \frac{1}{2}\int_t^T \hat{\sigma}_s^2 ds\right)$$

where $\tilde{W}_t$ is a $\mathbb{P}'$-Brownian motion, we have

$$\hat{v}_t = \mathbb{E}^{\mathbb{P}'}\left(\left(\hat{S}_t \exp\left(\int_t^T \hat{\sigma}_s d\tilde{W}_s - \frac{1}{2}\int_t^T \hat{\sigma}_s^2 ds\right) - K\right)^+ \middle| \mathcal{F}_t\right). \tag{13.31}$$

Remark Comparing with v_t in (13.15), $\hat{v}_t$ in (13.31) is not given by a closed form and hence, a numerical method is needed for its evaluation. Nevertheless, $\hat{v}_t$ depends only on the stock price and the observation process Y instead of on the unobserved σ_0.

13.5 PDE when S is observed

In this section, we assume that S_t is observed. As σ_t is $\mathcal{F}_t^S$-measurable, there is a function $b: [0,T] \times \mathbb{C}([0,T]) \to \mathbb{R}$ such that $\sigma_t = b(t,S)$. We write (13.2) as

$$dS_t = a(t,S_t)S_t dt + b(t,S)S_t dW_t. \tag{13.32}$$

Let $Z_t = S_t/B_t$. Then

$$dZ_t = \frac{S_t}{B_t}(a(t,S_t) - r_t)dt + \frac{b(t,S)S_t}{B_t}dW_t.$$

Let

$$\eta_t = \exp\left(\int_0^t \gamma_u dW_u - \frac{1}{2}\int_0^t |\gamma_u|^2 du\right)$$

where $\gamma_t = \frac{r_t - a(t,S_t)}{b(t,S)}$. By Assumptions (B), it is easy to show that η_t is a $\mathbb{P}$-square-integrable-martingale. Let

$$\frac{d\mathbb{P}^*}{d\mathbb{P}} = \eta_T, \qquad \mathbb{P} - a.s.$$

By Proposition 10.2.1 in (Musiela and Rutkowski, 1997), we see that the market with stock price given by (13.32) and bond by (13.2) is complete. Hence, $\mathbb{P}^*$ is the unique EMM for this market.

For any European contingent claim X settled at time T, the arbitrage price is

$$v_t' = B_t \mathbb{E}^{\mathbb{P}^*}(XB_T^{-1}\big|\mathcal{F}_t).$$

Since S_t is not Markovian, a PDE of the form of (13.14) is not available. We shall use the historical process to derive a PDE satisfied by the price process (we refer the reader to (Dawson and Perkins, 1991) for an introduction to historical processes). Let S_t' be $\mathbb{C}([0,T])$-valued process given by

$$S_t'(r) = S_{t\wedge r}, \qquad \forall r \in [0,T].$$

Then S_t', the historical process of S_t, is a Markov process. Let $\mathcal{A}_t'$ be the generator for S_t'.

As in Section 2, we assume that $r_t = r(t,S_t)$ and $X = g(S_T)$. We consider the following PDE for $(t,s') \in [0,T] \times \mathbb{C}([0,T])$:

$$\begin{cases} \frac{\partial}{\partial t}V'(t,s') + \mathcal{A}_t'V'(t,s') = r(t,s'(t))V'(t,s') \\ V'(T,s') = g(s'(T)). \end{cases} \tag{13.33}$$

Similar to Proposition 13.10, we have

Proposition 13.13 *Suppose that (13.33) has a solution V' such that*

$$|V'(t,s')| \le K(1 + \|s'\|^n)$$

where K and n are constants and $\|\cdot\|$ is the sup norm on $\mathbb{C}([0,T])$. Then

$$v_t' = V'(t,S_t').$$

14
The Russian Options

14.1 Introduction and background

In the European and American options of option pricing theory, the time period between the time the option is purchased and the time at or before which the option has to be exercised is fixed and known. If the purchase time is taken to be $t = 0$ and the exercise time $t = T$, then the European option pricing theory requires the option to be exercised at $t = T$ (the date of maturity); under the American option, you can exercise it at any time up to T, and moreover, the exercise time can be random (cf. (Karatzas and Shreve, 1988)).

A third approach to option pricing has recently been proposed by L.A. Shepp and A.N. Shiryaev and named the Russian options. In fact, they study the put option and, in a later work, the call option from this point of view (Shepp and Shiryaev, 1993a; Shepp and Shiryaev, 1993b). In both options, the period before the option is exercised can be indefinitely long and cannot be predicted in advance. Also, the option can be exercised at a random time. Another feature of both the put and call options is that an explicit expression for the fair price of the option is obtained. By "fair price" we mean the optimal expected present value based on the option. The optimal strategy is obtained and shown to be unique.

To our knowledge, the Russian options, treated in detail in this chapter, are not traded in any market. Some authors regard it as an "exotic" option (see, for instance, Musiela and Rutkowski, 1997). Nonetheless, the optimal stopping time problem which it solves is of considerable probabilistic and statistical interest. As in the European and American options, the Russian option assumes that the asset fluctuation follows the geometric Brownian motion model.

The Russian options differ from European and American options in that they are basically optimal stopping problems and hence of considerable probabilistic and statistical interest although (to our knowledge), they are not actually traded on any market. In this context, it is worth considering similar stopping problems that have been studied in other statistical contexts. A common thread that binds the Russian options and other examples about to be considered is the fact that all of them are related (albeit in a heuristic way) to Stefan-like or free boundary problems.

In 1961, H. Chernoff (Chernoff, 1961) studied sequential tests for the mean of a normal distribution which he concluded by relating his result to a free boundary problem. A detailed investigation in 1966 by B. Grigelionis and A.N. Shiryaev of optimal rules for Markov processes showed the relevance of Stefan's problem in this context.

As a final example, we mention the optimal stopping problem solved by Shepp in 1969 (Shepp, 1969). Shepp considers the continuous analog of the following optimal stopping problem: $X_i, i = 1, 2, \ldots$ are i.i.d. random variables which are being viewed, and if you stop viewing at n, you get the payoff $n^{-1}(X_1 + \ldots + X_n)$. What is required is the optimal stopping rule and a formula for the expected payoff. The continuous version, somewhat generalized, is to observe continuously, $\frac{u+W_t}{b+t}$ $(t > 0)$ where $-\infty < u < \infty$ and $b > 0$. If sampling is stopped at a random stopping time τ, the expected payoff is given by $V(u, b; \tau) := E\frac{u+W_\tau}{b+\tau}$. Let $V(u, b) = \sup_\tau E\frac{u+W_\tau}{b+\tau}$ (over those τ for which the expectation is defined). Shepp proved the following result:

$$V(u, b) = \begin{cases} (1 - \alpha^2) \int_0^\infty e^{\lambda u - \frac{\lambda^2 b}{2}} d\lambda & \text{if } u \le \alpha b^{1/2} \\ \frac{u}{b} & \text{if } u > \alpha b^{1/2}, \end{cases} \tag{14.1}$$

where α (not dependent on u, b) is the unique real root of the equation

$$\alpha = (1 - \alpha^2) \int_0^\infty e^{\lambda\alpha - \lambda^2/2} d\lambda. \tag{14.2}$$

The above problem (as noted by Shepp) is related to a free boundary problem (FBP) whose solution would furnish an alternative proof of the following theorem: Suppose C is the "continuation set", i.e., sampling is continued as long as $(t, W_t) \in C$ and is stopped at $t = \tau(C)$, the first exit from C. This formulation leads to the Stefan-like problem:

(i) $\frac{\partial V}{\partial b} + \frac{1}{2}\frac{\partial^2 V}{\partial u^2} = 0$ if $(u, b) \in C$ (differential operator of the process);

(ii) $V = g$ if $(u, b) \notin C$ (agrees with the complement of C);

(iii) $\frac{\partial V}{\partial u} = \frac{\partial g}{\partial u}, \frac{\partial V}{\partial b} = \frac{\partial g}{\partial b}$ if $(u, b) \in \partial C$.

Here $C = \{(u, b); u \le g(b)\}$ where g is the unknown free boundary. The existence of a unique solution (V, g), to the best of our knowledge, has not been proved.

In most of these optimal stopping problems, the FBP is used more as a guide to intuition and not as a technique for establishing the existence of a unique solution.

14.2 The Russian put option

In this section we generally follow Shepp and Shiryaev's work and show that their solution is unique. We do this by proving that the corresponding free boundary problem (FBP) has a unique solution. In the papers of the above authors, the FBP is indeed alluded to but is not exploited to obtain uniqueness.

We first consider the so-called put option. Here you, the buyer of the option, can exercise your option at any time; in other words, the *exercise* time is up to you; it can be random and the time period between buying the option and exercising it can be indefinitely long. We assume the Black and Scholes GBM model for the stock price

$$dX_t = \mu X_t dt + \sigma X_t dW_t, t > 0, X_0 = x, \tag{14.3}$$

or

$$X_t = xe^{(\mu - \frac{1}{2}\sigma^2)t + \sigma W_t} \tag{14.4}$$

where $\mu \in \mathbf{R}$ and $\sigma > 0$ are known constants. Let

$$S_t = \max\left\{s, \sup_{0 \le u \le t} X_u\right\}, t \ge 0. \tag{14.5}$$

If you stop at τ(τ being a finite stopping time) you receive the payoff S_τ discounted by $e^{-r\tau}$, i.e., $e^{-r\tau}S_\tau$ where $r > 0$ is the discount rate. You, as the owner of the option, want to seek a strategy that will maximize $E_{x,s}e^{-r\tau}S_\tau$. This quantity, maximized over all finite stopping times, may be regarded as a "fair" price for buying the option. Thus we have to find

$$V^*(x, s) := \sup_\tau E_{x,s}(e^{-r\tau}S_\tau). \tag{14.6}$$

We shall assume that $r > \mu$ (otherwise as will be seen below, $V^*(x, s)$ will be infinite).

14.3 A free boundary problem for the put option

At this stage, we introduce the following free boundary problem associated with the Russian option:

$$\frac{1}{2}\sigma^2 x^2 \frac{\partial^2 V}{\partial x^2} + \mu x \frac{\partial V}{\partial x} - rV = 0 \tag{14.7}$$

if $g(s) < x \leq s$.

Conditions on the free boundary: (14.8)

g is continuous and differentiable, $g(s) > 0$ if $s > 0$, and $g(0) = 0$;

(i) $V(g(s), s) = s$,

(ii) $\frac{\partial V}{\partial x}(g(s), s) = 0$.

Conditions on the known boundary: (14.9)

$$V(x, s) \geq s; \frac{\partial V}{\partial s}|_{x=s} = 0.$$

Conditions (14.8) (i) and (ii) are conditions of smooth fit that go back to A.N. Kolmogorov in the 1950s.

One way to arrive at the differential equation (14.7) is to follow the principle due, presumably, to Mikhailov (see Shepp's comments in (Shepp, 1969)). The generator of the Markov process (t, X_t) is given by

$$L_t = \frac{\partial}{\partial t} + \mu x \frac{\partial}{\partial x} + \frac{1}{2}\sigma^2 x^2 \frac{\partial^2}{\partial x^2}.$$

$V(x, s)$ is then chosen so as to satisfy $L_t\{e^{-rt}V\} = 0$ in the continuation region $g(s) \leq x \leq s$. This immediately yields (14.7). Another way to derive (14.7) is given in (Shepp and Shiryaev, 1993a). To solve the FBP (14.7)–(14.9), we have to find V and the free boundary $g(s)$. Our aim is to show that there is a unique solution (V, g) and then use it to obtain the optimal stopping strategy and the fair price.

Shepp and Shiryaev guess at a solution to the above free boundary problem (FBP) and obtain an expression for $V^*(x, s)$ for $0 < x \leq s$. We will take a somewhat different approach and concentrate on the single phase Stefan-like problem (14.7)–(14.9).

Observe that (14.7) is really an ordinary differential equation. The so-called indicial equation (for $V = x^m$) is given by

$$\frac{1}{2}\sigma^2 m^2 + (\mu - \frac{1}{2}\sigma^2)m - r = 0.$$

The roots of this quadratic, denoted by γ_1 and γ_2, $\gamma_1 < 0 < 1 < \gamma_2$, are given by

$$\frac{(\frac{1}{2}\sigma^2 - \mu) \pm \sqrt{(\mu - \frac{1}{2}\sigma^2)^2 + 2\sigma^2 r}}{\sigma^2}.$$

The general solution

$$V(x, s) = A(s)x^{\gamma_1} + B(s)x^{\gamma_2}.$$

From conditions (14.8) we have

$$\begin{aligned} Ag^{\gamma_1} + Bg^{\gamma_2} &= s, \\ A\gamma_1 g^{\gamma_1} + B\gamma_2 g^{\gamma_2} &= 0. \end{aligned}$$

Hence

$$\begin{aligned} A &= \frac{s\gamma_2}{\gamma_2 - \gamma_1} g^{-\gamma_1}, \\ B &= -\frac{s\gamma_1}{\gamma_2 - \gamma_1} g^{-\gamma_2}, \end{aligned}$$

and we obtain

$$V(x,s) = \frac{s}{\gamma_2 - \gamma_1} \left\{ \gamma_2 \left(\frac{x}{g(s)} \right)^{\gamma_1} - \gamma_1 \left(\frac{x}{g(s)} \right)^{\gamma_2} \right\} \tag{14.10}$$

if $g(s) < x \le s$. For purposes of a later comparison, we write $V(x, s; g)$ for $V(x, s)$. We need a number of lemmas whose proofs will be given later.

Lemma 14.1 *If* $g(S_t) < X_t \le S_t$, *then*

$$V(X_t, S_t; g) \le V(S_t, S_t, g). \tag{14.11}$$

Lemma 14.2 *Assume that* s *is in a bounded closed interval* $[0, \bar{s}]$ *where* $\bar{s}$ *is arbitrary. Then*

$$g(s) \ge \frac{s}{H}, \tag{14.12}$$

where H *is a constant greater than 1.*

Denote the function $V(s, s; g)$ *by* $V(s, s; H)$ *with* $g(s)$ *replaced by* $\frac{s}{H}$.

Lemma 14.3 $V(S_t, S_t; g) \le A.s$ *where*

$$A = \frac{\gamma_2 H^{\gamma_1} - \gamma_1 H^{\gamma_2}}{\gamma_2 - \gamma_1} > 0.$$

Lemma 14.4 *If* $X_t < S_t$, *then* $dS_t = 0$.

Apply Itô's formula to the process

$$\begin{aligned} Y_t &:= e^{-rt} V(X_t, S_t; g). \qquad (14.13) \\ dY_t &= e^{-rt} \left\{ -rV(X_t, S_t)dt + \frac{\partial V}{\partial x}(X_t, S_t)dX_t \right. \qquad (14.14) \\ &\left. + \frac{1}{2}\frac{\partial^2 V}{\partial x^2}(X_t, S_t)d\langle X, X\rangle_t + \frac{\partial V}{\partial s}(X_t, S_t)dS_t \right\}. \end{aligned}$$

Then, recalling that we always work in the region $g(S_t) < X_t < S_t$, the last term in the curly brackets is zero if $X_t = S_t$, $\frac{\partial V}{\partial s}(S_t, S_t, g) = 0$ by (14.8) and if $X_t < S_t, dS = 0$ by Lemma 14.4. Since $d\langle X, X\rangle_t = \sigma^2 X_t^2 dt$, (14.14) becomes

$$\begin{aligned} dY_t &= e^{-rt} \left\{ -rV + \mu X_t \frac{\partial V}{\partial x} + \frac{1}{2}\sigma^2 X_t^2 \frac{\partial^2 V}{\partial x^2} \right\} dt + e^{-rt}\sigma X_t \frac{\partial V}{\partial x} dW_t. \\ &= \sigma e^{-rt} \frac{\partial V}{\partial x} X_t dW_t, \qquad (14.15) \end{aligned}$$

and

$$Y_{t\wedge\tau} = Y_0 + \int_0^{t\wedge\tau} \sigma e^{-ru} \frac{\partial V}{\partial x} X_u dW_u,$$

where τ is any finite stopping time. Letting $\sigma_n \uparrow \infty$ be a sequence of stopping times such that

$$\int_0^{t\wedge\tau\wedge\sigma_n} \sigma e^{-ru} \frac{\partial V}{\partial x} X_u dW_u$$

is a martingale, we see that $Y_{t\wedge\tau\wedge\sigma_n}$ is a positive martingale and

$$E_{x,s} Y_{t\wedge\tau\wedge\sigma_n} = E_{x,s} Y_0.$$

Making $\sigma_n \uparrow \infty$ and then $t \to \infty$, and applying Fatou's lemma, we have

$$E_{x,s} Y_\tau \leq E_{x,s} Y_0.$$

From the first condition in (14.9) and the above inequality, we obtain

$$E_{x,s} e^{-r\tau} S_\tau \leq V(x,s), \tag{14.16}$$

and taking the sup over τ yields the inequality

$$V^*(x,s) \leq V(x,s). \tag{14.17}$$

To show the opposite inequality, we first define the stopping time

$$\tau^* = \inf\{t > 0; X_t = g(S_t)\}, \tag{14.18}$$

which is the time of the first exit from the continuation set. Our aim is to show that

$$E_{x,s} Y_{t\wedge\tau^*} = V(x,s;g). \tag{14.19}$$

It should be noted that, though $Y_{t\wedge\tau}$ is a local martingale for any τ (hence for τ^*), it does not automatically follow that (14.19) holds. Here again we prove the necessary lemmas.

Lemma 14.5 $P[\tau^* < \infty] = 1.$

Lemma 14.6 *For* $a > 0, b > 0,$

$$P\{W_t \leq at + b, \quad 0 \leq t < \infty\} \geq 1 - e^{-2ab}. \tag{14.20}$$

From Lemmas 14.1 and 14.3, if $g(S_t) < X_t \leq S_t$,

$$Y_t = e^{-rt} V(X_t, S_t; g) \leq e^{-rt} V(S_t, S_t; H),$$

where A is the constant in Lemma 14.1.

It will be shown in Lemma 14.7 below that $\{Y_t\}_{0<t<\tau^*}$ is uniformly integrable for which it suffices to show that $E_{x,s}(\sup_{0<t<\infty} Y_t) < \infty$ which in turn follows if

$$E_{x,s} \sup_{0<t<\infty} e^{-rt} S_t < \infty. \tag{14.21}$$

From $E_{x,s} \sup_{0<t<\infty} Y_t < \infty$, it follows that for $T > t$,

$$\begin{aligned} E_{x,s} Y_{t\wedge\tau^*} &= E_{x,s} Y_{t\wedge\tau^*} 1_{[\tau^* \leq T]} + E_{x,s} Y_t 1_{[\tau^* > T]} \\ &< \infty. \end{aligned}$$

Thus, the local martingale $Y_{t\wedge\tau^*}$ is actually a martingale so that

$$E_{x,s} Y_{t\wedge\tau^*} = E_{x,s} Y_0. \tag{14.22}$$

Since $Y_{t\wedge\tau^*} \to Y_{\tau^*}$ a.s. as $t \to \infty$, by the uniform integrability established above, (14.22) yields

$$E_{x,s} Y_{\tau^*} = E_{x,s} Y_0. \tag{14.23}$$

Now

$$\begin{aligned} Y_{\tau^*} &= e^{-r\tau^*} V(X_{\tau^*}, S_{\tau^*}) \\ &= e^{-r\tau^*} V(g(S_{\tau^*}), S_{\tau^*}) \\ &= e^{-r\tau^*} S_{\tau^*}, \end{aligned}$$

the last equality following from (14.8) (i). Hence from (14.23), we have

$$E_{x,s} e^{-r\tau^*} S_{\tau^*} = V(x, s; g). \tag{14.24}$$

(14.24) immediately gives

$$V^*(x, s) \geq V(x, s; g). \tag{14.25}$$

Combining with (14.17) we finally obtain the desired equality

$$V(x, s; g) = V^*(x, s). \tag{14.26}$$

(14.24) and (14.26) further show that τ^* is the optimal stopping time.

It will be seen from the proof of Lemma 14.2 that $g'(0) = \frac{1}{c}$, where $c = \left(\frac{1-1/\gamma_1}{1-1/\gamma_2}\right)^{1/(\gamma_2-\gamma_1)}$, $c > 1$ and that $c \leq H$ since $\frac{g(s)}{s} \geq \frac{1}{H}$ for all $s > 0$. Letting $g_c(s) = \frac{s}{c}$, it is easily verified that g_c satisfies conditions (14.8) (i) and (ii) required of the free boundary.

We may now repeat the above procedure by replacing g by g_c and obtain, exactly as above, the relation

$$V^*(x, s) = V(x, s; g_c) \quad \text{for} \quad \frac{s}{c} < x \leq s. \tag{14.27}$$

Comparing (14.26) and (14.27) we have

$$V(x, s; g_c) = V(x, s; g) \quad \text{for} \quad \max\{\frac{s}{c}, g(s)\} < x \le s, \tag{14.28}$$

i.e.,

$$\frac{s}{\gamma_2 - \gamma_1}\left\{\gamma_2(\frac{cx}{s})^{\gamma_1} - \gamma_1(\frac{cx}{s})^{\gamma_2}\right\} = \frac{s}{\gamma_2 - \gamma_1}\left\{\gamma_2(\frac{x}{g})^{\gamma_1} - \gamma_1(\frac{x}{g})^{\gamma_2}\right\}.$$

Putting $x = s$, we have

$$\gamma_2 c^{\gamma_1} - \gamma_1 c^{\gamma_2} = \gamma_1\left(\frac{s}{g}\right)^{\gamma_1} - \gamma_1\left(\frac{s}{g}\right)^{\gamma_2}. \tag{14.29}$$

Differentiating (14.29) with respect to s at $s > 0$ we get

$$\left\{\left(\frac{s}{g}\right)^{\gamma_1 - 1} - \left(\frac{s}{g}\right)^{\gamma_2 - 1}\right\}\frac{d}{ds}\left(\frac{s}{g}\right) = 0.$$

Since $g(s) > s$, we cannot have $(\frac{s}{g(s)})^{\gamma_1} = (\frac{s}{g(s)})^{\gamma_2}$. We thus have $\frac{d}{ds}(\frac{s}{g}) = 0$ or $\frac{g'(s)}{g(s)} = \frac{1}{s}$ which gives $g(s) = \frac{s}{k}$ for $s > 0$, k being a constant of integration.

Since $g(0) = 0$, we have $g(s) = \frac{s}{k}$ for all s. To find k, substituting in the expression for $g'(s)$,

$$g'(s) = \frac{\frac{1}{\gamma_1}(\frac{s}{g})^{\gamma_1} - \frac{1}{\gamma_2}(\frac{s}{g})^{\gamma_2}}{(\frac{s}{g})^{\gamma_1+1} - (\frac{s}{g})^{\gamma_2+1}},$$

we find that

$$\frac{1}{k} = (\frac{1}{\gamma_1}k^{\gamma_1} - \frac{1}{\gamma_2}k^{\gamma_2})/(k^{\gamma_1+1} - k^{\gamma_2+1})$$

which gives the value

$$k = \left(\frac{1 - \frac{1}{\gamma_1}}{1 - \frac{1}{\gamma_2}}\right)^{\frac{1}{\gamma_2 - \gamma_1}}. \tag{14.30}$$

The constant k is the same as α, which is Shepp and Shiryaev's value (and also $= c$). We have thus proved that the free boundary problem (14.7)–(14.9) has the unique solution $(V(x, s), g(s))$ where $g(s) = \frac{s}{\alpha}$.

Remark It is interesting to note that the unique solution of our free boundary problem is obtained without having to use $V^*(x, s)$ in the region $0 < x \le \frac{s}{\alpha}$.

We can complete the option problem as follows. From

$$V^*(x, s) = \sup_{\tau} E_{x,s} e^{-r\tau} S_\tau,$$

since $S_t \ge s$ for all t,

$$E_{x,s} e^{-r\tau} S_\tau \ge s E_{x,s} e^{-r\tau}.$$

Now choose the stopping time $\tau_\epsilon = \epsilon\tau$, where τ is a finite stopping time and $\epsilon > 0$. Then $V^*(x, s) \geq sE_{x,s}e^{-r\tau_\epsilon} = sE_{x,s}e^{-\epsilon r\tau}$ for every $\epsilon > 0$. Making $\epsilon \to 0$, since $E_{x,s}e^{-\epsilon r\tau} \to 1$ we have

$$V^*(x, s) \geq s. \tag{14.31}$$

(Alternately, we get (14.31) by taking $\tau \equiv 0$ if the latter is allowed to be a stopping time for the problem).

Next, writing

$$Z_t = e^{-rt}S_t \quad \text{in} \quad 0 < X_t \leq \frac{S_t}{\alpha},$$

by Ito's formula,

$$\begin{aligned} dZ_t &= -re^{-rt}S_t dt + e^{-rt}dS_t \\ &= -rZ_t dt \end{aligned}$$

since $dS_t = 0$ by Lemma 14.4.

$$Z_t = Z_0 - \int_0^t rZ_u du \leq Z_0 \quad \text{since} \quad Z_u \geq 0.$$

Therefore,

$$E_{x,s}Z_{t\wedge\tau} \leq E_{x,s}Z_0 = s,$$

where τ is any finite stopping time. By Fatou's Lemma as $t \to \infty$,

$$E_{x,s}e^{-r\tau}S_\tau = E_{x,s}Z_\tau \leq s. \tag{14.32}$$

Taking sup over τ of the left hand side, we get

$$V^*(x, s) \leq s. \tag{14.33}$$

This proves $V^*(x, s) = s$ if $0 < x \leq s/\alpha$. □

We have proved the following result:

Theorem 14.8 *The free boundary problem (14.7)–(14.9) has the* unique *solution* $(V(x, s), g(s))$ *with* $g(s) = \frac{s}{\alpha}$. *Furthermore,*

$$V(x, s) = V^*(x, s) \quad \textit{in} \quad \frac{s}{\alpha} < x \leq s.$$

It has also been shown that

$$V^*(x, s) = s \quad \textit{in} \quad 0 < x \leq \frac{s}{\alpha}.$$

Note that $V(x, s) \to s$ *on* $x \searrow \frac{s}{\alpha}$.

Remark Shepp and Shiryaev claim that the Russian option has "reduced risk" in the sense that you don't have "to worry about missing a good price in the recent past" because you get the "best price up to settlement time." The *reduced* risk formulation is needed since, if one considers not S_τ but the current value X_τ, the problem has no interest, as is shown by Shepp and Shiryaev in their second paper. Here is their result.

Let $V(x) = \sup_\tau E_x e^{-r\tau} X_\tau$.

Theorem 14.9 *(i) If $r \geq \mu$, then $V(x) = x$;*
(ii) If $r < \mu$, then $V(x) = \infty$.

Proof Write $Z_t = e^{-rt} X_t$. Then

$$dZ_t = e^{-rt} X_t \{(\mu - r)dt + \sigma dW_t\}.$$

(i) Proceeding as in the proof of Theorem 14.8, we have $E_x dZ_t \leq 0$ and $E_x Z_\tau \leq E_x Z_0 = x$ for any τ, so that $V(x) \leq x$. Taking $\tau = 0$ or using the stopping time τ_ϵ as in (14.31) we get $V(x) \geq x$.
(ii) Now $r < \mu$. Taking the fixed stopping time $\tau \equiv t$, we have

$$V(x) \geq E_x e^{-rt} X_t = e^{(\mu - r)t} \to \infty$$

as $t \to \infty$ giving $V(x) = \infty$. □

14.4 Proofs of the lemmas

Lemma 14.1 Differentiating $V(x, s; g)$ with respect to x in $g(s) < x \leq s$, we get

$$\frac{\partial V}{\partial x} = \frac{s\gamma_1\gamma_2}{\gamma_2 - \gamma_1} x^{-1} \left\{ (\frac{x}{g})^{\gamma_1} - (\frac{x}{g})^{\gamma_2} \right\} > 0$$

since $\gamma_1\gamma_2 < 0$ and $(\frac{x}{g})^{\gamma_1} - (\frac{x}{g})^{\gamma_2} < 0$. □

Lemma 14.2 For $s > 0$ we have from condition (14.8) (ii)

$$g'(s) = \frac{\frac{1}{\gamma_1}(\frac{s}{g})^{\gamma_1} - \frac{1}{\gamma_2}(\frac{s}{g})^{\gamma_2}}{(\frac{s}{g})^{\gamma_1+1} - (\frac{s}{g})^{\gamma_2+1}} \equiv h(s),$$

$$\text{so that } \frac{1}{h(s)} = \frac{(\frac{s}{g})^{\gamma_2+1} - (\frac{s}{g})^{\gamma_1+1}}{\frac{1}{\gamma_2}(\frac{s}{g})^{\gamma_2} - \frac{1}{\gamma_1}(\frac{s}{g})^{\gamma_1}}.$$

The denominator is positive for $s > 0$ (and of course, continuous). Since $g(0) = 0$ by assumption,

$$\lim_{s\downarrow 0} \frac{s}{g(s)} = \lim_{s\downarrow 0} \frac{1}{g'(s)} = \frac{1}{g'(0)} = c,$$

where c is a positive finite number. In fact, from the expression for $g'(s)$ it is easily seen that

$$c = \left(\frac{1 - \frac{1}{\gamma_1}}{1 - \frac{1}{\gamma_2}} \right)^{\frac{1}{\gamma_2 - \gamma_1}}.$$

Hence, restricting s to an arbitrary compact interval $[0, \bar{s}]$ (it will turn out that this is no restriction at all), we get $\frac{1}{h(s)} \leq H$ for all s in $[0, \bar{s}]$. We then get

$$g(s) = \int_0^s h(u)du \geq \frac{s}{H}$$

where $H > 1$ since $g(s) < s$. This proves the lemma.

Lemma 14.3 Let $0 < y < s$. Then

$$\begin{aligned}
V(s, s; y) &= \frac{s}{\gamma_2 - \gamma_1} \left\{ \gamma_2 (\frac{s}{y})^{\gamma_1} - \gamma_1 (\frac{s}{y})^{\gamma_2} \right\}. \\
\frac{dV(s, s; y)}{dy} &= \frac{s}{\gamma_2 - \gamma_1} \left\{ -\gamma_2 \gamma_1 s^{\gamma_1} y^{-\gamma_1 - 1} + \gamma_2 \gamma_1 s^{\gamma_2} y^{-\gamma_2 - 1} \right\} \\
&= \frac{s}{\gamma_2 - \gamma_1} \frac{\gamma_1 \gamma_2}{y} \left\{ (\frac{s}{y})^{\gamma_2} - (\frac{s}{y})^{\gamma_1} \right\} < 0
\end{aligned}$$

since $\gamma_1 \gamma_2 < 0$ and $(\frac{s}{y})^{\gamma_2} > (\frac{s}{y})^{\gamma_1}$. The latter holds because $(\frac{s}{y})^{\gamma_2 - \gamma_1} > 1$ since $0 < y < s$.

Now since from Lemma 14.2, $g(s) \geq \frac{s}{H}$, we have that

$$V(s, s; g) \leq \frac{s}{\gamma_2 - \gamma_1} (\gamma_2 H^{\gamma_1} - \gamma_1 H^{\gamma_2}).$$

Note that we cannot write the right hand expression as $V(s, s; g_H)$ where $g_H = \frac{s}{H}$ because g_H has not been shown to satisfy condition (14.8) (i). □

Lemma 14.4 From

$$\begin{aligned}
S_t &= \max\{s, \sup_{0 \leq u \leq t} X_u\}; \\
\text{we have for } h > 0, \quad S_{t+h} &= \max\{S_t, \sup_{t \leq u \leq t+h} X_u\}. \\
\text{For } u \geq t, \quad X_u &= X_t e^{(\mu - 1/2\sigma^2)(u-t) + \sigma(W_u - W_t)}.
\end{aligned}$$

From the limsup part of Lévy's version of the law of the iterated logarithm, we have

$$\limsup_{0 < t_2 - t_1 \to 0} \frac{W(t_2) - W(t_1)}{\left(2(t_2 - t_1) \log \frac{1}{t_2 - t_1}\right)^{1/2}} = 1 \quad \text{a.s.}$$

Hence, for $\delta > 0$,

$$W_u(\omega) - W_t(\omega) \leq (1+\delta)\big(2(u-t)\ \log(\frac{1}{u-t})\big)^{1/2}$$

for $0 < u - t \leq h_0(\delta, \omega)$ a.s.

Define Ω_1 to be the set of ω's for which the above inequality holds. Then $P(\Omega_1) = 1$ and for $\omega \in \Omega_1 \cap \{X_t < S_t\}$ and $0 < h \leq h_0$,

$$\begin{aligned} S_{t+h} &= \max\big\{S_t, X_t \sup_{t<u\leq t+h} e^{\mu-\frac{1}{2}\sigma^2)(u-t)+\sigma(W_u-W_t)}\big\} \\ &\leq \max\big\{S_t, X_t \sup_{0\leq u-t\leq h} e^{\mu-\frac{1}{2}\sigma^2)(u-t)+\sigma(1+\delta)(2(u-t)\ log\frac{1}{u-t}))^{1/2}}\big\}. \end{aligned}$$

By choosing δ and hence h_0 sufficiently small, and noting that $X_t(\omega) < S_t(\omega)$, we obtain for $0 < h \leq h_0$ and noting that $X_t(\omega) < S_t(\omega)$,

$$\begin{aligned} & S_{t+h} \leq S_t. \\ \text{But} \quad & S_{t+h} \geq S_t \quad \text{a.s.} \\ \text{Hence} \quad & S_{t+h} - S_t = 0 \quad \text{a.s. on the set } \Omega_1 \cap \{X_t < S_t\}. \\ \text{That is,} \quad & dS_t = 0 \quad \text{a.s. on } \Omega_1 \cap \{X_t < S_t\}. \end{aligned}$$

(In some of the above steps, ω has been suppressed for convenience.)

Lemma 14.5 For $T > 0$, by the definition of τ^*,

$$\begin{aligned} \{\tau^* > T\} &= \{X_t > g(S_t), \quad 0 \leq t \leq T\} \\ &\subset \{X_t > \frac{S_t}{H}, \quad 0 \leq t \leq T\}, \end{aligned}$$

the last inclusion, a consequence of Lemma 14.2. Hence

$$P_{x,s}(\tau^* > T) \leq P_{x,s}(X_t > \frac{S_t}{H}, \quad 0 \leq t \leq T).$$

Now for $0 \leq t \leq T$, $X_t > \frac{S_t}{H}$ implies $\log X_t > \log S_t + \log \frac{1}{H}$, i.e.,

$$(\mu - \frac{1}{2}\sigma^2)t + \sigma W_t > (\mu - \frac{1}{2}\sigma^2)u + \sigma W_u + \log\frac{1}{H}$$

for $0 \leq u \leq t \leq T$. Hence

$$P_{x,s}(\tau^* > T) \leq$$

$$P_{x,s}\left\{\sigma(W_t - W_u) + (\mu - \frac{1}{2}\sigma^2)(t-u) \geq \log\frac{1}{H}, 0 \leq u \leq t \leq T\right\}.$$

The right hand side is

$$\begin{aligned}
&\le P_{x,s}\left\{W_t - W_u \ge \frac{-\log H}{\sigma} - \frac{|\mu - \frac{1}{2}\sigma^2|}{\sigma}(t-u), 0 \le u \le t \le T\right\}\\
&= P_{x,s}\left\{W_u - W_t \le \frac{\log H}{\sigma} + \frac{|\mu - \frac{1}{2}\sigma^2|}{\sigma}(t-u), 0 \le u \le t \le T\right\}\\
&\le P_{x,s}\left\{W_d \le b, W_{2d} - W_d \le b, \dots, W_{nd} - W_{(n-1)d} \le b\right\}.
\end{aligned}$$

Here $T = nd$ and $b = \frac{\log H}{\sigma} + \frac{|\mu - \frac{1}{2}\sigma^2|}{\sigma} d$. From the above chain of inequalities we get

$$P_{x,s}(\tau^* > T) \le \left(\frac{1}{\sqrt{2\pi d}}\int_{-\infty}^{b} e^{-\frac{x^2}{2d}}dx\right)^n \to 0$$

since $n \to \infty$ as $T \to \infty$. Hence $\tau^* < \infty$ a.s. □

Lemma 14.6 If $a > 0, b > 0$, then

$$P_{x,s}\{W_t \le at + b, 0 \le t < \infty\} > 1 - e^{-2ab}.$$

This is a well known inequality and the proof is given below.

Let $Z_t = e^{2aW_t - 2a^2 t}$. Then Z_t is a martingale with $EZ_t = 1$.

$$\begin{aligned}
&P_{x,s}\{W_s \le as + b, 0 \le s \le t\}\\
&\quad = P_{x,s}\left\{\sup_{0\le s\le t}(2aW_s - 2a^2 s) \le 2ab\right\}\\
&\quad = 1 - P_{x,s}\left\{\sup_{0\le s\le t} Z_s > e^{2ab}\right\} \ge 1 - e^{-2ab}.
\end{aligned}$$

□

Lemma 14.7 $E\left(\sup_{t>0}, e^{-rt}S_t\right) < \infty$ *or*

$$\int_0^\infty P\left[\sup_t e^{-rt}S_t > y\right]dy < \infty.$$

Let $y > s, y > x$.

$$\sup_t e^{-rt}S_t = \sup_t e^{-rt}\max\left\{x, \sup_{0\le u\le t} X_n\right\}.$$

Therefore

$$\begin{aligned}
\sup_t e^{-rt} S_t > y \quad &<=> \quad \sup_t e^{-rt} \sup_{0 \le u \le t} X_u > y \\
&<=> \quad \sup_t \left[-rt + \log x + \sup_{0 \le u \le t} \{(u + \sigma W_u\} \right] > \log y \\
&<=> \quad \sup_t \left[\sup_{0 \le u \le t} \left\{ (\mu - \frac{\sigma^2}{2})u + \sigma W_u \right\} > \log \frac{y}{x} + rt \right].
\end{aligned}$$

Hence

$$\begin{aligned}
& P\left\{ \sup_t e^{-rt} S_t > y \right\} \\
&\quad = P\left\{ \sup_t \left(\sup_{0 \le u \le t} \left[(\mu - \frac{\sigma^2}{2})u + \sigma W_u \right] > \log \frac{y}{x} + rt \right) \right\} \qquad (14.34)
\end{aligned}$$

Writing

$$\begin{aligned}
a &= (r - \mu + \frac{1}{2}\sigma^2)/\sigma \quad (r > \mu) \\
b &= \frac{1}{\sigma} \log \frac{y}{x},
\end{aligned}$$

we have that if $W_t \le at + b \quad \forall t$, then

$$\begin{aligned}
& \sup_{0 \le u \le t} \left\{ (\mu - \frac{\sigma^2}{2})u + \sigma W_u \right\} \\
&\quad \le \sup_{0 \le u \le t} \left\{ (\mu - \frac{\sigma^2}{2})u + \sigma (au + b) \right\} \\
&\quad = \sup_{0 \le u \le t} \left\{ \log(\frac{y}{x}) + ru \right\} = \log(\frac{y}{x}) + rt, \quad \forall t. \qquad (14.35)
\end{aligned}$$

Noting that $ab = \frac{1}{\sigma^2}(r - \mu + \frac{1}{2}\sigma^2) \log \frac{y}{x} = (\frac{1}{2} + \frac{r-\mu}{\sigma^2}) \log \frac{y}{x}$, it follows from (4.1) and (4.2) that

$$\begin{aligned}
& P\left\{ \sup_t e^{-rt} S_t > y \right\} \\
&\quad \le 1 - P\{W_t \le at + b \quad \forall t\} \\
&\quad \le e^{-2ab} = e^{-(1 + \frac{2(r-\mu)}{\sigma^2})} \log \frac{y}{x} \\
&\quad = \left(\frac{y}{x}\right)^{-(1 + \frac{(r-\mu)}{\sigma^2})}.
\end{aligned}$$

Hence

$$\int_1^\infty P\left[\sup_t e^{-rt} S_t > y\right] dy \leq \int_1^\infty \left(\frac{y}{x}\right)^{-\{1+\frac{2(r-\mu)}{\sigma^2}\}} dy < \infty$$

if $r > \mu$. □

14.5 The Russian call option (or the option for selling short)

In this section and the next, we consider a dual to the Russian put option. This is a *call* option for "selling short," studied by Shepp and Shiryaev. The idea is that the seller pays the *minimum* price (in inflated as opposed to discounted dollars) of the share or asset during the time period between the *selling* time ("now") and the *delivery* time, the latter to be determined by the seller using a stopping rule. The seller gets the best (i.e., minimum) price up to settlement time. The problem is to find the optimal settlement time and to obtain an exact formula for the optimal expected fair price.

The stock price process is, as before, the geometric Brownian motion model

$$X_t = xe^{(\mu-1/2\sigma^2)t+\sigma W_t} \tag{14.36}$$

as in the previous section. Let $r > 0$ and $y \leq x$ be given and let τ be a finite stopping time which the seller chooses as the time of delivery. Let

$$Y_t := \min\{y, \inf_{0\leq u\leq t} X_u\}, t \geq 0. \tag{14.37}$$

That is, Y_t is the minimum value, starting at y, for X. The mathematical problem of our "short selling" option is, starting with the initial values x and y, to minimize $E_{x,y}e^{r\tau}Y_\tau$ over all finite stopping times τ, i.e., to calculate

$$V^*(x, y) := \inf_\tau \mathbf{E}_{x,y}e^{r\tau}Y_\tau \tag{14.38}$$

and, further, to find the optimal stopping time $\hat{\tau}$ to achieve this:

$$\mathbf{E}_{x,y}e^{r\hat{\tau}}Y_{\hat{\tau}} = V^*(x, y). \tag{14.39}$$

14.6 The F.B.P. for the call option

We shall first dispose of a trivial case when $V^* = 0$. Here we give Shepp and Shiryaev's proof.

Proposition 14.10 *$V^* = 0$ if any of the following conditions is satisfied:*

$$r < 0, \tag{14.40}$$

$$r = 0, \quad \text{and} \quad \mu \le \frac{\sigma^2}{2}, \tag{14.41}$$

$$0 < r \le \frac{1}{2\sigma^2}\left(\mu - \frac{\sigma^2}{2}\right)^2, \mu < \frac{\sigma^2}{2}. \tag{14.42}$$

Proof If either (14.40) or (14.41) holds, the result follows since e^{rt} is bounded and $Y_T \to 0$ as $T \to \infty$. By taking $\tau \equiv T$, V^* can be made arbitrarily small by taking T large enough.

Suppose (14.42) holds. Now use the formula

$$P(W_t \le at + b, 0 \le t \le T) = \Phi\left(a\sqrt{T} + \frac{b}{\sqrt{T}}\right) - e^{-2ab}\Phi\left(a\sqrt{T} - \frac{b}{\sqrt{T}}\right) \tag{14.43}$$

if $b > 0$, where Φ is the standard normal distribution function. Then it can be verified that

$$\begin{aligned}
&\lim_{T\to\infty} e^{rT} E \inf_{0\le t\le T}\left\{e^{\sigma W_t + (\mu-\sigma^2/2)t}\right\} \\
&= \lim_{T\to\infty} e^{rT}\int_0^1 e^{-x} P\left[\sigma W_t + \left(\mu - \frac{1}{2}\sigma^2\right)t \ge -x, \quad 0 \le t \le T\right] dx \\
&= \lim_{T\to\infty} e^{rT}\int_0^1 e^{-x} P\left[W_t \le \left(\frac{\mu}{\sigma} - \frac{1}{2}\sigma\right)t + \frac{x}{\sigma}, \quad 0 \le t \le T\right] dx \\
&= \lim_{T\to\infty} e^{rT}\int_0^1 e^{-x}\left\{\Phi\left[\left(\frac{\mu}{\sigma} - \frac{1}{2}\sigma\right)\sqrt{T} + \frac{x}{\sigma\sqrt{T}}\right]\right. \\
&\qquad \left. - e^{-\frac{2x}{\sigma}(\frac{\mu}{\sigma} - \frac{1}{2}\sigma)}\Phi\left[\left(\frac{\mu}{\sigma} - \frac{1}{2}\sigma\right)\sqrt{T} - \frac{x}{\sigma\sqrt{T}}\right]\right\} dx \\
&= 0
\end{aligned} \tag{14.44}$$

since $\mu < \frac{1}{2}\sigma^2$ and $r - \frac{1}{2}(\frac{\mu}{\sigma} - \frac{1}{2}\sigma)^2 < 0$. □

The above proof follows from Shepp and Shiryaev's second paper (Shepp and Shiryaev, 1993b).

For the remaining values of r, μ and σ we consider the FBP:

$$\frac{1}{2}\sigma^2 x^2 \frac{\partial^2 V}{\partial x^2} + \mu x \frac{\partial V}{\partial x} + rV = 0 \quad \text{in} \quad 0 < y < x \le g(y) \tag{14.45}$$

with the following conditions

$$V(x, y) \le y, \quad \frac{\partial V}{\partial y}(x, y)|_{x=y} = 0; \tag{14.46}$$

The conditions on the free boundary are

(a) $V(g(y), y) = y$; (14.47)

(b) $\frac{\partial V}{\partial x}(g(y), y) = 0$.

Further, g is assumed continuous and differentiable with $g(s) > 0$ for $s > 0$ and $g(0) = 0$. We also assume that g has a right hand derivative at $s = 0$. These conditions do not make sense when $g(y) \equiv \infty$. As we shall see, this case arises when $r = 0$. Conditions (14.47)(a) and (b) then have to be replaced by

(a′) $\lim_{x\to\infty} V(x, y) = y$ and (14.47)

(b′) $\lim_{x\to\infty} \frac{\partial V}{\partial x}(x, y) = 0$.

The indicial equation corresponding to (14.45) is the quadratic

$$\frac{1}{2}\sigma^2\gamma^2 + (\mu - \frac{\sigma^2}{2})\gamma + r = 0. \tag{14.48}$$

Several cases arise which have to be treated separately.

Case 1 $0 < r < \frac{1}{2\sigma^2}(\mu - \frac{\sigma^2}{2})^2, \mu > \frac{1}{2}\sigma^2$. Equation (14.48) has distinct real roots $\gamma_1 < \gamma_2 < 0$. Writing $V(x, y; g)$ for $V(x, y)$ we have

$$V(x, y; g) = \frac{y}{\gamma_2 - \gamma_1}\left\{\gamma_2(\frac{x}{g})^{\gamma_1} - \gamma_1(\frac{x}{g})^{\gamma_2}\right\}. \tag{14.49}$$

Case 2 $r = \frac{1}{2\sigma^2}(\mu - \frac{\sigma^2}{2})^2, \mu > \frac{\sigma^2}{2}$. Now the indicial equation has equal roots γ and

$$V(x, y; g) = y\left\{(\frac{x}{g})^{\gamma} - \gamma(\frac{x}{g})^{\gamma}\log\frac{x}{g}\right\}. \tag{14.50}$$

Note that (14.50) can be obtained by applying L'Hôpital's rule to the right hand side of (14.49).

Case 3 $r > \frac{1}{2\sigma^2}(\mu - \frac{\sigma^2}{2})$. Now we have the complex conjugate roots $\gamma, \overline{\gamma}$ with $Im(\gamma) > 0$ and

$$V(x, y; g) = \frac{y}{\gamma - \overline{\gamma}}\left\{\gamma(\frac{x}{g})^{\overline{\gamma}} - \overline{\gamma}(\frac{x}{g})^{\gamma}\right\}. \tag{14.51}$$

Case 4 $r = 0, \mu > \frac{\sigma^2}{2}$. This is the case when $g(y) \equiv \infty$. To see this, observe that $Y_t \downarrow Y_\infty$ as $t \to \infty$, and it is clear that there is no finite stopping time τ such that $E_{x,y}Y_\tau = V^*(x, y)$. In fact, we have

$$V^*(x, y) = E_{x,y}Y_\infty. \tag{14.52}$$

Turning to the indicial equation with the modified conditions (14.47)(a′) and (b′) and solving the equation $\frac{\sigma^2 x^2}{2}\frac{\partial^2 V}{\partial x^2} + \mu x \frac{\partial V}{\partial x} = 0$, we obtain

$$V(x, y) = B(y) - \frac{A(y)}{\beta - 1} \cdot \frac{1}{x^{\beta-1}},$$

where A and B are constants of integration and $\beta = \frac{2\mu}{\sigma^2} > 1$. Since $V(x, y) = y$ for all x, $y = \lim_{x\to\infty} V(x, y) = B(y)$. Next,

$$\frac{\partial V}{\partial y}|_{x=y} = 1 - \frac{A'(y)}{\beta - 1} \cdot \frac{1}{y^{\beta-1}} = 0,$$

(A' being the derivative of A). Hence $A(y) = (\beta - 1)\frac{y^\beta}{\beta}$. We thus have

$$\begin{aligned} V(x, y) &= y - \frac{y^\beta}{\beta} \cdot \frac{1}{x^{\beta-1}} \\ &= y\left\{1 - \left(\frac{y}{x}\right)^{2\mu/\sigma^2-1} \cdot \frac{\sigma^2}{2\mu}\right\}; \quad y \le x < \infty. \end{aligned} \tag{14.53}$$

The proof that $V^*(x, y) = V(x, y)$ will be deferred until after the first three cases have been discussed. Cases 1–3 are essentially non-trivial cases. We shall show that the FBP (14.45)–(14.47) has a unique solution with $g(y) = \theta y$ where θ is the value obtained by Shepp and Shiryaev for the three cases. We shall derive the result only for Case 2 since the other two cases can be treated similarly.

Case 2 Differentiating V with respect to y and using the second condition in (14.46), we get

$$g'(y) = \frac{\gamma(\frac{y}{g})^\gamma \log(\frac{y}{g}) - (\frac{y}{g})^\gamma}{\gamma^2(\frac{y}{g})^{\gamma+1}\log(\frac{y}{g})}, \quad 0 < y < g(y). \tag{14.54}$$

As $y \downarrow 0$ it follows that
$\frac{g(y)}{y} \to g'(0) = k$, where k is finite from (14.54),

$$k = \frac{\gamma \cdot \frac{1}{k^\gamma}\log\frac{1}{k} - \frac{1}{k^\gamma}}{\gamma^2(\frac{1}{k})^{\gamma+1}\log\frac{1}{k}} = e^{\frac{1}{\gamma(\gamma-1)}}.$$

Note that $k > 1$ since $\gamma < 0$. Thus from (14.54) we conclude that $g'(y)$ is continuous and bounded in a closed interval $[0, y_0]$ where y_0 is arbitrary. We then have

$$g(y) = \int_0^t g'(u)du \le Ly, \tag{14.55}$$

where L is an upper bound for g'. Much of the ensuing argument is similar to that in the put option case. If we define

$$\hat{g}(y) = ky \tag{14.56}$$

(Note that $k = \theta$ in Shepp and Shiryaev's notation), then $\hat{g}$ satisfies all the conditions for it to be a free boundary and is indeed the free boundary obtained by them. Our aim is to show that $\hat{g}$ is the unique boundary, which will then imply the uniqueness of the optimal stopping time for our option pricing problem. Let us denote by g, any free boundary which will solve the FBP. Our aim is to show that

$$V^*(x, y) = V(x, y; g). \tag{14.57}$$

Write

$$Z_t = e^{rt} V(X_t, Y_t; g). \tag{14.58}$$

In the continuation region $0 < Y_t \le X_t \le g(Y_t)$ (arguing as in the put option case), Y_t grows only when $X_t = Y_t$, that is, $dY_t = 0$ a.s. on the set $\{Y_t < X_t\}$.

Let $\sigma_n \uparrow \infty$ be a sequence of stopping times such that the integral below is a martingale. We then have

$$Z_{t\wedge\sigma_n} = Z_0 + \int_0^{t\wedge\sigma_n} \sigma e^{rs} \frac{\partial V}{\partial x}(X_s, Y_s) X_s dW_s,$$

where (by choice of σ_n), $Z_{t\wedge\sigma_n}$ is a positive martingale. Since the integral on the right hand side is a martingale, we obtain upon replacing t by any finite stopping time τ and taking expectations

$$E_{x,,y}(Z_{\tau\wedge\sigma_n}) = Z_0 = V(x, y; g). \tag{14.59}$$

Making $n \to \infty$, by Fatou's lemma,

$$E_{x,y} Z_\tau \le V(x, y; g). \tag{14.60}$$

Now define the stopping time

$$\hat{\tau} := \inf\{t \ge 0 : X_t = g(Y_t)\}. \tag{14.61}$$

Since $g(Y_t) \le LY_t$,

$$\begin{aligned} P(\hat{\tau} > T) &\le P\Big\{\sigma(W_t - W_u) + (\mu - \frac{\sigma^2}{2})(t - u) \\ &\le \log\frac{g(y)}{x} \le \log\frac{Ly}{x}, \quad 0 \le u \le t \le T\Big\} \\ &\to 0 \quad \text{as} \quad T \to \infty, \end{aligned}$$

we have that $P(\hat{\tau} < \infty) = 1$. Hence

$$\begin{aligned} Z_{\hat{\tau}} &= e^{r\hat{\tau}} V(X_{\hat{\tau}}, Y_{\hat{\tau}}) = e^{r\hat{\tau}} V(g(Y_{\hat{\tau}}), Y_{\hat{\tau}}; g) \\ &= e^{r\hat{\tau}} Y_{\hat{\tau}}, \end{aligned}$$

the last equality following from (14.47)(a). From (14.61) we then have

$$E_{x,y} e^{r\hat{\tau}} Y_{\hat{\tau}} \leq V(x, y; g),$$

from which it follows that

$$V^*(x, y) \leq V(x, y; g). \tag{14.62}$$

To obtain the reverse inequality we use the first condition in (14.46) to get

$$\begin{aligned} E_{x,y} e^{r(t\wedge\tau)} Y_{t\wedge\tau} &\geq E_{x,y} e^{r(t\wedge\tau)} V(X_{t\wedge\tau}, Y_{t\wedge\tau}; g) \\ &= E_{x,y} Z_{t\wedge\tau}. \end{aligned} \tag{14.63}$$

Since Z is a continuous local martingale, $Z_{t\wedge\tau\wedge\sigma_n}$ is a martingale for some stopping time sequence $\sigma_n \uparrow \infty$ and we have

$$E_{x,y} Z_{t\wedge\tau\wedge\sigma_n} = Z_0 = V(x, y; g).$$

Now (14.63) is true for any finite stopping time τ. So replacing τ by $\tau \wedge \sigma_n$, we have

$$E_{x,y} \left\{ e^{r(t\wedge\tau\wedge\sigma_n)} Y_{t\wedge\tau\wedge\sigma_n} \right\} \geq E_{x,y} Z_{t\wedge\tau\wedge\sigma_n} = V(x, y; g). \tag{14.64}$$

By the definition of Y_t, $Y_{t\wedge\tau\wedge\sigma_n} \leq y$, for all n and $t \wedge \tau \wedge \sigma_n$ so that for all n,

$$e^{r(t\wedge\tau\wedge\sigma_n)} Y_{t\wedge\tau\wedge\sigma_n} \leq e^{rt} y.$$

The left hand side of the above inequality $\to e^{r(t\wedge\tau)} Y_{t\wedge\tau}$ a.s. As $n \to \infty$, by the dominated convergence theorem we have, using (14.64)

$$E_{x,y} \left\{ e^{r(t\wedge\tau)} Y_{t\wedge\tau} \right\} \geq V(x, y; g), \quad \forall t \geq 0.$$

Again using the dominated convergence theorem as $t \to \infty$, we obtain from the above inequality

$$E_{x,y} \left\{ e^{r\tau} Y_\tau \right\} \geq V(x, y; g).$$

Hence

$$V^*(x, y) \geq V(x, y; g). \tag{14.65}$$

From (14.62) and (14.65), the desired equality follows:

$$V^*(x, y) = V(x, y; g), \quad y < x \leq g(y). \tag{14.66}$$

Since the free boundary $\hat{g}$ is also a solution of our FBM,

$$V^*(x, y) = V(x, y; \hat{g}), \quad y < x \le \hat{g}(y). \tag{14.67}$$

From (14.66) and (14.67),

$$V(x, y; \hat{g}) = V(x, y; g), \quad y < x \le \min\{g(y), \hat{g}(y)\}. \tag{14.68}$$

Substituting from (14.50) in the above equation we have

$$y\left\{(\frac{x}{ky})^{\gamma} - \gamma(\frac{x}{ky})^{\gamma}\log\frac{x}{ky}\right\} = y\left\{(\frac{x}{g})^{\gamma} - \gamma(\frac{x}{g})^{\gamma}\log\frac{x}{g}\right\}.$$

Making $x \to y$,

$$\frac{1}{k^{\gamma}} - \gamma\frac{1}{k^{\gamma}}\log\frac{1}{k} = \left\{\frac{y}{g(y)}\right\}\left\{1 + \gamma\log\frac{g(y)}{y}\right\}.$$

Writing $h(y) = \frac{g(y)}{y}$ for convenience,

$$\frac{1}{h^{\gamma}}(1 + \gamma\log h) = \frac{1}{k^{\gamma}} + \frac{\gamma}{k^{\gamma}}\log k = \text{ constant.}$$

Differentiating with respect to y at $y > 0$, we have

$$\frac{\gamma^2}{h^{\gamma+1}}\log h \cdot \frac{dh}{dy} = 0,$$

which gives $\frac{dh}{dy} = 0$ since $\log h(y) \ne 0$ because $h(y) > 1$ for $y > 0$. Hence $\frac{g(y)}{y} = c$, a constant for all $y > 0$. From the steps immediately following (14.54), we immediately see that $c = k$. Thus we have shown that if $g(y)$ is any other free boundary solving the FBP (14.45)–(14.47), then $g(y) = ky$. In other words, the solution obtained by Shepp and Shiryaev is unique. The stopping time $\hat{\tau}$ defined in (14.60) then becomes the stopping time obtained by these authors. We have thus proved the following result.

Theorem 14.11 *The FBP (14.45)–(14.47) has a unique solution (V, g) for the three cases considered. The expression for V has already been given above.*

For Case 1,

$$g(y) = \theta y \quad \textit{where} \quad \theta = \left\{\frac{1 - \frac{1}{\gamma_2}}{1 - \frac{1}{\gamma_1}}\right\}^{1/(\gamma_2 - \gamma_1)};$$

For Case 2,

$$g(y) = \theta y \quad \textit{where} \quad \theta = e^{\frac{1}{\gamma(\gamma-1)}};$$

For Case 3, $g(y) = \theta y$ where $\theta = \exp(\frac{\phi}{\beta})$; ϕ is given by $1 - \frac{1}{\gamma} = re^{i\phi}$ and $\beta = Im(\gamma) > 0$.

In all three cases, $\theta > 1$ and the unique optimal stopping rule is given by

$$\hat{\tau} := \inf\{t \geq 0 : X_t = \theta Y_t\}.$$

The value of the option is given by the formula

$$\begin{aligned} V^*(x,y) &= V(x,y) \quad y < x \leq \theta y \\ &= y \quad \textit{for} \quad x > \theta y. \end{aligned} \tag{14.69}$$

To obtain the value of the option claimed in (14.69) when $x > \theta y$, we proceed as follows:

$$\begin{aligned} V^*(x,y) &= \inf_\tau E_{x,y} e^{r\tau} Y_\tau \\ &\leq y \inf_\tau E_{x,y} e^{r\tau} \leq y \end{aligned}$$

by arguing as in the put option case and taking the stopping time $\epsilon\tau$ and then making $\epsilon \to 0$. When $X_t > \theta Y_t$, we have $dY_t = 0$. Hence, if $Z_t = e^{rt} Y_t$,

$$dZ_t = re^{rt} Y_t dt, \quad Z_t = Z_0 + \int_0^t re^{rs} Y_s ds \geq Z_0.$$

We then have, replacing t by $t \wedge \tau$,

$$E_{x,y} Z_{t\wedge\tau} \geq E_{x,y} Z_0 = y.$$

Note once again that

$$E_{x,y} Z_{t\wedge\tau} \leq E_{x,y} e^{r\tau} Y_{t\wedge\tau}$$

and that $e^{r\tau} Y_{t\wedge\tau} \downarrow e^{r\tau} Y_\tau$ as $t \to \infty$. We get

$$E_{x,y} e^{r\tau} Y_\tau = \lim_{t\to\infty} E_{x,y} e^{r\tau} Y_{t\wedge\tau} \geq y.$$

Taking the inf over τ on the left hand side we obtain the reverse inequality

$$V^*(x,y) \geq y,$$

thus showing that

$$V^*(x,y) = y \quad \text{if} \quad x > \theta y. \tag{14.70}$$

It remains to complete the proof for Case 4. Since $r = 0$, $Z_t = V(X_t, Y_t)$. From the argument preceding (14.63) we get

$$Z_{t\wedge\sigma_n} = Z_0 + \sigma \int_0^{t\wedge\sigma_n} \frac{\partial V}{\partial x}(X_s, Y_s) X_s dW_s$$

where $\sigma_n \uparrow \infty$ are stopping times such that the integral on the right is a martingale. Then

$$E_{x,y} Z_{t\wedge\sigma_n} = V(x,y).$$

From the first condition in (14.46),

$$E_{x,y}Y_{t\wedge\sigma_n} \geq E_{x,y}Z_{t\wedge\sigma_n} = V(x,y).$$

By the monotone convergence theorem applied twice to the first quantity in this inequality, we have

$$V^*(x,y) = E_{x,y}Y_\infty \geq V(x,y) \tag{14.71}$$

Next,

$$E_{x,y}V(X_{t\wedge\sigma_n}, Y_{t\wedge\sigma_n}) = E_{x,y}Z_{t\wedge\sigma_n} = V(x,y).$$

Hence making $\sigma_n \to \infty$, noting that V is continuous and positive, we have

$$E_{x,y}V(X_t, Y_t) \leq V(x,y). \tag{14.72}$$

Now,

$$X_t = xe^{t\{(\mu-\frac{1}{2}\sigma^2)+\frac{\sigma W_t}{t}\}} \to \infty$$

as $t \to \infty$ since $\mu > \frac{1}{2}\sigma^2$ by assumption. From the formula (14.53) for $V(x,y)$, $V(X_t, Y_t) \to Y_\infty$ as $t \to \infty$ because

$$\frac{Y_t}{X_t} \leq \frac{y}{X_t} \to 0.$$

Hence, again by Fatou's lemma and (14.72),

$$E_{x,y}Y_\infty \leq V(x,y). \tag{14.73}$$

Hence

$$V^*(x,y) \leq V(x,y) \tag{14.74}$$

from (14.52). Thus we have proved that $V^*(x,y) = V(x,y)$ for Case 4. □

References

J.P. Ansel and C. Stricker (1990). Quelques remarques sur un theorème de Yan, *Séminaire de Probabilités*, Lecture Notes in Mathematics, Springer-Verlag, Heidelberg, 226–274.

A.G. Bhatt, G. Kallianpur and R.L. Karandikar (1995). Uniqueness and robustness of solution of measure-valued equations of nonlinear filtering, *Ann. Probab.* **23**, 1895–1938.

R. Billingsley (1968). *Convergence of Probability Measures*, John Wiley & Sons, New York.

L. Breiman (1968). *Probability*, Addison-Wesley, Reading, MA.

H. Chernoff (1961). Sequential tests for the means of a normal distribution, *Proc. Fourth Berkeley Symp. Math. Statist. Prob.* **1**, Univ. of California Press, 79–91.

T. Choulli and C. Stricker (1996). Deux applications de la décompositon de Galtchouk-Kunita-Watanabe, *Séminaire de Probabilités XXX*, Lecture Notes in Mathematics **1626**, Springer-Verlag, Heidelberg.

J.M.C. Clark (1979). The representation of functionals of Brownian motion as stochastic integrals, *Annals of Math. Stat.* **41**, 1282–1295; correction 1978.

N.J. Cutland, P.E. Kopp and W. Willinger (1995). Stock price returns and the Joseph effect: a fractional version of the Black-Scholes model, *Progress in Probab.* **36**, 327–351.

A. Dasgupta (1997). Fractional Brownian motion: its properties and applications to stochastic integration, Ph. D. Thesis, Department of Statistics, University of North Carolina, Chapel Hill.

A. Dasgupta and G. Kallianpur (1998). Arbitrage opportunities for a class of Gladyshev processes, to appear.

D. Dawson and E. Perkins (1991). *Historical Processes*, Mem. Amer. Math. Soc. **454**.

F. Delbaen and W. Schachermayer (1994). A general version of the fundamental theorem of asset pricing, *Math. Ann.* **300**, 463–520.

C. Dellacherie and P.A. Meyer (1978). *Probabilities and Potential*, North-Holland, Amsterdam.

D. Duffie (1992). *Dynamic Asset Pricing Theory*, Princeton University Press, Princeton, NJ.

H. Föllmer and M. Schweizer (1991). Hedging of contingent claims under incomplete information, in *Applied Stochastic Analysis*, M.H.A. Davis and R.J. Elliott, eds., Stochastic Monographs **5**, Gordon and Breach, 389–414.

J.-P. Fouque, G. Papanicolaou and K.R. Sircar (1998). Asymptotics of a two-scale stochastic volatility model, in *Equations aux derivées partielles et applications en l'honneur de Jacques-Louis Lions*, Gauthier-Villars, 517–525.

A. Friedman (1964). *Partial Differential Equations of Parabolic Type*, Prentice Hall, Inc., New York.

E.G. Gladyshev (1961). A new limit theorem for stochastic processes with Gaussian increments, *Theory of Probability and its Applications* **6**, 52–61.

P.R. Halmos (1950). *Measure Theory*, P. Van Nostrand Company, Inc., New York, NY.

M. Harrison and S. Pliska (1981). Martingales and stochastic integrals in the theory of continuous trading, *Stochastic Processes Appl.* **11**, 215–260.

N. Hofmann, E. Platen and M. Schweizer (1992). Option pricing under incompleteness and stochastic volatility, *Math. Finance* **2**, 153–187.

N. Ikeda and S. Watanabe (1981). *Stochastic Differential Equations and Diffusion Processes*, North-Holland, Amsterdam.

K. Itô (1961). Lectures on stochastic processes, Tata Institute of Fundamental Research, Bombay.

J. Jacod (1979). *Calcul Stochastique et Problème de Martingales,* Lecture Notes in Mathematics **714**, Springer-Verlag, Berlin.

S. Kakutani (1948). On equivalence of infinite product measures, *Ann. of Math.* **2**, 214–224.

G. Kallianpur (1980). *Stochastic Filtering Theory*, Springer-Verlag, Berlin.

G. Kallianpur (1995). Stochastic Finance, Lecture Notes, University of North Carolina at Chapel Hill.

G. Kallianpur and J. Xiong (1998). Asset pricing with stochastic volatility, to appear.

G. Kallianpur and R.L. Karandikar (1988). *White Noise Theory of Prediction, Filtering and Smoothing*, Gordon and Breach, London.

R.L. Karandikar (1983a). A remark on paths of continuous martingales, *Expositione Mathematicae* **1**, 67–69.

R.L. Karandikar (1983b). On quadratic variation process of a continuous martingales, *Illinois Journal of Mathematics* **27**, 178–181.

I. Karatzas (1997). *Lectures on the Mathematics of Finance*, CRM Monograph Series No. 8, American Mathematical Society.

I. Karatzas and S.E. Shreve (1988). *Brownian motion and stochastic calculus*, Springer-Verlag, Berlin.

N. Kazamaki (1976). The equivalence of two conditions on weighted norm inequalities for martingales, *Proc. of the International Symp. on Stochastic Differential Equations*, Kyoto.

M.A. Krasnoselskii and J.B. Rutickii (1961). *Convex functions and Orlicz spaces*, P. Noordhoff, Groningen.

D. Kreps (1981). Arbitrage and equilibrium in economies with infinitely many commodities, *J. Math. Econ.* **8**, 15–35.

S. Kusuoka (1993). A remark on arbitrage and martingale measure, *Publ. RIMS, Kyoto University* **29**, 833–840.

T.G. Kurtz and D. Ocone (1988). Unique characterization of conditional distributions in nonlinear filtering, *Ann. Probab.* **16**, 80–107.

S. Levental and A. Skorohod (1995). A necessary and sufficient condition for absence of arbitrage with tame portfolios, *Ann. Appl. Probab.* **5**, 906–925.

S.J. Lin (1995). Stochastic analysis of fractional Brownian motions, *Stochastics and Stochastics Reports* **55**, 121–140.

R.S. Liptser and A.N. Shiryayev (1989). *Theory of Martingales,* Kluwer Academic Publishers, Dordrecht, Netherlands.

B.B. Mandelbrot and J.W. Van Ness (1968). Fractional Brownian motions, fractional noises and applications, *SIAM Review* **10**, 422–437.

R. Merton (1990). *Continuous-Time Finance*, Basil Blackwell, Oxford.

M. Metivier (1982). *Semimartingales*, de Gruyter, Berlin.

P.A. Meyer (1966). *Probability and Potentials*, Blaisdell, Waltham.

H.P. McKean (1969). *Stochastic Integrals*, Academic Press, New York, NY.

M. Musiela and M. Rutkowski (1997). *Martingale Methods in Financial Modelling*, Springer-Verlag.

J. Neveu (1965). *Mathematical Foundations of the calculus of Probability*, Holden Day, San Francisco, CA.

P. Protter (1990). *Stochastic Integration and Differential Equations*, Springer-Verlag, Berlin.

L.C.G. Rogers (1997). Arbitrage with fractional Brownian motion, *Mathematical Finance* **7**, 95–105.

W. Rudin (1974). *Functional Analysis*, TMH edition, New Delhi.

W. Schachermayer(1994). Martingale measures for discrete time processes with infinite horizons, *Math. Finance* **4**, 25–56.

L.A. Shepp (1969). Explicit solutions to some problems of optimal stopping, *Ann. Math. Statist.* **40**, 993–1010.

L.A. Shepp and A.N. Shiryaev (1993a). The Russian option: Reduced regret, *The Annals of Applied Probability* **3**, 631–640.

L.A. Shepp and A.N. Shiryaev (1993b). A dual Russian option for selling short, DI-MACS Technical Report, 93–126.

C. Stricker (1990). Arbitrage et lois de martingale, *Ann. Inst. Henri Poincare* **26**, 451–460.

D.W. Stroock and S.R.S. Varadhan (1978). *Multidimensional diffusion processes*, Springer-Verlag, Berlin.

J.A. Yan (1980). Characterization d'une classe d'Ensemble Convexes de L^1 ou H^1, in *Séminaire de Probabilités*, Lecture Notes in Mathematics **784**, Springer-Verlag, Berlin, 220–224.

Index